Order from Randomness
A Scientific Refutation of Intelligent Design

Daniel Logue, Ph.D.

Copyright © 2021 by Daniel Logue. All rights reserved.

Steve - Thanks for all those conversations that helped make this book what it is.

Contents

1	**Introduction**	**5**
2	**Is our intuition of reality accurate?**	**9**
	2.1 Complexity in nature	12
	2.2 Quantum mechanics	14
	2.3 Special relativity	22
	2.4 Probability is counterintuitive	25
	2.5 Infinite sets	28
	2.6 Exponential growth	31
	2.7 Conclusion	33
3	**Physical systems and mathematics**	**35**
	3.1 What is a theory?	36
	3.2 The difference between faith and evidence	44
	3.3 How does science model reality?	48
	3.4 Chaotic systems	54
	3.5 Fractals	66
	3.6 Cellular automata	70
	3.7 Randomness and information	75
	3.8 Bell's theorem and indeterminism	82
	3.9 The end of reductionism	89
	3.10 An uncaused cause	93
	3.11 You weren't there	96
	3.12 Conclusion	101
4	**Optimization**	**103**
	4.1 What is optimization?	104
	4.2 Optimization by random perturbation	108

	4.3 Optimization using persistent excitation	123
	4.4 Genetic algorithms	129
	4.5 Conclusion	133

5 Evolution 135
5.1	What is the theory of evolution?	136
5.2	Random mutation and natural selection	138
5.3	From small changes to large	141
5.4	The probabilities are too small	146
5.5	All forms are transitional forms	153
5.6	There are no superior forms	155
5.7	Complexity and convergence	158
5.8	The fractal nature of biological systems	167
5.9	Extending dimension of the function space	169
5.10	Evolutionary pressure	172
5.11	Paleontological evidence for evolution	177
5.12	Genetic evidence for evolution	179
5.13	Conclusion	183

6 Intelligent Design 185
6.1	Is intelligent design a legitimate theory?	188
6.2	Probability	192
6.3	Irreducible complexity	195
6.4	Specified complexity	206
6.5	The design inference	215
6.6	The generation of new information	220
6.7	Intervention by an intelligent designer	228
6.8	Can intelligent design provide any answers?	235
6.9	Is ID a valid scientific theory?	240
6.10	The Anthropic Principle and fine-tuning	250
6.11	Is there a supernatural?	257
6.12	Is intelligent design unfairly rejected?	260
6.13	Conclusion	263

7 The energy basis of life 267
7.1	Feedback systems	270
7.2	Entropy	276
7.3	Self-assembling systems	282

7.4	Self-replicating systems	293
7.5	Evolution one more time	297
7.6	Dissipative systems	299
7.7	Life as a dissipative structure	305
7.8	Maxwell's demon	315
7.9	Is there design in nature?	317
7.10	Conclusion	319

8 Intelligence and free will — 321
8.1	Intelligence and the brain	322
8.2	Free will	341
8.3	Artificial intelligence	349
8.4	Intelligent evolution	355
8.5	Simulated reality	360
8.6	Conclusion	364

9 The universal method — 365
9.1	The evolution of technology and science	366
9.2	The financial system and the economy	370
9.3	Religion	373
9.4	Conclusion	377

10 Last thoughts — 379

Index — 397

Chapter 1

Introduction

Many people would argue that humanity has been split between two different competing world views for a very long time and, in recent years, as advancements in technology and science have given us the ability to peer further back in time and unwind the complex intricacies involved in life's development on planet Earth, the chasm has become ever more pronounced between religion on one side and science on the other. This fact is attested to by the numerous books published on the subject, the voluminous YouTube debates and discussions, the distrust of the scientific community during the 2020 Covid-19 pandemic, and the morphing of the creationist movement into what has come to be known as the *intelligent design* (ID) hypothesis.

The conflict between these two worldviews can be observed from many different vantage points, of which the most conspicuous is that religion is formed from the top-down and science from the bottom up. Religion is based upon infallible pronouncement. Science comprises potentially fallible conclusions based on observational discoveries in physical reality. The former is immutable (supposedly), and the latter is always subject to revision with additional data. The former is not correctable, hence to effect change, one is forced to modify one's interpretation of divine pronouncement. The latter is self-correcting due to no one individual having ultimate authority; the natural world is open to everyone to explore, and science forces one's understanding to align with what is real and observable.

Religion is based on tradition and absolutes. It is a primary function of science to break through traditions. Religion does not require evidence and, in some cases, provides admonishment for even asking for proof. Science requires evidence. Religion splits reality into the natural and supernatural. Science considers all to be part of the natural (there simply is no separate container called supernatural). Science makes no assumptions of what is (except for maybe the rules of logic). Religion presupposes everything about reality.

I will stop there as this is not a book about religion. It is, however, important to provide the impetus for each of these points of view and how they differ in gross terms from one another. To be fair, in some regards, it is impossible to decouple ID completely from its religious roots. Still, I shall attempt to do so for the majority of this book, except where not doing so enhances clarity.

As time passes and the population becomes more informed, ideas with no evidential backing become more challenging to defend. The trial of Kitzmiller v. Dover [1] is a prime example of society, in general, rejecting a concept, specifically ID, which has no scientific basis. I suspect that the rejection of creationism as a viable alternative to science is the ultimate reason for the ID movement. It is a repackaging of creationism into an ostensibly credible argument that many people see as legitimate science.

Creationism is the belief that the Universe, Earth, and all the life on Earth were divinely created and are not the consequences of natural processes [2, 3]. Most creationists believe the Earth is about 6,000 years old, that evolutionary theory is wrong, and that the stories in the Book of Genesis in the Bible are an accurate account of creation. There is no evidence for these claims other than what is contained in the Bible and other religious texts. Intelligent design attempts to provide creationism with more scientific credence.

Intelligent design does differ from creationism in that a few of its contemporary advocates are skilled in mathematics, logical argument, and the scientific method, meaning these ID proponents can express their views in a way that is addressable by all scientists. Discussing ideas like the formation of the diversity of life on Earth is facilitated by both sides' ability to clearly and objectively present their respective views. There is no way to argue with someone who asserts without evidence that a god created and guided life on Earth. The best that can be done is to lay out the arguments and evidence so those who are willing to be objective can understand them.

Virtually all the arguments for ID are based on probability estimates and an argument from incredulity (the argument that since one can't think of a better answer, the answer one has in hand must be correct). The main premise of ID is that certain events, such as abiogenesis, the diversity of life on Earth being brought about by physical processes, the suitability of the fundamental constants of the universe to allow life to exist, etc., are so improbable as to indicate their impossibility and, therefore, an intelligent designer is undoubtedly responsible. As we shall see throughout this book, although these probabilities are insurmountable using a pure brute-force assault, that is not how nature goes about solving them. I'm sure that nearly every reader has heard the platitude that biological evolution is based solely on random modification of the deoxyribonucleic acid (DNA) chains in the cells of organisms. However, biological evolution not only uses random mutation, but several other non-random mechanisms as well, such as selection, positive feedback, and the spontaneous generation of complexity.

Random mutation of DNA nucleotides is just one factor that comes into play in life's evolution, and I might argue it is behind only self-replication in importance. Random events, in general, are essential precisely because they are indeterministic. Every other process proceeds in a deterministic fashion, be it a simple linear system, a nonlinear chaotic system, or cellular automata. I will argue in this book that new and novel information can only be constructed in the presence of indeterminism. Although immensely valuable and the primary contributor to life's complexity, deterministic processes alone cannot introduce anything novel within a closed system.

The second important component of evolution is natural selection. One can frame natural selection as an optimization process; however, it is not used to make one life form superior to another. As we'll see, there are no superior life forms. The modifier "superior" is an anthropomorphic concept and does not apply here. All organisms are adapted to fit their environment, including fitting with other organisms. It is not as if the survival of the fittest means that one organism wins over another; instead, fitness measures how well an organism *fits* within an ecosystem. All entities in the ecosphere are contingent on the others. A predator is not superior to its prey. One without the other will not survive. Evolution is continually working to find a better fit. As we will see, natural selection can be modeled as a simplified gradient ascent method that optimizes fitness. However, it is not a fixed-step or adaptive-step method. The steps it takes are random in size, but the selection, or optimization direction, is not random.

It is becoming clear that complexity plays a significant role in generating the diversity we see in life. For a long time, it has been known that mathematics is deeply embedded in the structures found in nature. From the Fibonacci series showing up in the arrangements of seeds and petals of plants to the fractal properties of trees, rivers, and lightning, nature seems to a great extent to be written in mathematics. Complexity and chaos theory have only come on the scene in full force in the last half-century and are striking in their effect. As we'll see in Chapter 3, a considerable amount of information can be stored within a small mathematical container, and these containers can correspond directly to reality. For example, the ratio of a circle's circumference to its diameter is pi. In base ten, this is an irrational number whose decimal portion, as far as we can tell, never repeats—in one context, it contains an infinite amount of information. But pi is not just an abstract mathematical constant. It is absolutely essential in nature's functioning. Due to being a transcendental number, we can consider calculating the ratio of a circle's circumference to its diameter a natural generator of complexity. (Note the word transcendental is used in the mathematical sense.) We will see many other examples of complexity in later chapters.

Chaos also plays a large role in physical reality. From the Lorenz equations that describe weather patterns, to cellular automata, to the functioning of electronic devices, chaos is an integral part of reality. Chaotic systems express infinite sensitivity to initial conditions, meaning that an infinitesimal change in the initial conditions of a dynamic system can result in large differences in its time evolution. This is the so-called butterfly effect, where the flapping of a butterfly's wings in South America can cause a tornado in Kansas. The dynamics of biological systems fall into this category. Small changes in an organism's genome can lead to large changes in that organism's shape, capabilities and behavior.

Self-assembly and replication are the last two contributors to evolutionary biology we will discuss in this book.

Self-assembling structures put themselves together autonomously; they do not require an external builder. It seems counterintuitive that a system could assemble itself, but it must be remembered that it is just a process that, once started, continues until it is complete. It is very akin to tipping over a string of dominoes; once the first domino has been tipped, the rest will follow suit without further assistance from an external agent. In the chemistry of molecular assembly, large molecules can be assembled by autocatalysis. A catalyst is a molecule that speeds up or assists a reaction but is not consumed by that reaction. In autocatalysis, the molecule that is being assembled assists itself. In the domino example, gravity can be loosely characterized as a catalyst; it is not needed since a human hand could turn the dominoes over. However, with the assistance of gravity, the process of dominoes falling is rapid and becomes self-sustaining. Self-assembly is very important to life's formation, and we will consider it in detail later.

Self-replication, or reproduction, is an entity's ability to copy itself and is a hallmark of life. Lifeforms do not exist in perpetuity; they die and are replaced by reproductions of themselves, although these reproductions are not exact copies. There is always some margin of error in the replication. It is this error that drives the evolutionary process. If the reproduction of lifeforms were exact, evolution would not proceed; life would be static. The fact that these reproduction errors lead to differences in organisms' ability to compete and survive in the real world modulates the reproduction rates of the organisms. Errors that lead to greater reproductive efficiency are propagated throughout subsequent populations. Errors that lead to lesser reproductive efficiency die out through successive generations. Therefore, errors in a population's genome modulate the reproductive dynamics of that population through competitive natural selection. We will discuss this topic at length in later chapters.

This book has two purposes. The first is to determine if the claims from the ID movement possess any ability to explain biological complexity. The second purpose is to explain how evolutionary theory succeeds at solving the biological complexity problem. Comparing the two viewpoints is inevitable as ID is really only a rejection of the claim that evolutionary biology can explain the diversity of life on Earth. Intelligent design makes no positive claims of its own.

It is important to note that this book is not concerned with religious concepts or arguments. It is only concerned with the scientific rationale behind the topics of ID and evolutionary biology.

Through the course of the book, we will take a trip through the various fields of physics, mathematics, optimization theory, complexity and chaos, biology, neuroscience, and philosophy. It will take all of these areas to explain the success of evolutionary theory and the failure of intelligent design.

Chapter 2

Is our intuition of reality accurate?

> "The greatest enemy of knowledge is not ignorance, it is the illusion of knowledge.",
> Stephen Hawking.

We humans are pattern-seeking animals, a skill honed over many eons to the point it acts without our conscious oversight. This ability kept our ancestors alive many thousands of years ago while living in the wild by allowing predators and food to be characterized and recognized almost instantly. The fact that many of the neural-processing routines enabling this ability are embedded into our subconscious is a testament to how important they were and are. And although still very useful to us today, these unconscious (and conscious) pattern-seeking tools we all possess can go awry as the world has changed dramatically. Not only has it changed, but we are delving into intellectual depths that never needed to be investigated and understood in our ancestral youth.

I want to use this chapter to introduce several disjoint topics that are not necessarily important to the explicit thesis of this book but instead illustrate the problem with a naive approach to drawing conclusions about the functioning of reality. The spirit or essence of the issue of rationalizing incorrectly due to our biological machinery being ill-suited for the job, or simply being misled by our preconceived notions, is a major problem with the ID argument. An argument from ignorance is indeed a logical fallacy and cannot be used to come to a reliable or rational answer. This chapter intends to give the reader a feel for what is to come and why it is crucial in an earthy, common-sense kind of way.

As human beings, our intuition of how reality functions is typically not just off; it more often than not is entirely wrong. The examples in this chapter are chosen to give a taste of this issue before moving into the more complex subjects of evolutionary biology and intelligence. In other words, this chapter intends to put you on your toes in advance of the actual game. For those of you who are familiar with this material, it is perfectly safe to skim the section headings and skip to the next chapter.

Except for the last two sections that deal with infinite sets and exponential growth, I am going to leave the mathematical oddities for the next chapter, where they will be discussed in conjunction with concepts in physics.

The fundamental argument of ID states that the complexity demonstrated within nature, particularly in biological structures, is beyond what nature could have produced on its own. I think most of us at times in our lives have marveled over the intricate design of a spider's web glistening in the early morning, the mesmerizing structure of a crystal geode, the rigid meandering of a splay of lightning in the western sky, or the patterning of the petals of a flower. As I grew up in the woods, I don't remember ever tumbling to the conclusion that these things must be made by someone or something; they were always just there. However, it is not hard to imagine how it could cross someone's mind that design must have been involved. After all, there is a great deal of perfection and precision in many structures within nature. That said, if a rational, objective hypothesis that is backed up by solid evidence can be constructed that explains the origins of these structures, it must take precedence over our intuitions. Otherwise, we will find ourselves attributing intelligent design to objects we shouldn't.

In a YouTube video presented by ID proponent Douglas Axe titled 'Author Douglas Axe presents his book "Undeniable"' [4], Axe provides an overview of what he calls the *universal design intuition* that he defines in his book implicitly with the statement, "Tasks that we would need the knowledge to accomplish can be accomplished only by someone who has that knowledge," and goes on further to explain that the ability to intuit design is a common human faculty we all possess. In the video, he shows a folded, paper origami swan and claims that a child who has never seen one before and comes across the swan will innately *know* that someone must have designed it as it could not have formed on its own. Further, he claims that this intuition, when properly used, is reliable and can even be used to gather evidence necessary to refute Darwin's theory of evolutionary biology.

This first claim of Dr. Axe, that there is a reliable means of detecting design, is one of the primary contentions of ID theory that we will explore in detail in later chapters.

I don't deny that we as humans search for explanations of things we don't understand, like how an origami swan is made, but time and again, our intuitions about the origins or functions of entities and processes within the Universe have been wrong. It was apparent to adherents of the Catholic Church in Galileo's time that the Sun revolved around the Earth—only a fool would think otherwise as anyone could see it happen every day. (This is not entirely correct as the Sun and Earth actually rotate about the centroid of their masses, but it is clear this was not the issue in Galileo's time.) The Church turned out to be quite wrong.

Yet, we today still demonstrate our innate intuitions in common expressions such as the Sun "comes up every morning" or "the Sun goes down in the west". There's nothing wrong with making these types of colloquial statements, but they are factually incorrect. It takes a very insightful person, without being first told, to conclude that the apparent motion of the Sun is derived from the rotation of the Earth about its axis and not from the revolution of the Sun about the Earth. It is just not a common-sense thing.

Because a clear understanding of how design is determined is a central theme of ID, in this chapter, we will look at several examples within the physical world that *only appear to be designed*. Seeing examples of what is and is not designed and where our intuition is wholly inadequate will aid those who haven't spent some time thinking about these questions. All of us have some intuition about what constitutes design, and we may be quite sure that our categorical judgment ability is accurate. Really, how hard could it be to differentiate between the naturally occurring and that which has been designed? In addition, the rules of nature should adhere to common sense. As we'll see in this and the following chapters, design is not always easy to intuit, and the laws of physics often run counter to our intuitions.

A second point that Axe makes in the video and his book *Undeniable* [5] is that the complexity a paper swan exhibits does not simply show up by accident. He drives the point home by emphasizing that paper origami swans are not known to randomly assemble themselves within the paper scrap piles of paper factories. It is not a new notion to equivocate upon the idea of evolution making use of random processes and claim that it is akin to a fully functional 747 airliner forming in midair flight as a tornado tears through a junkyard.

This represents a flawed analogy as the evolutionary process is not entirely random. The only persons who claim that natural complexity is generated from purely random processes are ID proponents and the religious. It is a straw-man fight. There are random events that drive evolution, but natural selection is certainly not random.

Throughout this chapter, we are going to look at several examples of spontaneous generation of complexity and the dubious claim that it is trivial to discern whether said complexity is designed.

In the interest of being fair to intelligent design advocates, I am not going to say at this point that biological complexity is not designed nor include any examples of such in this chapter. Rather, I am going to consider only examples that are obviously driven by unintelligent processes in an attempt to address claims like those from Axe that design is not only present in nature, but it can be intuited correctly by virtually anyone, including children.

The term "design" will not be defined just yet. Instead, for now, I am going to start with the colloquial, naive assumption that complexity indicates design and demonstrate that this conclusion is not valid when it comes to the natural world and mathematics.

2.1 Complexity in nature

At a young age, many children are taught how and become wholly engrossed with constructing paper snowflakes at Christmastime. They find it almost magical to repeatedly fold the paper along diagonal lines, cut out triangles, squares, and half-circles along one edge, then finally unfold the paper to reveal a series of symmetrically-placed, geometric-shaped holes in their newly created snowflake. For a week or so, or until the enchantment wears off, parents will find the floors of their home continuously covered with tiny flakes of paper of every color.

Of course, Dr. Axe is correct in his assertion in the YouTube video that origami swans do not show up spontaneously from the waste clippings on the floors of paper factories. The implication is that similar levels of complexity cannot be generated randomly in nature. This is true, but it is a false equivocation. If a paper swan were to assemble itself spontaneously from a pile of trash paper clippings, it would need to do so using pure random luck. However, this is not the case in nature, and this is where the ID legerdemain is attempted. *Complexity in nature does not stem from random chance alone.* To claim it does is to mislead. Let's look at some examples of complexity in nature and show they are not created by mere happenstance.

Consider the image of a snowflake under magnification in Figure 2.1. It certainly shows all the complexity of an origami swan or a paper snowflake constructed by a first-grader, but its formation relies not at all on intelligence. The hexagonal geometry derives from the angle of the covalent bonding of hydrogen and oxygen atoms in the water molecule, and the complex, fractal extrusion is driven by intrinsic optimization of heat flow out of the center of the structure.

What is the probability of a snowflake being formed by chance versus an origami swan? Axe comments that no one ever sees a swan form from the cuttings on a paper mill floor. Given that there could be tens of precise folds to form an origami swan, it would undoubtedly be a surprise to see one formed accidentally from random paper clippings. However, the number (10^{18}) of water molecules in a typical ice crystal, 100 of which make up a typical snowflake, is astronomically larger than the number of folds and cuts in an origami swan. But here's the thing, and it is very important: *neither the origami swan nor the snowflake form by random chance.* An intelligent agent indeed designs the swan and the snowflake forms via the deterministic physical laws of nature. No intelligence is needed to form a snowflake. The hexagonal shape is a self-assembling emergence that stems from the 104.5° angle of polarization of the water molecule. Spines grow outward from the center in a hexagonal pattern as a consequence of the minimization of energy. Heat finds the optimal path to leave the body of the snowflake, and the result is a hexagonal shape.

The assumption that all of us, children and adults alike, can discriminate design from the action of natural processes is colored by the fact that most people have generally only been introduced to the inner workings of one of these processes; that is, design. We are all aware of the complexity present in designed

2.1. COMPLEXITY IN NATURE

Figure 2.1: A typical snowflake structure under magnification.

objects like books, cell phones, and automobiles because we are all aware that these things are created or manufactured by other human beings. On the other hand, we are all not aware of the mathematics, physics, and biology associated with life on Earth. Let's look at another example.

A vertical temperature gradient is generated when a thin layer of viscous liquid is placed between two horizontal surfaces held at different temperatures. If the bottom plate is at a higher temperature than the top plate, with small temperature differences, a linear gradient is produced from top to bottom. A corresponding density gradient is also created. The hotter liquid at the bottom becomes less dense (for example, a specific volume of water is larger at a high temperature than it is at a low temperature) than the cooler liquid at the top. As long as the temperature gradient is not too large, conduction is enough to manage the heat flow. But, as the heat injection at the lower plate crosses a specific value, convection takes over and circulating convection cells called Bénard cells form. An example of Bénard cells is shown in Figure 2.2.

The formation of Bénard convective cells could easily be mistaken as design if one were not aware of how they actually form. They are not a derivative of intelligent behavior but are simply the physical laws of nature being used to solve an energy-minimization problem. Convective rolls form because they are the most efficient means of transferring energy from the bottom, hot plate, to the top, cooler plate. These convective rolls appear because only with convection can enough heat be transferred to meet the boundary conditions.

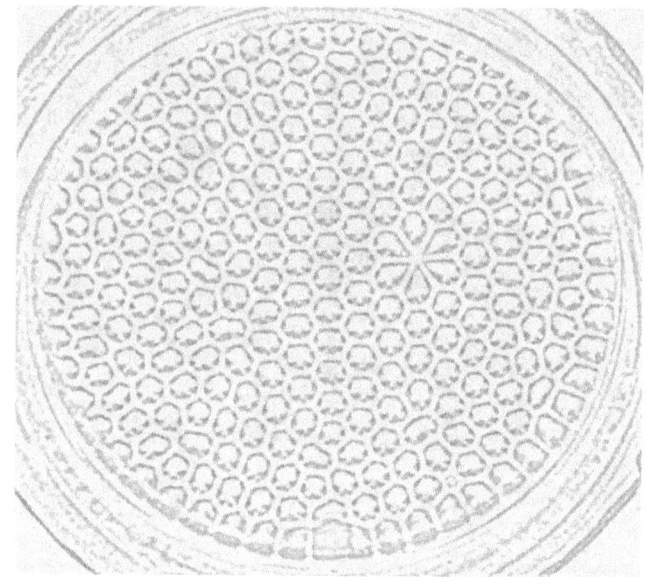

Figure 2.2: Above a particular temperature gradient, conductive heat transfer spontaneously changes to hexagonal, convective rolls that optimize heat transfer [6].

2.2 Quantum mechanics

Perhaps the topic in physics most infamous for violating our intuition is quantum mechanics. Richard Feynman's quintessential quote concerning quantum mechanics, "If you think you understand quantum mechanics, you don't understand quantum mechanics," perfectly illustrates the problem with Axe's assertion that we can trust our intuition about the operation of the physical world. Feynman's meaning was that the physics of the Universe often runs completely counterintuitive to our notions of how it should operate. In this section, I want to examine several examples of quantum mechanical behavior and question the validity of the notion of whether the intuition of the majority of people would predict this behavior.

It is a well-known property of physical matter that it spontaneously emits radiation when heated. In the late 19th century, the *Rayleigh-Jeans law* was used to predict the spectral radiance from a *blackbody* as

$$B_\nu(T) = \frac{2\nu^2 k_B T}{c^2}, \qquad (2.1)$$

which expresses the power emitted per-unit-area at a given frequency ν and temperature T [7]. The constants k_B and c are Boltzmann's constant and the speed of light, respectively. A blackbody radiator is basically a highly non-

2.2. QUANTUM MECHANICS

reflective emitter that is in temperature equilibrium with its environment. In other words, the only radiation it emits is due solely to its temperature and does not include reflections from outside sources.

The Rayleigh-Jeans equation can be integrated with respect to frequency to obtain the total energy density produced by the blackbody. However, the expression has a square of frequency in the numerator, leading to a problem when the integration is performed across wide frequency ranges. In fact, the integral diverges or goes to infinity as the frequency of the light is allowed to increase without bound. An answer that explodes to infinity can be tolerated in some instances; in this particular case, by making the assumption that the frequency of light could not be infinite, maybe things would turn out fine. That was not to be as data collected in the late 19th and early 20th centuries showed that the Rayleigh-Jeans expression deviated from experimental results. Specifically, there was divergence as the frequency of the light neared the ultraviolet region. At the time, this problem was termed the *ultraviolet catastrophe* because it pointed toward an irreconcilable problem with the overall model of physics.

The ultraviolet catastrophe showed that a change in the understanding of light energy would be necessary and led Max Planck in 1900 to propose a radically different formulation for blackbody radiation, one based on discrete energy levels. He modeled the energy contained within the blackbody as an ensemble of harmonic oscillators that emit electromagnetic radiation at discrete energy levels. A classic example of a harmonic oscillator is a guitar string fixed at both ends. If the string is plucked, the lowest mode of oscillation is half-moon in shape and vibrates up and down. The first harmonic mode is shown in Figure 2.3 as the dashed line. The dashed-line, half-cycle sinusoid is only shown as positive in the figure, but it actually oscillates between positive and negative; the plot only shows the shape of the motion, not the motion itself.

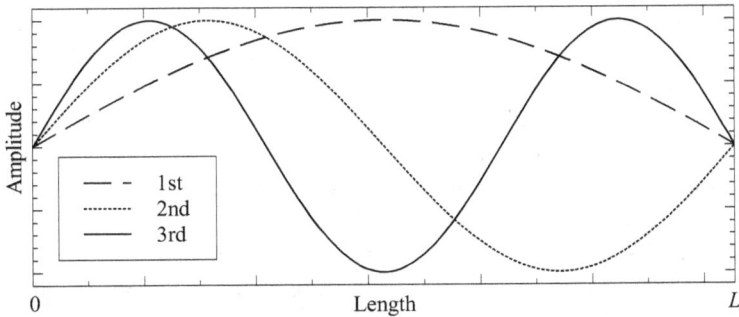

Figure 2.3: The first three modes of a harmonic oscillator.

Also shown in Figure 2.3 are the second and third modes, represented by dotted and solid lines, respectively. However, the guitar string can sustain an infinite number of modes, each higher in frequency than the former (of course, I am talking about an idealized string here). The string does not allow non-integral modes because the boundary conditions of no motion at the endpoints would not be satisfied. A sine wave goes through zero amplitude at the angles 0 and π, and this meets the boundary conditions for this problem as shown by the dashed line in the figure. A three-quarter sine wave with endpoints at 0 and $3\pi/2$ is not allowed because $\sin(3\pi/2) = -1$ and, hence, does not satisfy the right boundary condition.

Max Planck took a giant leap forward in the understanding of physical reality by reasoning that electromagnetic radiation could similarly only assume integral energy (frequency) values. The expression that Planck used for energy was $E_n = nh\nu$, where n is the mode number and $h = 6.626 \times 10^{-34}$ J·s and has come to be called Plank's constant. This equation explicitly defines the energy as an integral multiple of the frequency scaled by Planck's constant. I'm not going to go through a derivation here, but when a discrete solution is assumed, Plank's alternative to the Rayleigh-Jeans law is given by equation (2.2) and is referred to as Planck's law [7–9]. Planck's equation does not diverge when integrated to infinity because the exponential in the denominator increases faster than ν^2 in the numerator, thus resolving the ultraviolet catastrophe.

$$B_\nu(T) = \frac{2h\nu^5}{c^3} \frac{1}{e^{h\nu/(k_B T)} - 1} \qquad (2.2)$$

The energy densities predicted by Rayleigh-Jeans and Planck's laws are shown in Figure 2.4 as functions of frequency. Note that whereas the Rayleigh-Jeans result diverges to infinity, Planck's law predicts the energy density rising to a maximum and then falling off as the frequency increases, resulting in a finite value when integrated across all frequencies. The disparity results from the former law being based on the equipartition of energy and the latter on statistical mechanics, specifically the Boltzmann distribution law. Equipartition of energy assumes that for a system in thermal equilibrium, energy is shared equally across all degrees of freedom, that is, all frequencies. As we have seen, this assumption leads to a prediction of infinite energy as the temperature increases, which is obviously incorrect. On the other hand, once the constraint that energy must only exist in discrete packets of $h\nu$ is added, the equipartition of energy does not hold, and Boltzmann's distribution dictates that there are fewer high-energy than low-energy modes. In other words, as the frequency increases, the probability that these discrete modes exist decreases; i.e., there is not an equal number of high- and low-energy states; there are fewer at high energy.

The mathematical details are not the most important takeaway from the above discussion. Rather, it is vastly more important to recognize the fact that energy is quantized, that is, comes in discrete packets, was not anticipated by the brightest scientific minds of the 19th century, much less being intuitively obvious. Those of us in the 21st century are familiar with energy quantization due to news articles and science fiction movies, so we underestimate the magnitude of Planck's

2.2. QUANTUM MECHANICS

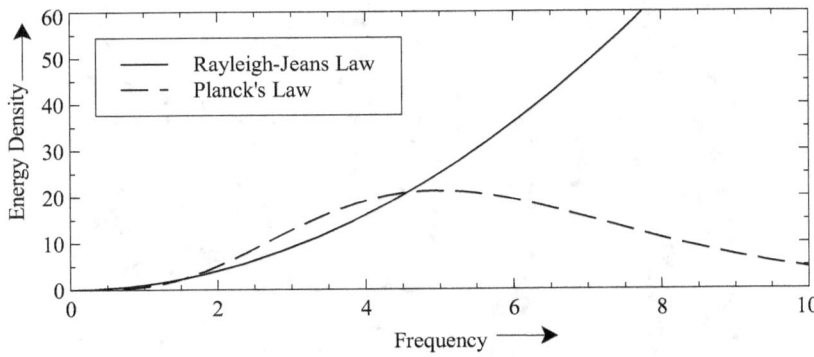

Figure 2.4: Energy density predicted by the Rayleigh-Jeans and Planck's laws as a function of frequency.

insight. Planck, who was a genius in his own right, spent half a decade working on this problem and still had not solved it. It is hard to imagine the solution could have been just intuited upon first consideration by the majority of us.

Surprisingly, Planck's law and the resolution of the ultraviolet catastrophe is one of the least strange of the strange things encountered within quantum mechanics. The double-slit experiment is a setup that many people, including those not associated with the field of physics, have heard of today. Some of them have actually witnessed such an experiment in high-school or college physics. It involves firing particles, typically photons or electrons, at a screen with two distinct parallel, vertical openings. (In this description, we are going to assume that electrons are being used to remove any concern that photons possess some special characteristic that aid in the outcome.) A detector panel behind the screen records the positions of the impact of the particles that make it through the slits. (See Figure 2.5.)

Classical physics, and also our intuition, insists that the particles go through the slits and form two concentrations of particle impacts behind the screen. This is obvious, right? Let's say, at a target range, a shooter repeatedly fires at a target that is behind a wall with two six-inch circular holes in it. After she fires a few hundred rounds, it would be expected that there would be two distinct, circular patterns in the target. Our intuition tells us this would be the case for the shooting-range target and that it should also hold true for the dual-slit experiment.

Now, consider throwing rocks into a lake. We have all done this at some time as kids and watched the circular ripples grow outwardly. If a friend throws a rock into the same pond at the same time you throw yours, two circular wavefronts will be created, and eventually, the two wavefronts will collide and interfere with one another. We all know what that looks like. At certain locations, the waves constructively interfere to produce high waves or low troughs, and at

Figure 2.5: The dual-slit experimental setup. An emitter of photons, electrons or other particles is aimed at an opaque screen with two vertical slits. Behind the screen is a detector panel that records the impact positions of the particles. The pattern of two vertical lines on the detector is what one would expect for classical particles.

certain points, they cancel each other completely out. Observing these interfering wavefronts hitting the vertical concrete wall of the lake, you and your friend would see bands of high and low water levels spread horizontally across the face. There's nothing surprising here. It is just how waves behave.

So, what do water waves have to do with electrons? Waves are waves, and electrons are particles. However, when the experiment of firing electrons through a dual slit is performed, what is imaged on the backdrop of the screen is not two distinct groupings of electron impacts. Instead, there appears an interference pattern exactly matching that of the wavefront on the wall of the lake above, meaning the electrons are in some way behaving as waves as they pass through the slits in the screen. This is a shocking revelation and was first proposed by the French physicist Louis de Broglie in the early 20th century after working on x-ray wave mechanics with his older brother Maurice [10].

The notion of photons or electrons acting as both waves and particles is aptly called the *wave-particle duality of nature*. But what exactly does this mean? This discussion has been around for a very long time, stretching back even to Newton and Huygens arguing over the corpuscular and wave properties of light. In Newton's time, these were viewed as physical properties of light; quantum mechanics takes a slightly different viewpoint on what precisely the wave and particle behavior represent.

Quantum mechanics says that the properties of position, momentum, energy, etc., of a particle, are probabilistic. For example, when a single electron is moving through space, and it is *not being observed*, its position is not well-defined. Its precise position is not explicitly defined at all. Only when the electron is interacted with, "observed", does it appear at a given location. However, it

2.2. QUANTUM MECHANICS

does not just appear anywhere. The probability of it appearing in a particular location is defined by its associated *wave function*. The wave function represents the probability distribution describing where the electron will be found when the wave function *collapses* due to interaction or observation.

Observation became a primary point of discussion in Niels Bohr's Copenhagen Interpretation of quantum mechanics in the early 20th century. It is still controversial today just what an observation means, and justifiably so. Here, observation will be synonymous with an interaction that causes the collapse of a quantum wave function.

Whether a particle is moving or not, and so long as it is isolated from disturbance, its wave function continually expands. The wave function cannot be used to identify exactly where a particle will appear when it is interacted with, but it can be used to determine the probability of finding the particle in a particular location. More specifically, the square of the wave function provides the requisite probability. For example, if one wishes to calculate the probability of finding a particle at $x = 2$ and the value of the wave function at this location is $\phi = 0.5$, then the probability of finding the particle at this location when measuring is $P = \phi^2 = (0.5)^2 = 0.25$, a one-in-four chance. (I am being a bit loose with the mathematics here. Actually, we would integrate the square of the wave function over a region.) Let's see how the wave function is calculated.

The wave equation is a linear partial differential equation (2.3) and was heuristically formulated by Erwin Schrodinger in 1925. Just like any other wave, the quantum mechanical wave needs to have a wave equation. However, it is different from those of water or electromagnetic waves in that it is complex (note the complex operator $i = \sqrt{-1}$). The fact that it is complex implies that solutions to the equation do not correspond to real quantities. For example, because the wave equation for currents on an electrical transmission line has real coefficients, the solutions correspond directly to real currents on the line. This is not so for the Schrodinger equation, and the natural way to convert them to real quantities is to square them.

$$i\hbar \frac{\partial \psi(x,t)}{\partial t} = -\frac{\hbar^2}{2m} \frac{\partial^2 \psi(x,t)}{\partial x^2} \qquad (2.3)$$

I won't go through the solution of the equation here but will plot the result instead; see Figure 2.6. The solution comprises two multiplicative components, one that generates the shape of the envelope and a second that oscillates. Note that the dashed line is not part of the wave function; it is just there to emphasize the envelope. The wave in the figure demonstrates the meaning of wave-particle duality: the wave function is a wave, but it is a wave that is localized in an area specified by its envelope.

Figure 2.7 shows a particle moving to the right of the page and its continually expanding wave function. It continues expanding as long as there is no interaction with the particle. The wave part of the wave function produces the interference pattern in the dual-slit experiment above. It needs to be understood that this

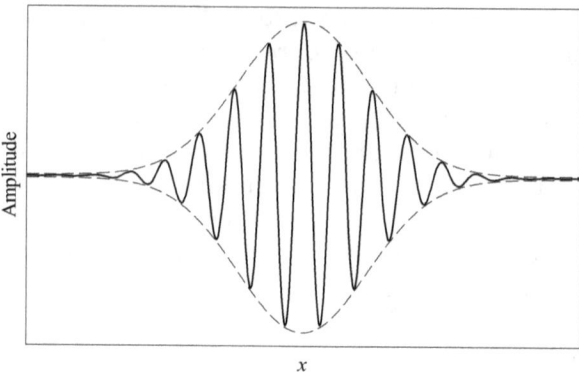

Figure 2.6: A quantum-mechanical wave packet corresponding to a particle. It is a localized wave.

wave is not a physical thing. Instead, it is a probability wave, if anything. It is a representation of the probability of finding a property of the particle in a particular state. For example, it provides information on where there is a likelihood of finding the particle.

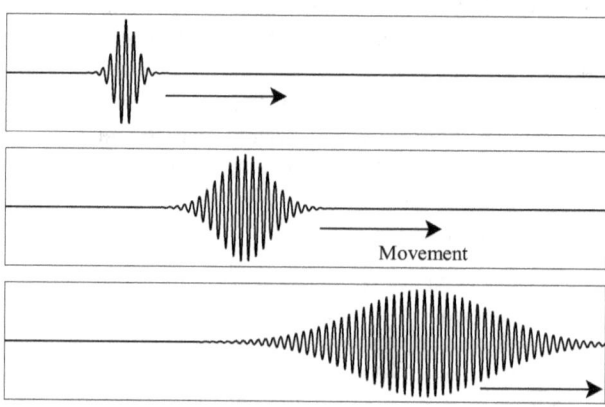

Figure 2.7: The expanding wave function representing a particle's position as it travels from left to right.

In the Copenhagen Interpretation of quantum mechanical theory, this wave packet collapses to a delta function of zero width when the particle is observed or measured, revealing its particle nature. Light, for example, is not physically wave-like but is composed of photons. The wave nature of light is apparent only because the probability of finding the locations of the photons that comprise the

2.2. QUANTUM MECHANICS

light is dictated by a wave-shaped distribution. In other words, light presents itself as a wave because the probability of finding the photons in a wave-shaped pattern is high. This is a very subtle point that requires some thought to grasp. It is not as if photons or other particles are both particles and waves simultaneously; instead there are particles that are probabilistically distributed in the shape of a wave.

Figure 2.8 shows the results of the dual-slit experiment. The two peaks, one on each side of the plot and drawn in a dashed line, represent the incorrect classical, particle-only solution. If the electrons were nothing more than physical particles, two localized collections of impacts would be the expected outcome. However, this is not what is seen when the experiment is performed; instead, the interference pattern in the center of the plot is imaged on the backdrop behind the slitted screen. If one of the slits is then blocked, a single peak will appear.

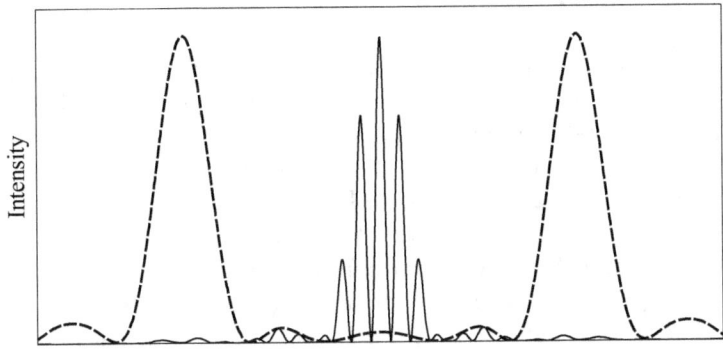

Figure 2.8: The actual detected pattern for the dual-slit experiment. The real result, and the one predicted from quantum-mechanical theory, is shown as the solid line and exhibits wave interference *even if the particles are fired one at a time*. Classical physics, where photons and electrons are simple particles, predicts the double-peaked, dashed-line result, which is incorrect.

Interestingly, the interference pattern is obtained even if the rate of electron firing is reduced to guarantee that at most, there is only one electron in flight at any given time. In stark contrast to the way our minds believe reality functions, *a single electron can interfere with itself*. One electron essentially passes through both slits. This result is indeed strange and completely flies in the face of classical theory. And, as de Broglie showed, the phenomenon of mass behaving as a wave does not apply only to electrons. It applies to all massive bodies. There is no absolute certainty of the location of a baseball, a person, or a mountain. The truth of reality runs entirely orthogonal to our intuitions.

When an electron is in flight and not being observed, it seems it only exists in potentiality and does not physically exist. The electron does not actually pass through both slits; rather, its *potential* for doing so is what interferes with itself to generate the resulting interference pattern on the backdrop. It is only when the electron arrives at the backdrop and must show itself that the wave function collapses and the electron resolves itself as a particle at a specific location. Moreover, this is a general result: quantum mechanical properties such as position, momentum, etc., only exist in potentiality until they are interacted with or observed, and thus forced to reveal themselves.

The behavior of particles is not a result that can just be intuited. It took many brilliant physicists many decades to reach these conclusions.

There are other counterintuitive examples in quantum mechanics that we could examine, such as nonlocality, or action at a distance, but I am going to leave this topic for now as the point has been made, and I'm going to switch to another area of physics that has also generated many surprising results.

2.3 Special relativity

One of the most perplexing revelations in physics in the early 20th century was Albert Einstein's development of special relativity. It produced revolutionary changes in the way we view the world and concreted Einstein's reputation as one of the most remarkable geniuses of all time. The story of *special relativity* begins with two assumptions made by Einstein [8]: (1) the laws of physics are the same in any inertial reference frame, and (2) the speed of light in a vacuum is the same for all observers in any inertial reference frame. An inertial reference frame is any reference frame that is not experiencing acceleration, or stated differently, where Newton's first law is obeyed; that is, an object moving at constant velocity in an inertial reference frame continues to move at that velocity unless acted upon by a force. It turns out that these assumptions, especially the second one, lead to some startling implications.

Consider the situation depicted in Figure 2.9 of a train car traveling to the right of the page at velocity v. The car has a light emitter (the triangle) fastened to the floor and facing the roof. The emitter produces a pulse of light that travels to the car's roof, and its time-of-flight is measured by an observer on the train car as shown in (a) and by an observer on the ground next to the car as shown in (b). Since the observer in the car is traveling with the light source, the light simply takes a straight path to the ceiling. On the other hand, the light source is moving relative to the observer on the ground, so what he sees is the light moving in a diagonal line to the roof of the car (note (b) shows the car at two different times, the first when the light pulse is initiated, and the second when it hits the roof). If we assume, as Einstein did, that the speed of light is constant for all observers, then how could the light pulse travel the shorter distance observed by the individual on the train car and the longer distance seen by the stationary observer in the same amount of time? This question is the crux of the matter.

The time it takes for the light pulse to reach the roof as observed by the person on the train car is $\Delta t_m = h/c$, where c is the speed of light, and h is the height of the ceiling. Similarly, the time the stationary observer measures is

2.3. SPECIAL RELATIVITY

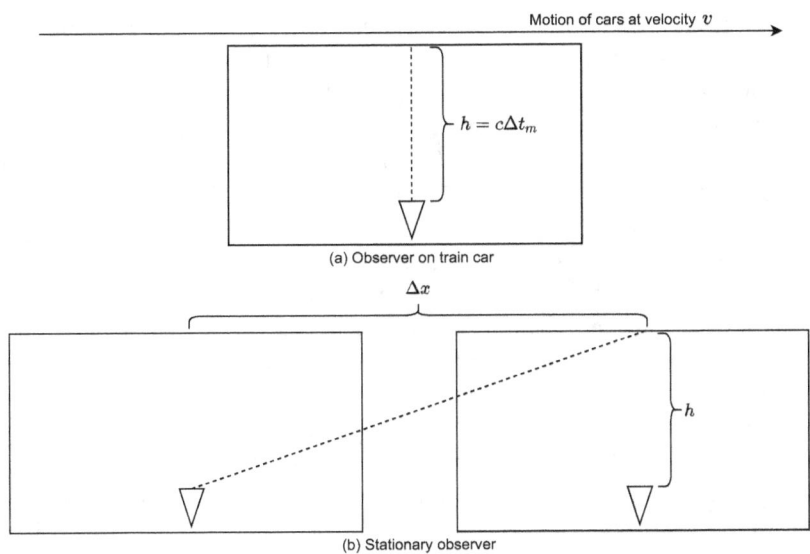

Figure 2.9: A train car with a light emitted from the floor to the ceiling. Figure (a) is the reference frame of an observer on the train, and (b) is what a stationary observer sees.

$\Delta t_s = (1/c)\sqrt{h^2 + (v\Delta t_s)^2}$. Solving the second equation for Δt_s yields $\Delta t_s = (h/c)\sqrt{1 - v^2/c^2}$, and substituting the expression for Δt_m into this equation, we have $\Delta t_s = \Delta t_m/\sqrt{1 - v^2/c^2}$. Since the denominator is always less than or equal to one, the answer, startling as it may be, is that time moves faster for the stationary observer than for the moving observer. In other words, the travel of the light beam progresses faster for the stationary observer than for the one moving, allowing a larger distance to be traversed.

As an example that demonstrates the implications of this facet of special relativity, assume that there is a pair of twins, one who is a futuristic astronaut and the other who is not. The astronaut intends to travel to Alpha Centauri, 4.3 light-years from Earth, while the second twin stays on Earth. An advanced starship will be used that is capable of traveling at 99% the speed of light. Using the expression we derived above, time will pass $\sqrt{1 - 0.99^2} = 0.141$ times slower for the astronaut than for the twin on Earth. From Earth's point of view, it will take $4.3/0.99 = 4.34$ years for the ship to make it to the destination. However, from the pilot's point of view, the trip will only take $4.34(0.141) = 0.612$ years, or about seven months. Essentially, the twin on Earth will be 3.7 years older than the other when the ship arrives at Alpha Centauri.

The change in the flow of time between the two observers is not a trick or a play on words. It is literal. The passage of time slowing for objects in motion has been confirmed repeatedly by experiment. But this isn't the only counterintuitive phenomenon that arises from Einstein's special relativity. Since both the astronaut twin and the homebound twin agree that the speed of the starship is 0.99c, and they measure different travel times, the conclusion is that they also measure different distances from Earth to the star. The stationary twin on Earth measures $(0.99c)(4.34 \text{ years}) = 4.3$ light-years. However, the pilot of the ship measures $(0.99c)(0.612) = 0.606$ light-years. Distance actually contracts for the traveler. Again, this is literal, not a play of mathematics, and has been validated by experiment many times.

These two aspects are enough to confirm the strangeness of special relativity, but I want to talk about one more implication before leaving this section. It has to do with what is probably the most famous mathematical equation of all time, written $E = mc^2$, and still seen on the t-shirts of college students today.

Einstein's mass-energy equivalence expression, $E = mc^2$, is one of the most surprising results to come out of special relativity theory. This formula states that matter and energy are transmutable, one into the other. In fact, one could say that matter is sort of a condensation of energy. The proportionality constant between matter and energy is the speed of light squared, or c^2, which is a very large number, meaning that a small amount of matter contains a large amount of energy. For example, the state of New York in 2019 consumed 145,600,356 mega-watt-hours of electricity, or 5.2×10^{17} J [11], which takes an incredible amount of energy-generating infrastructure to maintain. However, the amount of energy in the mass of *anything* that could be hauled in a half-ton pickup truck is $(1000 \text{ lB})(1 \text{ kg}/2.2 \text{ lB})(3 \times 10^8 \text{ m/s}^2)^2 = 4.1 \times 10^{19}$ J. Dividing this out, that half-ton of any material, if entirely converted into electrical energy, could power New York state for 78 years (at the same rate as in 2019). This answer gives some idea of the huge amount of energy that is stored within matter.

The mass specified in the example above is called the object's *rest mass*, and its equivalent energy is the amount of energy the mass contains at rest. That is, it is the amount of energy contained in the half-ton of material while it's not moving. If the mass is moving at a speed of v_m, its total energy is given by $E_t = (mc^2)/\sqrt{1-(v_m/c)^2}$. Note that c is a constant on the right of this expression, implying that the mass increases. Substitute $m' = m/\sqrt{1-(v_m/c)^2}$, and the energy equation becomes $E_t = m'c^2$, where $m' > m$.

The fact that an object's mass increases as its speed increases may not be shocking to many people as it shows up occasionally in science fiction, and it is an often-cited reason as to why massive objects cannot approach the speed of light (it would take infinite energy because the object's mass grows without bound as its speed increases). However, there are stranger aspects of this idea.

All forms of energy are subject to the mass-energy equivalence equation. Let's say that we have two identical weights that we have measured right down to the last atom as being equivalent. When they are placed on a set of magical scales that measures their masses precisely, the two are at dead zero—they are exactly the same mass. Now, we put one mass in a freezer that brings its temperature to -40C, and we place the other in an oven that brings it to 100C. Then we put them back on the magic scales and measure their masses while at these disparate temperatures. It will be observed that the hotter object has become more massive and tips the scale. This result is anything but intuitive. But, hold on, it gets stranger yet.

Consider two old-fashioned wind-up watches. While they are both in the wound-down state, we use our magical scales and measure their masses to be equal. Now, we wind one watch, which places torsional strain on the spring it contains, and measure their masses again. The scale will show that the wound watch has the greater mass. The potential energy in the wound watch's spring increases its total energy, resulting in an increase in its mass as dictated by the mass-energy equivalence formula.

Relativity theory does an excellent job of demonstrating just how much we cannot trust our common sense, even when it comes to aspects of reality that we see every day. There is a reason that many brilliant scientists have spent their entire lifetimes trying to figure out these things: these are difficult problems to solve and are not at all intuitive.

2.4 Probability is counterintuitive

Probability theory is an area of study that leads to many counterintuitive outcomes. Some of them are simple, and some are more difficult to reconcile. For example, if you flip a fair coin and it comes up heads ten times in a row, what is the probability it will come up heads on the eleventh flip? Obviously, for a fair coin, the probability of it coming up heads for a single flip is 50%, and the same is true for it coming up tails. This likelihood of one-half for coming up heads can be explained by understanding there are two possible outcomes for a single flip and only one actual outcome for each flip. Therefore, the probability of coming up heads is found by dividing the number of actual outcomes, one, by the number of possible outcomes, two, to arrive at one half, and this stands even after ten flips came up heads. (Of course, in real life, if you see this happen, you might want to assume that the coin has a propensity to come up heads. I have gotten around this issue by asserting that the coin is a "fair" coin.)

Our minds are poor judges of probability and even tend to reject accurate assessments when they are known. For instance, there are many people who well know, statistically speaking, traveling on a plane is far safer than driving, but regardless, they are afraid of flying. Similarly, many people play the lottery, all the while knowing that the odds of winning are astronomically small. These are straightforward instances of probabilistic calculations, and we still either get them wrong or tend to reject the results altogether. In the cases where the analysis is less clear, we are even more inept.

There used to be a game show on TV called *Let's Make a Deal* hosted by Monty Hall in which the final portion of the game involved having the contestant choose one of three closed doors (see Figure 2.10), behind one of which was a million dollars, and behind the other two, nothing of value. After the contestant chose a door, the host would always pick one of the remaining two doors and show that it had junk behind it, and then he would give the contestant the option of staying with the door she had chosen or switching to the last remaining door. The question here is, should she stay with the door she initially chose, should she switch to the remaining door, or does it matter?

The intuition of many people, including my own, is that the choice of remaining doors does not matter since both choices possess a one-third probability of being correct. This problem actually confused a good many people educated in mathematics, but it has a straightforward answer. One does not even have to resort to conditional probability formulations to do the calculation. I think we can all agree that if the million dollars has been placed behind a random door, and we have no knowledge of which, then each door has a one-third probability of being the correct one. Therefore, when the contestant chooses a door, they have a one-third chance of it being the winning door. The key to understanding this problem is that the probability that the million dollars is behind one of the remaining two doors is two-thirds as there are two doors, and each has a one-third chance of being correct.

Figure 2.10: The three-doors game, in which there is a million dollars behind one door, and the game's participant must guess which door it is.

The game starts with the three doors shut and the million dollars hiding behind one of them. Assume the contestant chooses door one; it is clear that she has a one-third chance of having selected the correct door. We also know there is a two-thirds chance that the prize is behind the combination of doors two and three; in other words, there is a two-thirds probability associated with two doors as $1/3 + 1/3 = 2/3$. Next, the host opens door three demonstrating the prize is not behind it.

At this point, the contestant is given the opportunity to switch to door two. So should she stand on door one or switch to door two? The probability that the prize is behind door one is, as we have calculated, one-third. Moreover, the probability of the combination of doors two and three is two-thirds. However, the probability it is behind door three is zero, as the host demonstrated by opening that door to reveal the prize was not behind it. Therefore, the probability the prize is behind door two must be the probability of the combination of doors two and three, two-thirds, minus the probability it is behind door three, zero, which happens to be two-thirds. So, the final answer is that the probability of the prize residing behind door one is one-third and the probability it is behind door two is two-thirds. Therefore, she should always switch from door one to door two because it is twice as likely that the million dollars is behind door two. This answer is certainly not intuitively obvious. Let's look at another example.

State lotteries often engender talk among people about the odds of winning. Of course, if there are a million tickets to be sold, and each person only buys one ticket, then the odds of having the winning ticket is one in a million. This result is easily calculated by dividing the number of successful outcomes, one winning ticket, by the total number of outcomes possible, one million. Hence, the odds of acquiring the winning ticket is one in a million. Conversely, if you are a loser, what is the odds of having the ticket you do? It is also one in a million due to the same reasoning. In fact, the probability of getting any one of the tickets is one in a million. We imbue the winning ticket with its significance, but it is just as likely to be chosen as any other ticket.

The only reason the winning ticket is special is that we define it to be so. Aside from that, all ticket numbers are equally likely to be chosen, including the winning number. That the chances are the same for obtaining any one of the million lottery tickets is an essential concept that will show up later. We think that the extant life on Earth is unique—that we are unique—and that life couldn't possibly have shown up in a different form. We imbue this particular form with significance.

As one last example, suppose there is a family with two children, and you are told one of them is a boy. What are the odds that the other is a boy? The immediate answer almost everyone gives is one-half. This answer makes perfect sense as the other child can only be a boy or a girl; one out of two possible choices yields a probability of one-half. But the probability is actually one-third. Seeing this involves counting up all the possible outcomes. It will be easier to accomplish this if we denote possible combinations in the form (x, x) where the first entry is child one, the second entry is child two, and the x's are placeholders for gender. The four possibilities are then (boy, boy), (boy, girl), (girl, boy), and (girl, girl). The confusion comes in because the question does not specify which child is the boy. The number of possible outcomes with at least one boy is three. The number of outcomes with two boys is one. Hence, the probability of the other child being a boy is one divided by three.

We'll talk more about probability in later chapters, but this section is sufficient to give a flavor as to why one cannot just make blind assumptions about the likelihood of specific outcomes.

2.5 Infinite sets

An area of mathematics that can seem strange when people first learn about it is the theory of infinite sets. Most of us are aware of what constitutes a set. The group of numbers $S = \{1, 2, 3, 4, 5\}$ is a set called S that contains the integers one through five. The cardinality, or size of this set, is five. We say that two sets are equal in size if they are of the same cardinality or, saying it another way, there is one-to-one mapping onto one set from the other. For example, consider the set $A = \{a, b, c, d, e\}$. It also has cardinality five. It can be mapped onto the set S as follows: $a \to 1$, $b \to 2$, $c \to 3$, $d \to 4$, $e \to 5$. No element from set A maps to multiple elements of S. Each element of A maps to one and only one element of S, and all the elements of S are mapped. These two sets have a one-to-one correspondence and are thus equal in size.

The set of natural numbers is given by $\mathbb{N} = \{0, 1, 2, 3, \ldots\}$. This sequence extends from zero to infinity. It is, however, in mathematical jargon, termed a *countable* set. Not that anyone could ever sit and count from zero to infinity, but rather because every number in the sequence can be enumerated (or known). Despite it being infeasible to count all the natural numbers, it is conceptually possible to start at zero, go to one, and so on to infinity. The function $n_{k+1} = n_k + 1$ with n starting at zero describes how all of these numbers can be arrived at. Furthermore, since they admit to such a process, the natural numbers are a countable set even though it is impossible to carry it out practically.

The top number line in Figure 2.11 represents all the natural numbers from zero to infinity, $\mathbb{N} = \{0, 1, 2, 3, \ldots\}$, while the bottom line shows only the even natural numbers (including zero), $\{0, 2, 4, 6, 8, \ldots\}$. Which of these sets is larger? One's intuition almost certainly argues that the first set is twice as big as the second, but is this true?

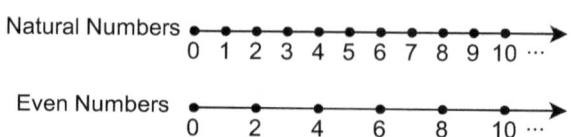

Figure 2.11: One set containing all of the natural numbers, and a second containing the even natural numbers. It appears that the top set is larger than the bottom, but this is not the case.

Let's check to see if there is a one-to-one correspondence between the two sets. To do this, we create a function that relates the two sets and maps only one element of the first to the second. This function is $f(x) = 2x$. We can plug each and every element of the set of natural numbers $\mathbb{N}$ into $f(x)$ and arrive at one and only one of the even numbers in the second set: $f(0) = 0$, $f(1) = 2$, $f(2) = 4$, and so on. For each element of $\mathbb{N}$ there is a corresponding element in the set of even numbers; therefore, they are the same size. The conclusion is the

2.5. INFINITE SETS

set of natural numbers is the same size as the set of even numbers.

This result is certainly not simply intuited by looking at the lists of the elements of each set and going with a gut feeling. It requires a bit of analysis to arrive at such a counterintuitive answer. If the reader does not buy it yet that our intuition is not a good judge of simply counting numbers, let's look at another example that may be even more surprising.

We have seen that the set of natural numbers is infinite in size. What is the size of the *real numbers*? The set of real numbers contains every possible number on the number line, including integers, rational numbers (numbers of the form p/q where p and q are integers), and irrational numbers such as π (these numbers cannot be expressed as a ratio of integers). In other words, the set of real numbers contains all natural numbers and is also infinite in size. However, the real numbers are *uncountable*, which was demonstrated by Georg Cantor in 1874 [12], and means that there is no functional synthesis that can represent them all like was done with the natural numbers. Although both the natural and real numbers are infinite sets, Cantor asked whether there were more real numbers between zero and one than contained within the entire set of natural numbers between zero and infinity. Intuitively one might assume that there is no way this can be the case.

In showing that it is indeed true that there are more real numbers between zero and one than natural numbers from zero to infinity, I'm going to follow Cantor's binary proof. We are all acquainted with decimal numbers that are represented in base ten. For example, 0.5 means the ratio of 5/10 and is equal to $5(1/10) = 0.5$. In the same way, 0.51 is found by $5(1/10) + 1(1/100) = 0.51$, and $0.513 = 5(1/10) + 1(1/100) + 3(1/1000)$. Each digit N positions to the right of the decimal point represents a value equal to that digit divided by 10^N. This is what it means to represent a decimal in base ten.

Decimal numbers can also be represented in base two, or binary. Whereas the digits to the right of the decimal point in base ten had the multiplier sequence $1/10, 1/100, 1/1000, \ldots$, binary decimals go $1/2, 1/4, 1/8, 1/16, \ldots$ with the denominator being 2^N where N is an integer corresponding to the decimal position. For example, $0.101 = 1(1/2) + 0(1/4) + 1(1/8) = (0.625)_d$ where the subscript d means the base-10 representation; that is, the binary decimal 0.101 is equal 0.625 in our conventional base-10 number system.

Now, let's do the same thing we did with the natural numbers above and create a one-to-one correspondence between the natural numbers and the binary decimals as shown in the array (2.4). There is a binary decimal sequence assigned to every subscript of the number s, which are denoted by the natural numbers $0, 1, 2, 3, \ldots$. So, for every natural number, we write a binary decimal that directly corresponds to it, and we have no danger of running out of binary decimals because we can always extend the sequence of zeroes and ones to the right. A binary decimal number has been written for each natural number.

$$s_0 = 0.(0,0,0,0,0,0,0,0,\ldots)$$
$$s_1 = 0.(1,1,1,1,1,1,1,1,\ldots)$$
$$s_2 = 0.(1,0,1,0,1,0,1,0,\ldots)$$
$$s_3 = 0.(0,1,0,1,0,1,0,1,\ldots)$$
$$s_4 = 0.(1,0,0,1,1,0,0,1,\ldots)$$
$$s_5 = 0.(0,1,1,0,0,1,1,0,\ldots)$$
$$s_6 = 0.(0,0,1,1,1,1,0,0,\ldots)$$
$$s_7 = 0.(1,1,0,0,0,0,1,1,\ldots)$$
$$\vdots$$
$$s_N = 0.(a_{N,0}, a_{N,1}, a_{N,2}, a_{N,3}, a_{N,4}, a_{N,5}, a_{N,6}, a_{N,7}, \ldots) \quad (2.4)$$

Let's modify the array as shown in (2.5) by inverting the first bit of the binary decimal corresponding to s_0, doing the same to the second bit of s_1, the third bit of s_2, and so on. Now, construct a new decimal number using these inverted digits in their respective bit positions and call this new number $x = 0.(x_0, x_1, x_2, x_3, x_4, x_5, x_6, x_7, \ldots)$. Note that this number is, by construction, different than any binary decimal in the list of (2.4). But, remember, we mapped every natural number to one of these decimals between zero and one and, yet, we have created a new decimal that does not belong to this set.

$$s_0 = 0.(\bar{0},0,0,0,0,0,0,0,\ldots)$$
$$s_1 = 0.(1,\bar{1},1,1,1,1,1,1,\ldots)$$
$$s_2 = 0.(1,0,\bar{1},0,1,0,1,0,\ldots)$$
$$s_3 = 0.(0,1,0,\bar{1},0,1,0,1,\ldots)$$
$$s_4 = 0.(1,0,0,1,\bar{1},0,0,1,\ldots)$$
$$s_5 = 0.(0,1,1,0,0,\bar{1},1,0,\ldots)$$
$$s_6 = 0.(0,0,1,1,1,1,\bar{0},0,\ldots)$$
$$s_7 = 0.(1,1,0,0,0,0,1,\bar{1},\ldots)$$
$$\vdots$$
$$s_N = 0.(a_{N,0}, a_{N,1}, a_{N,2}, a_{N,3}, a_{N,4}, a_{N,5}, a_{N,6}, a_{N,7}, \ldots) \quad (2.5)$$

This newly constructed number does not lie in the set of binary decimal numbers between zero and one that were constructed, and since every natural number has been mapped to one of these decimals, there must be at least one more decimal that does not correspond to one of the natural numbers. In fact, there is an infinite number of them. We have demonstrated that there are infinitely more real numbers between zero and one than total natural numbers, just as Cantor originally did.

Cantor's result is about as counterintuitive as things get. Couldn't we just add another natural number to account for any surplus of real numbers? The fact is, we tried and, although logic supports the conclusion reached, our intuition does not. It is wishful thinking to go against the reasoning used.

2.6 Exponential growth

Before leaving this chapter, I want to discuss one more topic that some find surprising and challenging to understand. It is explicitly included as another example of how our intuition betrays us and because the concept will be essential to understanding the material in later chapters.

Let's start with an old parable about a king in India around 3,000 BC named King Belkib [13]. The king is bored and challenges anyone in the kingdom to come up with a distraction that will keep him entertained. At this point in the story, the wise man Sissa enters and provides the game of chess he has just invented for the king's amusement. The king is entirely enthralled by the game and offers Sissa a large portion of gold as reward for his ingenuity. Sissa refuses the gold and, instead, asks for rice, but in a very special way.

Sissa asks for one grain of rice to be placed on the first square of the chessboard on the first day, two grains on the second square on the second day, four on the third square on the third day, and so on, until all 64 squares the board be completed. The king thinks that this is quite a deal compared to giving Sissa a large sum of gold, and he quickly agrees.

As the days go by, the king enjoys his game and ponders how a genius like Sissa, who invented such a fantastic game, could be so silly as to accept payment in terms of small quantities of rice.

The quantities arrived at each day are calculated by simple binary (power-of-two) arithmetic. The quantity on the first day is $2^0 = 1$, that of the second day is $2^1 = 2$, and that on the third day $2^2 = 4$. These are paltry sums of rice, and it is no wonder the king remained confident in his decision. This confidence waned on the sixteenth day when the sum of rice was $2^{15} = 32,768$. It ended on the twenty-fifth day when the number of grains of rice due became $2^{24} = 16,777,216$. At this juncture, the king knew he had been taken.

The number of grains of rice Sissa would have received by the 64th day is given by $2^{64} - 1 = 18,446,744,744,073,709,551,615$. According to Sylvain Saurel [13], this is greater than the world rice production in the year 2020.

The outcome of the story above is counterintuitive to many people, including King Belkib. It is called *exponential growth* because it is of the form a^n where $a = 2$ in the story and n is a parameter that is increased each day. In other words, exponential growth occurs when the exponent of such a function is incremented. The growth is hard for our minds to comprehend because it increases so fast, as shown in Figure 2.12. The top plot in the figure shows the first seven steps and is easy to keep up with mentally. The growth appears slow, even though it doubles at each step. When considering only a few steps like this, the behavior of the function makes sense. However, the bottom plot in the figure shows the first 30 steps of the function, and we can no longer conceptualize the magnitudes of the values. The latter steps, those after 25, are growing so fast that all the steps before 25 appear to be zero.

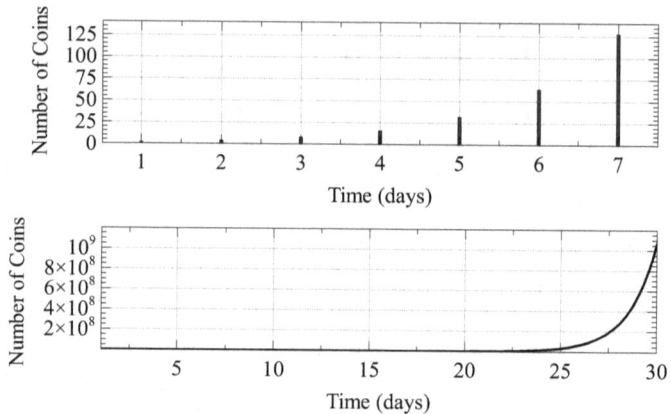

Figure 2.12: Exponential growth of the form 2^n. The top plot shows the first few steps of the growth, and the bottom plot shows the progression over 30 steps. In both cases, each successive step doubles the value of the previous step.

As a second example, consider folding an American dollar bill over and over. The dollar bill has a thickness of about 0.01 cm. If we fold this dollar once, the resulting thickness is $2(0.01 \text{ cm}) = 0.02$ cm. It is easy to see that if we then fold it again, there will be four dollar thicknesses overlapping, yielding a total thickness of $4(0.01 \text{ cm}) = 0.04$ cm. A third time will result in eight overlaps and a total thickness of 0.08 cm. Each time the dollar is folded its thickness doubles. This is exponential growth in the thickness of the dollar.

The Moon is 239,000 miles from Earth which, when translated into centimeters, is 38,240,000,000 cm. How many times would the dollar need to be folded such that the resulting thickness would reach from the Earth to the Moon? This value can be found as follows: $\log_2(38,240,000,000/0.01) = 41.8$. If the bill could be folded just 42 times (which it could not be), the result would be thick enough to reach from here to the Moon. This is an astonishing result.

Of course, a dollar bill could not be folded 42 times. Think about the cross-sectional area of the bill as it is folded; it decreases by 1/2 for each fold. So as the thickness increases as 2^n, its cross-sectional size decreases exponentially as $(1/2)^n$. Therefore, by the time the bill is at its 41st penultimate fold, its width would be $1/(4.55 \times 10^{13})$ of its original size. No fold could possibly be made with only this width to work with. However, the result of the thought experiment still holds.

This phenomenon is the same one that produces monetary growth with compounded interest. As an example, if you were to invest $1,000 at a return of 8% for ten years, the result would be $(\$1,000)(1.08)^{10} = \$2,159$ dollars. Furthermore, letting the investment grow for 40 years will result in a total of $21,725. With this knowledge, it is easy to see how compounding interest is a bit of a cheat that probably should not exist in the financial world. The reason is that the time over which one invests or borrows is linear, whereas the profits made or expenses incurred are exponential. This is hardly fair.

It is the fact that exponential growth is nonlinear that tricks our minds up. Our minds understand linear operations well, however, they have extreme problems visualizing nonlinear functions.

2.7 Conclusion

Our intuition in deciphering design or the outcomes of physical processes is only accurate when considering those instances where our thinking process has been explicitly trained to recognize them. The reason a child assumes a paper origami swan has been designed is that is her experience. Cars are designed, iPhones are designed, birthday cakes are designed, video games are designed, and the logical extrapolation is that paper origami animals are also designed. There is no innate god-like magic being invoked, and this is a perfect example of the problem with divinity-related speculations about reality. If God did create this Universe, and you are empowered to spot his handiwork, you need to be god-like enough to recognize such handiwork, and you're not. I am playing loosely with the phraseology here, but I think it comes across very clear what I mean: there is no god-like intuition granted to you in the form of a magical common sense. If that were the case, this would indeed be a very dull Universe, devoid of awe. Instead, here you find your own way, and no imaginary incarnations are going to help. Rather they will only hinder.

The mental machinery of humans tends to operate as if the world functions linearly. We are used to concepts such as two people weigh twice as much as one person, traveling to a place twice as far away takes twice as long, and so on. And this calibration works well at the speeds and distances that humans evolved to live within. It is when this space must be extrapolated that our built-in computational acumen starts to fail.

Handling nonlinear problems is so difficult for our conceptual mind that the body itself has evolved to isolate us from having to do so. For example, we think sounds that are twice as intense sound twice as loud to our ears. That is not the case. The biological circuits in our low-level hearing hardware actually perform a logarithm on the sounds around us to allow us to hear a wider range of sound levels. We do not even realize this is happening. A very similar thing occurs with our eyesight. Our minds automatically re-reference the ambient, background light level of the world around us so as to enhance our visual acuity. There is a vast amount of processing of our sensory perceptions that occurs before we ever have conscious access to the data, and even then, we can never see the raw data.

Stacked on top of the sensory preprocessing is a layer of pattern matching that developed during our evolution. This built-in processing is what makes you jump unconsciously away from a coiled rope in low-light conditions thinking it is a snake. Although very useful to keep from being bitten thousands of years ago, today, it is misleading and unnecessary for the most part. Our pattern-matching hardware leads us in many instances to see patterns that are not there.

And finally, there is a layer of assumptions that have been built into us through eons of sociological evolution. A simple example of this is the culinary differences observed across different parts of the globe. Food items that will make Americans puke are considered a delicacy in China or India, and vice-versa. This psychological training is so ingrained within us that we generally are not even aware of it.

The purpose of science and mathematics is to get around these personal and biological biases to see an accurate picture of reality. Claims that we can trust our intuition to come to a correct model of reality are simply ridiculous. We tried that for many centuries before the intellectual enlightenment, and it does not work well at all. One of the objectives of this book is to dispel the belief that we can trust our senses and intuition to arrive at factual conclusions about the Universe. Following that line of reasoning led us to the Spanish Inquisition, slavery, and the trying of purported witches.

The next chapter will introduce concepts in physics and mathematics that will be needed throughout the remainder of the book to provide a critical assessment of both ID and the mechanics of evolutionary biology. As we've seen here, it can be a difficult task to separate objectivism from our need to make intuitive judgments. Furthermore, since ID has an ulterior motive to favor preference over logic, digging into these subjects is made even more complicated. Nonetheless, it is going to be an interesting journey.

Chapter 3

Physical systems and mathematics

> "Theories are nets cast to catch what we call 'the world': to rationalize, to explain, and to master it. We endeavor to make the mesh ever finer and finer.", Karl Popper, *The Logic of Scientific Discovery*.

> "Another thing I must point out is that you cannot prove a vague theory wrong.", Richard P. Feynman.

This chapter covers a broad range of subjects within physics, mathematics, and dynamic systems to introduce the reader to the physical and mathematical tools necessary to understand the material in the following chapters. The central thesis of this book hinges upon the concepts of randomness, chaos, optimization, and cellular automata, and this chapter presents the background necessary to understand those concepts. Each will be covered in detail, and examples of their appearance in nature provided.

In addition, the concept of a theory or scientific model is introduced and defined in this chapter. Knowledge of what a theory is will be very important later in determining whether ID is valid science. The concept of a theory is given first, followed by practical examples of how the concept is used.

In addition, a few ancillary topics will be covered as well, including Bell's theorem as it relates to indeterminism, the Anthropic Principle, and the limits of reductionism.

It is essential to understand the overall objective of science and how it uses theories to build a complete and consistent, interdisciplinary framework. Science is not, as it is usually portrayed in Hollywood movies, a haphazard set of guesses based on one or two data points. Rather, it is a collection of observed behaviors of the Universe woven together to provide an internally consistent description of physical reality. Think about it this way. A sharpshooter could explain to a student that hitting the target is a simple matter of putting a bullet through the bullseye. Although this description is correct, it is certainly not complete. In fact, if someone claimed to be an expert marksman and this was the only bit of information she could provide about the sport and how it was performed, one would have a hard time believing that she was indeed an expert.

On the other hand, if she could provide a detailed description of sighting in the gun's scope, adjusting the crosshair to take into account lateral wind velocity and distance from the target, knew how to calculate muzzle velocity, the fall of the bullet, air friction drag on the bullet, etc., one might be more inclined to believe she was indeed a marksman because the entire picture would become more consistent. Muzzle velocity has to correspond with travel distance based on air drag and angle of the weapon, and the crosshair has to be matched up as well. If all of these variables are not suitably accounted for, successfully hitting the target is not likely. All these parameters must work together for good marksmanship. The same is true for a scientific theory.

For example, claiming that reality is based on the rules of quantum mechanics alone might not be very convincing. But when quantum mechanics is brought to bear to successfully explain the dual-slit experiment, diffraction, spectral lines, the operation of semiconductor junctions, and an almost uncountable number of other investigations, it becomes very compelling. A theory capable of explaining multiple physical phenomena carries much more weight because its validity can fail in numerous ways. In other words, the more parameters and behaviors a theory accounts for, the more confidence we have in it precisely because there are more ways it can fail.

The field of science as a whole is reinforced when interdisciplinary theories corroborate one another. For example, the foundations of chemistry can theoretically be derived from quantum mechanics. Only with consistency across and between both fields could this be the case.

In the chapters to come, many of the ideas presented here will be brought to bear, hopefully in a coherent manner, to explain the basis of biological complexity and diversity.

3.1 What is a theory?

The colloquial meaning of a theory is any hypothesis for the cause or explanation of an event or process. It could even be just an educated guess. For example, while driving home, your car starts making a thump-thump sound and loses power, forcing you to pull to the side of the road. You know a little bit about engines and that they have a crankshaft that rotates when the engine is run-

3.1. WHAT IS A THEORY?

ning, and you suspect that there is a problem with the crankshaft. In common parlance, this type of shallow inference is taken as a theory.

Science does not use the same definition for the word theory. In science, a theory is a model or framework that consistently links a set of observed facts to explain a feature of or a process within reality. It may contain only one or a few relational or explanatory statements such as Newton's laws of motion, or it may include many mathematical relations such as contemporary electromagnetic theory. There are many such theories within science, but the one thing that they all have in common is consistently explaining a wide variety of experimental data using a minimal set of statements. Note that a law is simply a theory that has withstood the test of time and experimentation.

As they are very simple, let's look at Newton's laws of motion for a moment. There are only three, and the first states that an object remains at rest or moves in a straight line unless acted upon by a net force. Second, the acceleration of an object is directly proportional to the net force applied to it and inversely proportional to its mass. Third, for every action (or force), there is an equal and opposite reaction (force). These three statements represent the totality of Newton's laws of motion.

The first law is self-evident: objects remain still unless they are acted upon. A ball placed in the middle of a flat kitchen floor will sit there motionless until a child kicks it. An object not subjected to the force of gravity will remain suspended motionless in space, as has been demonstrated many times by astronauts. Simply put, inanimate masses do not move on their own but do so only if a force acts upon them.

I am going to look at Newton's third law before coming back to the second. We experience the third law every day of our lives; we just don't usually think about it. The easiest way to understand it is by example. Maybe at some point in the past, you've pushed on a car that was stuck in slick snow in an attempt to help the unfortunate driver. If you had sound footing, such as on a dry portion of the asphalt road, you could push very hard on the car. However, if you were standing on ice, there was no way to exert force on the vehicle without simply falling. The reason for this is because any force you exerted against the car was *exerted back upon you*; you pushed the car, and it pushed back on you. This fact may not be apparent when standing on dry pavement, but it is painfully clear when you're standing on ice. The reaction force from the car is what pushes you off your feet.

Another way of looking at this example is using Newton's first law. When you push hard on the car, and it does not move, the first law says that since it remains stationary, the net force acting upon it must be zero. In other words, there is a force canceling out precisely the force you are applying with your hands. The force the car applies to your hands is the reaction force mentioned in the third law.

Whereas the first and third laws lay the groundwork for why objects in the Universe move, the second law describes the dynamics of this motion. The former two laws describe when and why objects move, and the second law describes how they move. The second law says if a force F is applied to a mass M, it will move with an acceleration of $a = F/M$. Due to the division on the right, the larger the force or, the smaller the mass, the faster the mass is accelerated. A car with a high-performance engine takes off much faster than one with a simple four-cylinder engine due to the larger force (torque) that the former's engine can supply to the wheels. On the other hand, an empty truck accelerates faster than one that is pulling a camper behind it. This is because the camper adds to the mass the truck must move, reducing the overall acceleration of the combination (mass is in the denominator of the equation above).

Do not assume Newton's laws are overly limited just because the statements that form them are simple. The second law describing the dynamics of moving bodies opens the door for mathematics to enter the picture. Understanding that acceleration is the time-rate-of-change of speed, the second law can be rewritten as

$$a = \frac{dv}{dt} = \frac{F}{M}. \tag{3.1}$$

Integrating this equation, the speed is $v = at + C_1$, where t is time and C_1 is a constant (it is zero if the object was initially at rest). This result says that the speed increases with time and a constant of proportionality of a. In other words, starting from a standstill, in one second, the velocity of the mass is $v = a$, in two seconds, it is $v = 2a$, and so on.

Understanding that velocity is the time-rate-of-change of distance as given by

$$v = \frac{dx}{dt} = at + C_1, \tag{3.2}$$

the distance the object travels can be found by integrating. This distance is $x = (1/2)at^2 + C_1 t + C_2$, where C_2 is another constant of integration. This equation says that the distance traveled under constant acceleration is proportional to time squared. That is, assuming the object started at zero speed and at $x = 0$ (so the constants C_1 and C_2 are zero), the distance the object travels in one second is $x = (1/2)a$. In two seconds it moves $x = 2a$. In three seconds it goes a distance of $x = (9/2)a$. And so on.

My point of going through the above mathematics was not just to put the reader to sleep. It was to show the complexity and consistency of typical scientific theories. And, remember, Newton's laws are one of the simplest of them, but using these laws, all the sophisticated motions of bullets, space capsules, baseballs, etc., can be unraveled, predicted, and understood. Once mathematics is linked in, consistency and predictive derivation from the original equations are possible. All of this has been tested over and over again throughout the past centuries and shown to hold. These traits are the hallmarks of a scientific theory.

Another way of saying that theories should be minimalistic and internally

3.1. WHAT IS A THEORY?

consistent is to say that they should be axiomatic. Newton's laws are repeated here for convenience:

1. All objects moving at a constant velocity continue to move at that velocity unless acted upon by a force.

2. The acceleration of an object is proportional to the net force acting upon it divided by its mass.

3. For every action there is an equal and opposite reaction.

As Karl Popper reasoned, a theoretical system can be called axiomatic if the set of statements that characterize it meet specific requirements [14]. The first of these requirements is that the statements must be consistent with one another; they cannot contradict one another or be self-contradictory. The three statements comprising Newton's laws of motion meet this criterion. The second requirement is that the statements be independent. In other words, there is no member of the set that can be derived from the others, meaning the set is minimalistic. The first law ensures that the system we are talking about resides within an inertial reference frame (Newton's laws do not apply in noninertial frames.) The second states how massive objects move when they are subjected to a net force. When a force is applied to a mass, it accelerates in the same direction as that force. The third statement sets a conservative constraint upon the boundary between objects. If one object exerts a force upon a second, they each must experience this force in opposite directions.

Finally, the statements that define a theory must be necessary and sufficient to effect the consequences of the theory. That is, each statement must be required to express the theory. None of the statements should be superfluous. In addition, the statements must be capable of delineating the functioning of the theory.

A second, less formal way to define a theory is that it is the best explanation of the data at hand. However, a theory is *not always the exact explanation of a phenomenon*. Newton's laws of motion are a perfect example of this. At extremely high speeds, these laws break down, and Einstein's theory of relativity takes over. When dealing with the extremely small, the statistical processes of quantum mechanics take over. However, in many situations, Newton's laws perform exceedingly well in describing the dynamics of massive objects.

When better explanations are discovered, they replace the older theories, such as relativity replacing Newton's laws. However, this does not mean that Newton was wrong. Furthermore, it does not mean that all the current theories, which will likely be replaced in the future, are necessarily incorrect. They are the best explanations of the observational data we have at present, and they perform better than any alternative in predicting nature's behavior, or else they would have been rejected in lieu of a better theory. In this way, science keeps evolving toward better explanations of the functioning of reality, but it does not make truth proclamations.

Theories typically are reductionist frameworks, meaning that they are (possibly) approximations (the idea of reductionism is taken up later in this chapter). Newton's laws are reductionist in nature because they encapsulate the behavior of all massive bodies moving within inertial frames in three succinct statements. The statements do not describe how each body as an individual behaves; rather, they describe in general how all massive bodies move. In other words, Newton's laws say nothing explicit about a specific object, but, instead, act as a template to model all massive bodies. Nevertheless, by manipulating the equations' initial conditions and the gravitational constant, the equations can be made to model the behavior of a specific mass. We will see a little later some processes that do not readily admit to reductionism.

So, we have seen what a good theory is. Let's take a look at what a not-so-good theory might look like.

In the mid-nineteenth century, a fundamental question concerning the propagation of light through a vacuum presented itself. James Clerk Maxwell was busy formulating the very famous, so-called Maxwell's equations that describe the physics of electromagnetic waves. Scientists understood that light was an electromagnetic wave that, unlike sound or water waves, could propagate through the emptiness of space. Maxwell had even developed a wave equation virtually identical to the wave equation that defines the motion of waves in fluids. These developments led to the crucial question that if other waves need a medium to travel, be it air, water, or whatever, how can it be that a light wave can travel in a vacuum where there appears to be no medium?

The consensus at the time was that some kind of medium did exist—they called it the *luminiferous aether*—that, for some reason, was incredibly difficult to detect. The question consumed many great scientists who developed numerous novel means by which to potentially answer it. The Michelson-Morley experiment, with Albert A. Michelson and Edward W. Morley being the two scientists from which the namesake is derived, is the most famous of the experiments performed during this time with the sole intent of finding this elusive aether.

If the aether pervaded the whole of the Universe, as it must if it is required for light to travel from far distant stars, then the Earth must be moving through it, and the Michelson-Morley experiment intended to measure the relative speed between it and the Earth. The Earth travels around the sun at a speed of 30 km/s. It would travel with the aether through half of the journey and against it during the other half. If light traveled like any other wave in a medium, then when the Earth was moving with the aether, the measured speed of light would be less than expected, and when the Earth was traveling against the aether, the apparent speed of light would be higher.

Conducting this experiment relied on coming up with an accurate means of measuring the speed of light. Once that was possible, two measurements would be made: one while the Earth was traveling in one direction, and a second while the Earth was traveling in the opposite direction. If a significant difference were to be found between these two measurements, it would be a strong indicator that

3.1. WHAT IS A THEORY?

the aether was real.

Michelson designed a two-path interferometer to precisely measure the speed of light. The device split a light beam using a half-silvered mirror, sending the split light down two paths at right angles to one another, and then brought them back together, allowing them to mix and create an interference pattern the operator could examine. Shifting of the lines that made up the pattern would indicate that the time it took for the light beams to complete their respective circuits was changing. It wouldn't work if both paths were affected identically; in that case, there would be no shift in the pattern lines. The device was designed to float on a pool of liquid mercury so the entire assembly could be rotated, allowing the measurement angle to be changed.

Michelson and Morley collected data from the experimental setup in the summer of 1887, and the results showed no shift in the fringe pattern (within the measurement error of the device), demonstrating conclusively that the aether did not exist. This significant result ultimately led to the correct understanding of light propagation developed later by Lorentz and Einstein.

I didn't bring this 19th century question and the hypothesis of the aether up to say that the hypothesis was bad. Neither was Michelson and Morley's experiment a bad example of a scientific attempt to answer a question. It *did* answer the question. The Michelson-Morley experiment is a real and illustrative example of how science values a correct explanation of reality above all else.

The scientists living 150 years ago had a strong inclination that there was no explanation for light propagation other than the presence of some kind of medium through which it traveled. Honestly, most scientists at the time assumed the converse was not only impossible but indeed ludicrous. How could any wave travel through *nothing*? But when evidence found this to be the case, it was accepted, and from that point on, the development of electromagnetic theory went in a different direction.

Now, think about what would have happened if, instead of accepting the new findings, these scientists had stuck with their preconceived notion that every propagating wave must have a medium in which to travel and had refused to accept the new evidence, asserting that it was impossible and that the only reasonable conclusion was that Michelson's interferometer had had a problem. Or, worse yet, they claimed that the aether existed, but its existence was impossible to confirm.

It was indeed a possibility that the scientists could have come to one or both of the latter conclusions. However, these incorrect assumptions would have been corrected as time went on and additional experiments were performed, including those using different means to make the measurement, and the aether would have been shown not to exist. Going with one's intuition in opposition to physical evidence is dangerous. If the aether were accepted as a part of a theory that was simply untestable and, thus, unfalsifiable, there would be no guarantee that the theory correctly modeled reality. All portions of a theory must be testable and falsifiable and, if data supports the theory to be false, it must be rejected. This is how science separates imagination from reality, and it is how theories continue to zero in on the truth of reality, always getting asymptotically closer.

Theoretical development relies on the testing of hypotheses. For instance, Einstein's theory of relativity predicts that the space-time around massive bodies is curved toward the mass. The bending of space-time was predicted by the mathematics of general relativity, but it needed to be verified, and that was done by Arthur Stanley Eddington in 1919 during a complete solar eclipse as he observed a shift in the positions of the stars near the blackened disk of the Sun. Of course, he was seeing the light from stars that were much farther away than the Sun, and they appeared to shift position because the light rays emanating from them and passing by the Sun to enter Eddington's eyepiece were being bent. The shift in the stars' apparent positions allowed the amount of bending to be calculated, and it matched very closely to what the theory of relativity predicted.

If there are aspects of a theory that cannot be verified, either directly or indirectly, through experimental testing, then there is no way of confirming that the theory accurately reflects reality. For example, one could theorize that there is an entity that stands outside of space and time, which is unseen and cannot be measured, that moves all of the particles in the universe appropriately such that they follow the known laws of physics. There is no way this hypothesis can be refuted, but there is also no way that it can be proven or demonstrated to be true. These kinds of claims can and have led to never-ending debates, especially regarding religious assertions about reality, but they serve no useful purpose in the realm of science. That they cannot be tested is another way of saying that they have no impact on the physical world, and therefore do not matter.

I am not ruling out the existence of meaningful, personal experiences; I am sure they do happen. However, they have no place in science, at least not until they can be measured. Science not only creates a consistent, understandable view of the physical world in which we reside, it also acts as a means of communication and collaboration between all people. All scientific theories are written in a language that is understandable by everyone. They also rely heavily upon mathematics (it seems that reality's operation is written in the language of mathematics) so their concepts can be written in an entirely consistent fashion that others can easily understand. Duplication of testing by others is also facilitated. For example, I don't believe Newton's laws or Maxwell's equations because I was dogmatically taught them at university. I don't have to; I know they are true because I have done the experiments myself. It is challenging to develop an optimal system for controlling a permanent-magnet synchronous motor without, in the process, verifying that Newton's second law is indeed correct. Similarly, designing a 24 GHz doppler radar relies heavily on Maxwell's equations and electromagnetic theory. These theories are verified every day by engineers and physicists making use of them. Science is not a dogma. It is open for everyone to verify independently. Newton's equations work the same in your apartment bedroom as they do in the Everett building at the University of Illinois.

3.1. WHAT IS A THEORY?

I want to consider one last important item that often comes up in the bandying about of hypotheses: the argument from ignorance. It generally involves a clause like "well, what else could it be?". For example, you walk up to a street card magician who asks you to pick a card, remember it, and put it back in the deck. He shuffles the deck for additional theatricality, holds it in one hand as he waves his other hand over it, and whispers the platitude abracadabra. He then tells you to check your jacket's breast pocket, and, low and behold, there your card is, hidden in your coat. How could this have been done? The correct answer is that you simply don't know. However, in situations like this, people tend to fall back on the rationalization that since it seems impossible, it must be magic.

I was a little bit simplistic in the above card story. Most people have seen plenty of card tricks like this on TV and know they are done by legerdemain and subterfuge and, after being part of such a card trick, are not convinced they just witnessed an act of the supernatural. But you might be amazed at just how often this kind of thing has happened in the history of humans trying to understand the world around them. The most prominent example in the societies of today is due to the Universe not having yet been explained completely. Couple our lack of knowledge of how the Universe began with the tendency of human beings to believe they are the pinnacle of intelligence, and the emergence of a belief in a god or intelligent entity that created the Universe is almost a certainty. As a second example and a prelude to the content of later chapters of this book, we know that the best minds in molecular biology, algorithmic design, physics, etc., cannot explain how abiogenesis occurred or exactly how observed biological complexity came to be. This leads some to the conclusion that it must have been designed. This is the argument from ignorance fallacy. Just because we can't yet fully explain how natural complexity came about does not mean we are justified in invoking a supernatural explanation.

Scientific theories are generally constructed from a set of positive assertions based on experimental evidence. They can make predictions. For example, Newton's law of gravitation is based on one experimental setup that measures the force that massive bodies exert on one another. Once this force's existence has been verified repeatedly, the original theory graduates to a law that can be used to make reliable predictions about the functioning of the world around us, and, in addition, it can be verified by anyone anywhere.

On the other hand, consider a negative assertion such as that by ID advocates, which states there is no way nature could have created the complexity apparent in the biological entities observed, and therefore they must have been designed. The only way to confirm that an intelligent designer is required to generate complex life is to literally show that there is no other way this complexity could have been generated. Painfully, there are a lot of different candidate ways that would need to be shown as inadequate. Turning ID into a positive assertion would require showing that an intelligent designer exists and indeed did engineer the biological complexity we observe. I do not foresee this happening, however.

To show this more clearly, consider that you know that there is an interstate between Decatur, IL and Springfield, IL, and you suspect it is IL-72. What is the quickest way to demonstrate the truth of the claim that IL-72 connects these two cities? That would be to merge your vehicle onto IL-72 at the edge of Decatur and drive to Springfield. But there is another way of determining the answer provided that all cities are connected by at least one interstate, and that is to drive the length of every interstate highway in the set excluding IL-72 between Decatur and Springfield and show that they do not connect the two cities. (I am assuming only interstate highways are considered, and only one is used to connect each pair of cities—no jumping between interstates is allowed.) In this case, it is actually possible to drive all these interstate highways. However, this approach does not work in general because the number of possibilities can approach infinity. This approach is called an exhaustive search because it exhausts all possibilities to demonstrate the contrary.

In physics, generally, a set of data tends toward a possible explanation. The approach of looking for a complete set of data that would rule out a possible theory is not typically employed. I need to clarify that I am not talking about disproving a specific aspect of a theory. The Michelson-Morley experiment was very appropriate to disprove the existence of an aether by exposing a contradiction. It was directed at a very specific point of a theory. This same logic should not be used in ruling out an entire space of theories. Again, the thesis of ID is that there is no natural theory at all that could explain the complexity of life as we know it. This is an entirely different thing.

Scientific theories are inferences (generally) based on data collected from real-world experiments. Their constituents must be falsifiable, admit to testing, and be expressible in a way that is understood by others. If this is not the case, they degenerate into mere dogmatic speculation. Science and mathematics and their implications are available to everyone. There is no central figure that dictates which theories or theorems are good and which are bad. And although this aspect may cause consternation from some who would rather their views supersede the truth of reality, no matter how painful the realization, the truth is always preferable to delusion.

3.2 The difference between faith and evidence

One of the most often used arguments for creationism or ID being on the same footing as scientific investigation and theorization is the erroneous idea that both systems have a reliance on one kind of faith or another. The hackneyed expression used by religious individuals to assert that they "do not have enough faith to believe in evolution" is a perfect example of this belief. The claim generally arises from a misguided understanding of what constitutes evidence and the assumption that the Bible or other book provides accurate information about reality whether or not it contradicts direct observation.

Before going further, a working definition of faith is necessary. The word faith has many different definitions, the simplest of which means confidence. If

3.2. THE DIFFERENCE BETWEEN FAITH AND EVIDENCE

one has faith in a particular endeavor being successful, this is nothing more than an expression of confidence that the event will occur. Another definition from a standard dictionary defines faith as a strong belief in a supernatural power or powers that control human destiny. Of course, the word faith can also refer to an institutionalized set of prescriptions, such as the Catholic faith.

However, the definition of faith I want to make use of in this book is the one that explains how the religious community approaches the concept of evidence. The book of Hebrews from the Bible expresses it as "being sure of what we hope for and certain of what we do not see" (Hebrews 11:1) [15]. Commenting on this verse, Mary Fairchild gives a succinct definition of what it means to have faith in the Biblical sense [16]:

> What do we hope for? We hope that God is trustworthy and honors his promises. We can be sure that his promises of salvation, eternal life, and a resurrected body will be ours someday based on who God is.
> The second part of this definition acknowledges our problem: God is invisible. We can't see heaven either. Eternal life, which begins with our individual salvation here on earth, is also something we do not see, but our faith in God makes us certain of these things. Again, we count not on scientific, tangible proof but on the absolute reliability of God's character.

The definition above relates very well to the inherent problem of using faith to discern the functioning of reality: it is a top-down approach with a dogmatic book at the top instead of direct observation followed by reasoning. It is almost impossible to form objective conclusions about observations within the world when one is bound to a definition like the one above. And don't get me wrong; I think most religious people reinterpret the words in ways to get around following the pronouncement. That is, these people are more reasonable than the text of the Bible. Even still, there are bound to be some remnants of such ideas left that affect one's view of the world; how could this not be the case? I have seen it in my own life growing up in a Christian community.

Creationism pushes this to an extreme by outright asserting that nature and history must follow the interpretations of a holy book. In other words, the book takes precedence over our own observations of reality. If the book states that the world was created in six days, 6,000 years ago, then, despite the mountains of physical evidence to the contrary, it must be so. Forced conclusions such as this are, to one degree or another, inevitable if one must adhere to proclamations in a book.

Science, on the other hand, is a bottom-up approach. It forms conclusions (that are unknown at the outset) from observed data. Reality is the ultimate source of what is and isn't, not a revelation in an ancient book. For example, if the evidence points toward evolution as the reason for the biological diversity we see in the world, then so be it. I would argue that if people thousands of years ago had been privy to the experimental evidence we have today, that they too would have come to the same conclusion in lieu of inventing supernatural powers. Supernatural entities stem from societies trying to understand processes beyond their ability to understand at the time instead of just admitting they don't know. Based on their lack of knowledge, we could even classify their suppositions as possible theories for the time. However, it is well past time for these theories to have been supplanted. Using faith as a means to truth completely stymies the advancement of our knowledge. It, by definition, sets up a roadblock that can't be bypassed without throwing out the dogmatic religious framework with which it is associated.

Many advocates of a creator point out that the laws of logic and rationality, which science uses to reach its conclusions, cannot be verified as the verification would require the laws of logic, which leads to circular reasoning. And this is true. Philosopher William Lane Craig claims that without god, the laws of logic would be merely human conventions [17]. How would one go about backing up such a position? I have not the slightest idea. But, in a sense, he is correct—just leave out the adverb "merely". From the perspective of human beings, the laws of logic are conventions that have, so far, been shown to always work; that is, they are consistent with our observations of reality. As many others have pointed out, there exists no means of proving their validity explicitly. Instead, we rely on implicit verification, but verification nonetheless.

Let's look at a little bit of math for a moment. The equation $x + 2 = 5$ has an explicit solution that can be found simply by subtracting 2 from both the left- and right-hand sides to get $x = 3$. Then again, the equation $(1/2)x = \sin(x)$ does not have an explicit solution—there is no way to manipulate this equation into a form that only has x on the left-hand side. A solution can only be verified implicitly as the value of x that makes one side equal to the other. This equation actually has three solutions at $x = 0$ and $x = \pm 1.8954$. When plugged into the original equation, these values implicitly solve it.

The solutions to some mathematical and engineering problems cannot be proven. They can be demonstrated, however. We cannot at this time prove that gravity should behave the way it does; we can only demonstrate that it does in every case. Should we throw out Newton's laws and relativity based on this fact, or sit around and opine that we do not have a theory of everything? This would be silliness. It is not necessary to be able to prove the laws of logic—there is a good chance that it cannot be done anyway. It is enough to repetitively demonstrate that these laws mesh with the functioning of reality. Throwing out logic because there is no provable basis for its consistency will not be the end of it. The next step is hard solipsism. There is no way to prove conclusively that we truly share a reality. Do we throw it all out at that point? What is more, conjuring a god into existence will not resolve this issue either. There simply is no concrete proof against hard solipsism.

Furthermore, the laws of logic cannot be explicitly peeled away from reality and verified as though they were an isolated piece. We know they are valid because their use continually provides consistent results within reality. Adding a god to the equation does not improve this situation but rather makes it more convoluted. We still would have no way of explicitly demonstrating the validity

3.2. THE DIFFERENCE BETWEEN FAITH AND EVIDENCE

of the laws of logic and would be required to point to a god and state that we take it on faith. This is silly. The only reason we continue to use them is that they work, not because a group of people claims a god upholds them. This god would be irrelevant. I could just as easily claim any other reason for the consistency of the laws, and it would have the same effect.

Once we have the laws of logic firmly in hand, we can describe and explain physical processes in reality in a consistent and predictive manner that is transparent to others. Even ID proponents are attempting to take this route in providing explanations that are grounded in the sciences. I am not saying that they are successful theories—in fact, I am going to claim just the opposite—but at least an effort is being made to approach the ID problem scientifically.

If faith, as I have defined it, leads one to truth, then it is by accident. It is a terrible way to come to conclusions about the functioning of reality. And if indeed it comes to a correct decision about a fact, this conclusion must necessarily agree with science at some point, as there can be only one truth of physical reality.

As I clarify elsewhere in this book, scientists, physicists, engineers, etc., do not take what they learn in textbooks on faith, and I believe all of them would be indignant about such a claim. I do not just believe Maxwell's equations are correct and apply them like a recipe. Each design or construction that requires their use demands the designer understand them and, implicitly, through their use, their validity and consistency are confirmed. Along the same line, I do not just take Newton's word for it that force equals the mass of an object multiplied by its acceleration. As a student, I confirmed it in the lab, and as a professional, I verify it implicitly every day.

Faith cannot be verified in the same way as scientific claims; in fact, faith cannot be verified at all. Science can demonstrate that physical aspects of reality are accurate and, if faith agrees, then faith is also confirmed, but it is not justified. What gets us to correct answers—what always gets us to the correct answers—is not faith; it is reason and science or a similar logical process for decoding the world around us.

As Ricky Gervais said on the Stephen Colbert show when he was accused of having faith in science (I am paraphrasing): if we burnt all of the holy books and the science books today, the religious books would be gone for good, never to come back, but all the content within the science books would reemerge after some time exactly as they were before [18].

The reemergence of all of the scientific material would happen because there is a single truth to reality. It does not matter when or where one makes observations of the Universe to derive the laws of nature, these laws will always be found to be the same. If two university professors half a world apart each write a textbook on quantum mechanics, for example, the wording and thoughts on the subject may be slightly different, but the mathematical description of the functioning of reality will be exactly the same despite the two individuals never having collaborated. This is not the case for religious books. Many different holy books across the globe have come into existence over time, and no two of them ever agree. This is very telling.

Science, at its core, is really the search for a singular truth that spans the experiences of all sentient creatures. On the contrary, religious experience is personal and, as evidenced by the many religions worldwide, has many truths. Science is unifying, whereas religion tends to be divisive. Religious affiliations are exclusive, whereas scientific discovery and understanding are all-inclusive.

3.3 How does science model reality?

As mentioned above, a scientific theory is not only a collection of experimental data. Instead, the purpose of a theory is to explain data in a coherent, complete way. So, what does this process actually look like? In this section, several examples of modeling physical processes will be presented to give the reader a feel of what it means to model physical reality.

The main objective in constructing a theory is to find rules or patterns that relate physical variables to one another. All scientific theories or laws are descriptive in nature. They do not explain why nature behaves the way it does. They only provide descriptions of how it behaves. When heat is applied to a mass of water, its temperature increases; for example, consider injecting 1,000 Joules (J) of energy into a bucket of water, causing it to climb one degree Celsius (C) in temperature. If, instead, 2,000 J of energy is provided to the same mass of water, its temperature will increase by 2 C. Similarly, a 3,000 J heat injection will lead to a 3 C climb in temperature. Converting these observations to a descriptive rule, the change in temperature of the water can be written as $\Delta T_w = k\Delta E/\Delta m_w$. In other words, the change in the temperature of mass Δm_w of water is directly proportional, with constant k, to the amount of energy, ΔE, the water absorbs.

It is essential to realize that the theory just constructed does not say anything about *why* the water changes temperature when heat is applied; it simply says that the water does change temperature. Physics is entirely silent when it comes to questions of why things happen and, instead, provides only descriptions of what does happen.

This does not mean that the story ends with one simple descriptive law. Scientific investigation and the subsequent generation of models are recursive; the further we investigate, the more levels we find. Take the above example of heating water. Temperature is actually a parameter related to the motion of the molecules within the water. As more energy is added to the water, it has to go somewhere. It is actually transformed to kinetic energy in the water molecules—the more energy added, the more agitated and violent the motion of these molecules becomes. When you burn your hand with hot water, the pain you feel is a consequence of the high-speed, erratic molecules of water transferring some of their kinetic energy to the molecules in the skin of your hand. The temperature of the water is a measure of the average kinetic energy of the molecules it contains.

3.3. HOW DOES SCIENCE MODEL REALITY?

Newton's law of gravity describes how objects composed of matter interact with one another in this Universe, specifically describing the force they exert upon one another. As obvious as it seems to all of us today that all massive bodies produce a gravitational field, it certainly was not apparent to those of past centuries. When Isaac Newton conceived the idea in the 18th century that the force that causes a cannonball to fall here on Earth is the same force that maintains the motions of the planets and Earth's moon, it was a wholly radical idea. It is hard for us living in modern times to understand the magnitude of this realization. In a real sense, and Newton was a religious man, this discovery was akin to finding God. Not the Biblical God, but rather the architect of the laws of motion within the Universe.

If we consider two point masses m_1 and m_2, separated by a distance r, as shown in Figure 3.1, Newton's universal law of gravity defines the force between them as

$$F_g = -G\frac{m_1 m_2}{r^2}. \tag{3.3}$$

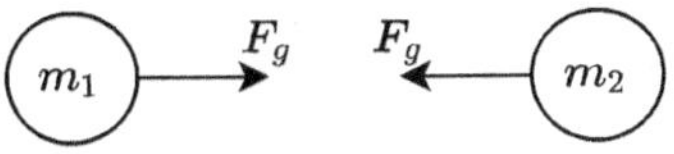

Figure 3.1: Two massive objects of masses m_1 and m_2, and the force between them, F_g.

It is required that the masses be point masses or possess the required symmetry, but let's leave those details aside. The way Newton arrived at (3.3) was taking data on the gravitational pull on massive objects at the surface of the earth. Then using the rough approximation that the density of the earth was five or six times that of water and having an estimate of the diameter of the earth, he was able to arrive at an estimate of the constant G of 6.7×10^{-11} m^{3}kg^{-1}s^{-2}. (The most accurate value in 2018 was 6.67430×10^{-11} m^{3}kg^{-1}s^{-2}, making Newton's estimate amazingly accurate for its time.)

As a second means of solving this problem, consider the plot in Figure 3.2. The dots represent actual data taken as a massive object is lifted above the Earth's surface and subsequently falls. The height from which the object is dropped is denoted on the y-axis. The time it takes for the object to impact the ground is given on the horizontal axis. Newton's second law of motion in the case of objects falling near Earth's surface is $F = mg$, where g is the gravitation constant at the Earth's surface.

The force on an object of mass m_o at the Earth's surface is given by Newton's law of gravitation as $F_g = (-Gm_e/r_e^2)m_o = gm_o$. This expression holds as a close approximation for objects near the Earth's surface; in other words, this is the force an object feels when it falls from a distance above Earth's surface

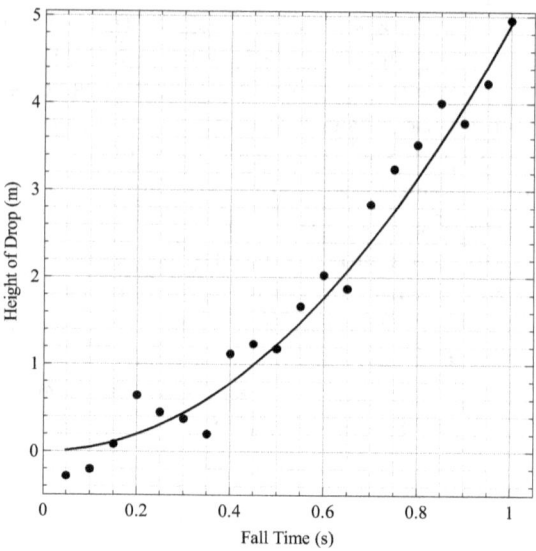

Figure 3.2: A plot of the time (x-axis) for a massive object to fall to the Earth from the specified height (y-axis).

much less than that of Earth's radius. The proportionality constant g is easily calculated from the Earth's mass and radius as 9.81 m/s^2.

Using this newly calculated constant and the equation we derived earlier for falling bodies near the surface of the Earth, $x = (1/2)gt^2$, we can perform a *least-squares fit* of the data provided in Figure 3.2 and verify that falling objects do indeed follow a square law. This experiment simultaneously corroborates Newton's universal law of gravitation and his second law of motion. Physical laws come directly from experiments like these, where an underlying descriptive law is proposed and verified against actual data.

For the next example, assume we have a coil of copper wire with ten looped turns. The wire has electrical insulation, such as polyvinyl chloride, that keeps the loops of the coil from shorting to one another. The insulation is stripped from the ends of the wire, and these are connected to a voltmeter. The coil is positioned horizontally, and a bar magnet, magnetized end-to-end, is held above the coil and dropped through it. As the magnet falls closer to the coil, passes through it, and then falls away from the coil, the voltmeter's needle registers a deflection. This story is similar to experiments Michael Faraday performed in the early 19th century [19]. Faraday's law of induction can be stated as

$$E = -k\frac{d\phi}{dt}, \qquad (3.4)$$

where E is the electric field induced between the coil ends, k is a constant of proportionality, and ϕ is the magnetic flux linking the coil. The voltage at the

3.3. HOW DOES SCIENCE MODEL REALITY?

coil terminals is proportional to the difference in the electric field across the terminals; this constant of proportionality is denoted k. The equation states that the voltage at the coil terminals is proportional to the time rate of change of the magnetic field passing through the coil. This was a fantastic discovery at the time and opened the door to the idea that electricity and magnetism were intimately linked, culminating later in the unification of these two forces and the realization that the magnetic field is simply a manifestation of the electric field.

It is not just this one experiment that led to a model of the electric and magnetic fields. Another experiment involved passing an electric current through a coil of wire and showing that it exhibited a magnetic field. This is the so-called electromagnet in which an electric field is used to produce a current in a coil of wire, which in turn produces a magnetic field. An electric current is a flow of electrons in the copper wire. Moving electrons or, more generally, moving electric charges produce a magnetic field. (We will see how before we leave this section.)

A final demonstration of the law of induction involves two coils. The coils are placed close together and concentric with one another. A rapidly changing voltage is applied to one coil to induce a changing current in it. (If the voltage is changing slowly enough, the current in the coil is determined by Ohm's law, $I = V/R$, where I is the current through the coil, V is the voltage applied across its terminals, and R is the resistance of the copper wire.) This changing current creates a magnetic field that changes in intensity. Since the coils are co-located, the changing magnetic field produced by the first coil passes through the second coil. Consequently, a voltage given by (3.4) is induced in the second coil. This is the last piece of the puzzle in unraveling the electromagnetic induction phenomenon. With the results of these experiments in hand, it was possible for Faraday (with help from the findings of Hans Oersted) to formulate the fundamental relationship between the electric and magnetic fields.

The last example we will look at is the unification of the electric and magnetic field theories. For many decades, the electric and magnetic fields were considered distinct from one another. In fact, they are in general still treated as distinct unless there is a reason not to do so. We have just shown that the discoveries of Faraday and others demonstrated a close connection between the two fields. This leads to the intuition that this connection might be closer than a casual glance reveals. Showing that the two fields are actually artifacts of one field will require that we pull from the theory of special relativity and draw upon at least one more result from electrodynamics, the expression for the Lorentz force [8,20].

The Lorentz force is the force experienced by a point charge moving in a magnetic field and is given by

$$\mathbf{F} = q\mathbf{v} \times \mathbf{B}. \tag{3.5}$$

Basically, this equation says that the force experienced by a positive charge q moving at a velocity $\mathbf{v}$ in a magnetic field $\mathbf{B}$ is proportional to both the speed of the particle and the magnitude of the magnetic field, and directed at a right angle to the particle's motion. (It is a little more involved than that, but this description captures the gist of what is going on.) This is the story from classical electrodynamics, an area of physics describing the behavior of moving charges and changing fields and pioneered by geniuses of the likes of Maxwell, Faraday, and Ampere.

Consider a current-carrying wire alongside which moves a unit charge q parallel to the wire at a speed v_q as shown in Figure 3.3. (This analysis is adapted from a description in David J. Griffith's *Introduction to Electrodynamics* [20].) We are going to look at this problem in a heuristic fashion and leave the mathematics out of it.

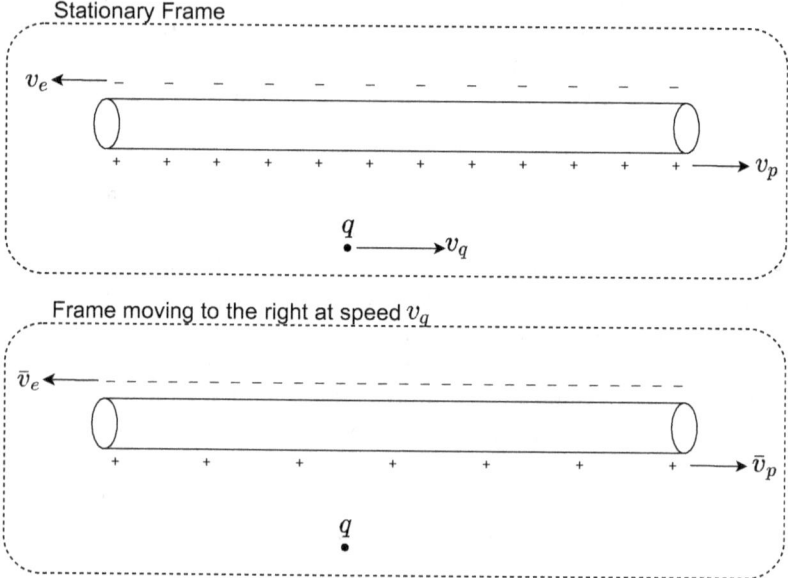

Figure 3.3: The charge q moves parallel to a current-carrying wire. The top figure shows the stationary reference frame, and the bottom is a reference frame moving to the right at the same speed as the charge.

In the stationary reference frame at the top of the figure, negative charges move to the left with speed v_n, and positive charges move to the right at v_p. This needs a little explanation. The electrons that move in a conductor to produce a current come from the electron shells of atoms that comprise that conductive metal. However, when an electron moves from one atom to another, the atom it leaves has lost one electron, a negative charge; hence that atom has a net positive charge. Every time an electron makes a jump to the left, effectively, an equivalent positive ion makes a jump to the right. This is why the figure shows a flow of negative charges to the left and positive charges to the right. Also, in the stationary frame, it is obvious that $v_n = v_p$.

Since there are just as many negative charges in the wire as there are positive

3.3. HOW DOES SCIENCE MODEL REALITY?

charges, the wire is electrically neutral and, thus, exerts no force on the charge q. However, experimental evidence shows that there is indeed a force on the charge q, which is described by the Lorentz equation (3.5).

There is another way of solving this problem that does not require the empirically derived Lorentz force equation. The lower half of Figure 3.3 shows a reference frame that is moving at the same speed as the charge q. The charged particle is stationary in this frame, and the wire moves to the left at speed v_q. It is valid for us to make this coordinate change because, from what we learned about special relativity in the previous chapter, observers must see the same physics in all inertial reference frames.

Due to relativistic effects (refer to Chapter 2), in the reference frame moving with the charge q, the densities of the charge carriers in the wire change. The relativistic speed of the electrons is $\bar{v}_e = (v_e + v_q)/\sqrt{1 + v_e v_q/c^2}$ and that for the positive charges is $\bar{v}_p = (v_p - v_q)/\sqrt{1 - v_p v_q/c^2}$. In the reference frame of the charge q, the speed of the negative charges in the wire is greater than the positive charges. Now, remembering from Chapter 2 that there is a length contraction as the speed of an object increases, it is observed that the effective length of the wire for the positive and negative charges is different. In fact, there is a length contraction for the faster negative charges compared to the slower positive charges. This means there are more negative than positive charges per unit length of the wire. Furthermore, here is the answer to the problem: in the reference frame of q, there is a net charge on the wire producing a net force on q. The *magnetic field* in the stationary reference frame is an *electric field* in the moving reference frame of the charge q. This is an amazing result.

It has just been shown that there are two different explanations for the effect of a magnetic field upon a moving charge, the first result coming from classical electrodynamics and the second being a consequence of relativistic effects. The result obtained is the same; only the interpretation of what is happening is different. In electrodynamics, the force on the particle is attributed to a newly conceptualized quantity called a magnetic field. However, the relativistic approach shows that the force is electrically derived from the different charge densities in the moving reference frame. This example demonstrates how theories can be correct and not capture the entire picture. The relativistic approach is simply a much deeper view of the reality at hand than the classic approach, but the two deliver the same results.

Despite the last few examples being woefully incomplete, the reader is getting an introduction to how a theory in a particular discipline is constructed. The various pieces of the theory must predict correct behavior on their own and also work together as a coherent whole. In addition, multiple theories must not conflict with one another if science is to accurately represent the overall picture of reality. Theories are not just a collection of disjoint experiments thrown together haphazardly. Taken together, the components of a theory consistently describe the workings of reality.

3.4 Chaotic systems

Chaotic systems will be discussed in this section. These are simple mathematical formulations that also appear in nature and are capable of producing an immense degree of complexity. Weather models, multi-body gravitational dynamics, electronic devices, and biological entities all exhibit chaotic behavior. Chaotic systems will be our first introduction to a class of systems for which future states are unpredictable. For example, the weather can only be predicted with any accuracy a few days in advance. This unpredictability does not arise because the sensors used for collecting temperature, humidity, wind speed, etc., are inaccurate, nor is it due solely to randomness in the system. The inability to predict future states of a weather system is instead caused by the very underlying mathematics that describes weather systems in general.

The notion that the behavior of simple mathematical systems would be unpredictable seems counterintuitive. It seems that the iteration of a single, second-order equation should result in an easy to predict behavior that is periodic at worst. On the contrary, the result can be chaotic to the degree that it appears at times to be random. I need to be clear here. In the ideal case, these systems certainly can be simulated to ascertain their future behavior. But there is no closed-form solution to predict their trajectories. These are *simulate and see* systems. This will be a very important concept throughout the rest of the book.

There is no need to develop or repeat formal definitions here, but a description is helpful before we go on. Chaos means that the output of a system is not predictable based on its initial conditions. Chaotic systems can generate utterly different ending states despite having been input virtually the same initial conditions. Complexity, on the other hand, is the emergence of order within the chaos. For example, the weather exhibits certain patterns such as storms, sunny days, hurricanes, windy days, etc., but there is no way to predict with certainty which of these patterns will show up on a given day. If one looks closely, the weather is chaotic, but this chaos is constrained within specific orderly patterns. In general, complexity emerges when a certain amount of order is imposed upon chaotic systems.

Consider the system described by the differential equation $\dot{x} = u(t) - x$, where the dot signifies differentiation with respect to time, t. It is easy to see that when the input u is larger than x, the value of x grows, and vice-versa. The variable x stops growing or shrinking when $u = x$ and $\dot{x} = 0$. Assume that u is stepped from zero to some finite value u_s; integrating this differential equation, we obtain $x(t) = u_s(1 - e^{-t}) + x_0 e^{-t}$. In the limit as t goes to infinity, x approaches u_s no matter the initial condition x_0. Not all systems are this well-behaved.

The solution of this differential equation always converges to the same value, $x = u_s$, no matter the system's initial condition. On the other hand, the solution to the second-order differential equation $\ddot{x} = -x(t)$ is of the form $x(t) = A\sin(t) + B\cos(t)$, where A and B are integration constants to be found from the initial conditions. This solution never settles down to a specific value but continuously

3.4. CHAOTIC SYSTEMS

varies in time with a sinusoidal response. It is an oscillator. However, it is very predictable as the sinusoid generated repeats forever with a constant amplitude and frequency.

Although the responses of linear dynamic systems such as those above can get quite complicated, involving many combinations of different modes, they exhibit nothing we would label as complexity. For example, given the amplitude and frequency of the sine wave above, it is a simple matter to find the value of the response at any point in the future *without* integrating the differential equation; the closed-form solution tells us everything about the system's response. The same is true for the single-order equation we looked at earlier.

Another example of a dynamic system is the simple pendulum similar to those used in old-fashioned clocks. A massless beam with a weight on its end is attached to a pivot point and allowed to swing back and forth. If the displacement is not too great, the pendulum approximates a linear oscillator and has a very regular period, making it suitable for use as a time-keeping mechanism. I will not go through the mathematics of how it is arrived at, but the pertinent differential equation is given by

$$\frac{d^2\theta}{dt^2} + \frac{g}{L}\sin(\theta) = 0, \qquad (3.6)$$

where θ is the angle of the pendulum off of the vertical, g is the gravitational constant at the Earth's surface, and L is the length of the beam connecting the mass to the pivot point. Note that the mass drops out and is not involved in the equation. It was usually a good idea to make the mass large to overcome friction in the pivot. To model an ideal pendulum, we will say the system is frictionless, and its energy is conserved.

Equation 3.6 is nonlinear because of the sine function on the right-hand side. The reason the pendulums of old clocks were allowed to swing only by a small amount is that as long the angle θ was small, the approximation $\sin(\theta) = \theta$ held, and the dynamics of the oscillating pendulum were nearly linear, leading to a consistent period of oscillation as the amplitude decreased due to the inevitable friction in the pivot point. This is easy to see because the characteristic roots of the small-angle approximate differential equation $\ddot{\theta} + (g/L)\theta = 0$ are $\pm i\sqrt{g/L}$ and thus result in pure sinusoidal oscillation at a fixed frequency. This is not the case when the displacement is large enough such that the previous approximation does not hold and the $\sin(\theta)$ term must be taken into account. In that case, the differential equation does not have a closed-form solution, and the pendulum's oscillation frequency becomes dependent upon its displacement.

The angular displacement of the pendulum, θ, and its angular speed, ω, are shown in Figure 3.4 for four different initial starting angles. At an initial displacement of 10° off the vertical, the resulting oscillation is very sinusoidal, demonstrating that our above solution found by approximating $\sin(\theta)$ as θ was accurate. This is the mode of operation that clockmakers sought.

With a 90° displacement, the sides of the sinusoidal response of the pendu-

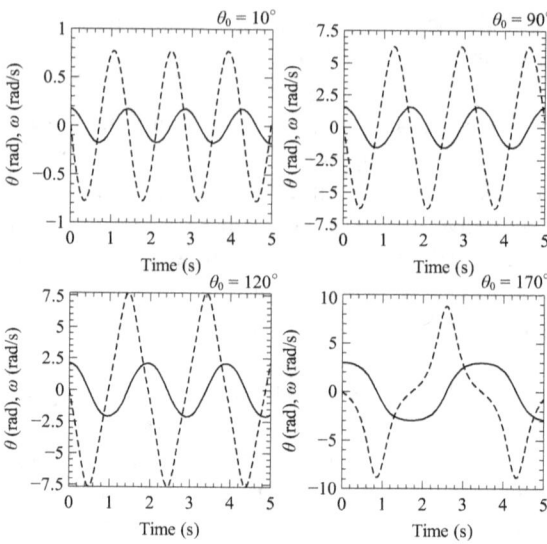

Figure 3.4: The dynamic motion of a simple pendulum started from four different initial angles of 10°, 90°, 120° and 170°. At small swing angles, $\sin(\theta) \approx \theta$, and the motion is nearly sinusoidal, but at larger displacement angles, the motion is quite nonlinear.

lum's angle become nearly straight lines. The nonlinearity is absolutely apparent when the initial angle reaches 120°. Finally, when the starting point is at 170°, the response becomes completely non-sinusoidal, indicating that the simplified linear differential equation does not at all represent the physical system. The pendulum swings up to 170° (the solid line is the angle in radians) and slows down as it does (the dashed line is the angular speed), then it falls with the angle changing very fast and the speed approaching a maximum.

This example of the simple pendulum demonstrates what happens as the dynamics of a system move from being linear to nonlinear. Although both of these systems appear to have very predictable responses, they are very different. As we discovered above, the frequency of the linear system is independent of displacement. However, that is not the case for the nonlinear system, as can be seen by comparing the upper-left plot with the lower-right plot. Clearly, the period of oscillation grows as the pendulum swing angle increases. We can easily predict the linear system's behavior in terms of its amplitude and frequency; this is evinced by it admitting a closed-form solution. There is no closed-form solution for the nonlinear case, and we cannot characterize it by the two parameters frequency and amplitude—it must be solved numerically or tested empirically to ascertain its behavior.

Although the nonlinear solution to the simple pendulum problem is intractable in comparison to the linear approximation, neither of these two approaches

3.4. CHAOTIC SYSTEMS

show signs of what we would call complexity. It is certainly the case that the solution to the nonlinear case likely requires the evaluation of an infinite sum for its solution, but it can be solved in theory (see [21]). However, some systems leave us no choice and must be integrated, that is, stepped through from the beginning, to see how they behave.

The layout of a double pendulum is shown in Figure 3.5. Instead of only one massless beam and one weight at its end, this pendulum has two beams and two weights. The two beams are connected end-to-end with the first mass at the connection point between the two beams and the second mass at the end of the second beam. Again, the fixed pivot point to which the first beam is attached is assumed frictionless, and the entire system is lossless. The point of the simulation is to lift the second mass and, since it's connected, the first mass as well, to a particular starting point and let it go and trace the motion of the second mass in the plane.

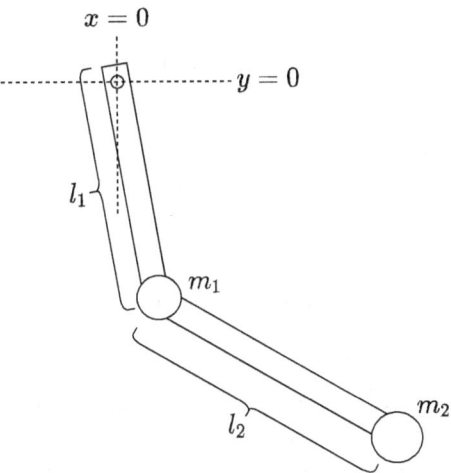

Figure 3.5: A double pendulum with beam lengths of l_1 and l_2. There is a mass, m_1, located where the two beams connect together at the first pivot point, and a mass, m_2, at the end of the second beam. Both pivot points are frictionless.

One might assume that the motion of the dual-beam pendulum, when released, although possibly nonlinear, will also be well behaved, but that is not the case. In some sense, it is reasonable to expect that the motion of the dual pendulum would be more complicated than that of the single pendulum simply because it is a more complex structure, but what we actually see when we experiment with these two setups is a qualitative difference. Figure 3.6 shows an example of the resulting motion when the second mass of the dual pendulum (and, by consequence, the first mass as well) is lifted and let go. The curve being traced in the figure is the position of mass two as the masses swing freely after being let go.

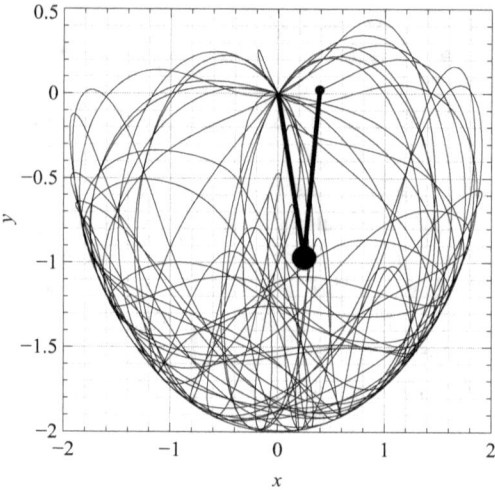

Figure 3.6: Plot of the position of the second mass of a two-mass pendulum. The larger mass (larger black circle) is three times more massive than the smaller. The rigid rods between the two masses are each one-unit long and there are two pivot points, one located at the origin and a second located at the larger mass. The second mass is placed at the starting position shown and let fall.

It is clear that the dual pendulum's resulting motion is qualitatively different from that of a single pendulum as it is no longer simply periodic but, instead, displays a significant degree of complexity. In addition, it is not immediately clear from where this complexity stems. The only difference between this setup and the previous one is that a second pendulum has been added to the first. The response of the single pendulum, even with a large initial angle, although nonlinear, has a well-defined periodic behavior (although the period changes with displacement). The dual-beam pendulum exhibits what is called chaotic behavior, where the trajectories starting from arbitrarily close initial conditions diverge by a large amount.

Assume that the time-evolution of a given dynamic system is given by $x(t) = H(t, x_0)$ where t is time and x_0 is the starting condition of the system. Now define $\Delta x_f = |H(t, x_0 + \Delta x_s) - H(t, x_0)|$. For a chaotic system, there is divergence in the time-evolution of trajectories starting arbitrarily close to one another; therefore, there exists the possibility that $\Delta x_f >> \Delta x_s$. An example of this is shown in Figure 3.7 where the same dual-pendulum system is started from *almost* the same initial condition and evolves a completely different trajectory. By almost, I mean that the two sets of initial conditions for the two simulations are virtually identical. Again, this is an example of the so-called butterfly effect, a term coined specifically in reference to chaotic systems. The butterfly effect describes how a tiny change in the atmosphere of South America, such as that

3.4. CHAOTIC SYSTEMS

caused by a butterfly flapping its wings, can propagate through the weather system to cause the formation of a tornado in Kansas.

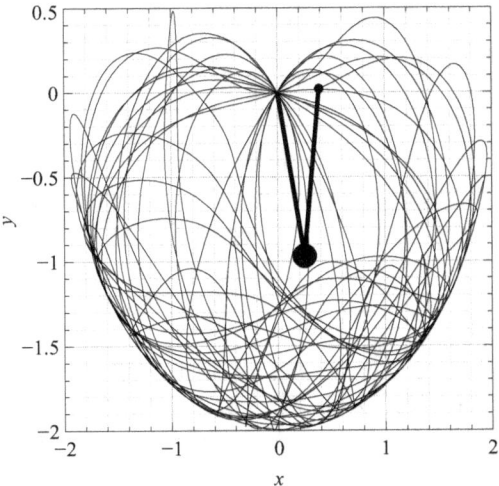

Figure 3.7: The same simulation of the double-pendulum system *except that the initial conditions changed by 0.001*.

Keep in mind that the unpredictable behavior progresses from any initial condition and that any point along the time-evolution of a dynamic system can be considered the initial condition for the evolution of the next portion. These systems are indeed deterministic and repeatable from the same initial condition, but they are not predictable as the initial condition varies. In other words, if the system is started from the initial condition $\mathbf{x} = \mathbf{x}_0$, it will follow a particular trajectory. If the system is started from that same initial condition again, it will follow precisely the same trajectory as before. However, if the initial condition changes even by an infinitesimal amount, the trajectory can differ. (Note that it is possible to make the initial condition the same and simulate the system to get the same trajectory on a deterministic computer. Nevertheless, it is impossible to do this with a real, physical system as even microscopic changes in the initial conditions will result in different trajectories.)

The angles with respect to the vertical for the two beams are plotted as functions of time in Figure 3.8. Compare the waveforms here with those of the single-beam pendulum in Figure 3.4. Whereas the angle of the first beam appears somewhat periodic, that of the second beam seems random. It is counterintuitive that the complexity contained within these waveforms is deterministic and derived from such a simple structure.

The chaotic dynamics of the dual pendulum are governed by a fractal-shaped region of stability. We will study fractals in the next section but, for now, just know they are complex shapes that are self-similar as they are zoomed into.

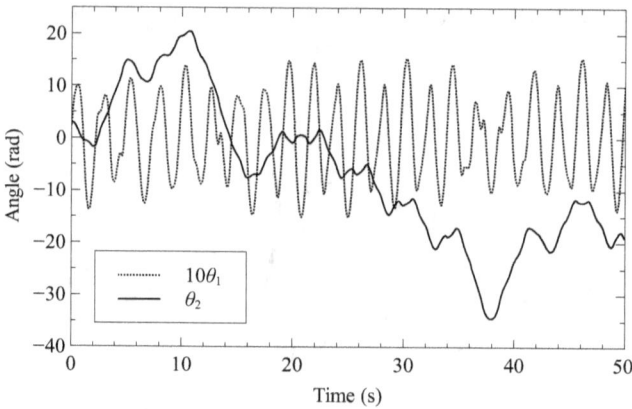

Figure 3.8: The angles of the two beams with respect to the vertical; θ_1 corresponds to the top beam and θ_2 to the lower beam. (Note that the angle of the top beam has been multiplied by ten for easier visualization.)

That is, the same level of complexity reappears no matter how closely they are examined. For example, a study case performed by Jeremy S. Heyl on the double pendulum demonstrated this clearly [22]. The research looked at all initial starting points in the (θ_1, θ_2) plane, and examined how long it took for either of the pendulum's beams to flip a full 360^{deg}. If the plot is color coded by how long it takes for a flip, the result is an astonishingly complex fractal. The color of a particular point within the fractal pattern represents the degree of stability of the resulting trajectory when the system's initial conditions lie at that point.

For reference, let's go back to a simple linear system for a moment and consider the solution of a dynamic system of the form $y(t) = A_0 e^{\alpha + i\beta} + A_1 e^{\alpha - i\beta}$ where i is the imaginary operator, A_k are constants, and $\alpha \pm i\beta$ are the characteristic roots of the system. By definition, this system is stable if and only if $\alpha < 0$; that is, the real part of the roots is in the left-half plane, resulting in the exponential terms decaying in time. If the real part of the roots are positive, the solution grows exponentially without bound and is unstable. Finally, if the real parts of the roots are zero, then the system is marginally stable, oscillating with a constant amplitude. It is very convenient when designing dynamic systems to say that the system is stable if the real part of the characteristic roots are negative. It is not that simple with chaotic systems.

Whereas the stability of linear systems is easily determined from the characteristic roots of the system, nonlinear chaotic systems can be much more complicated. Given a specific set of initial conditions, there is no straightforward way to determine that a nonlinear system is stable. Again, the region of stability for a linear system can be delineated by partitioning the state space of the systems in simple ways. For example, the stability of continuous-time linear dynamic

3.4. CHAOTIC SYSTEMS

systems is determined by whether the system's roots are in the left- or right-half plane. However, the corresponding regions of stability for chaotic systems are fractal in nature.

So far, everything we have looked at has been in the continuous-time domain. Let's now shift gears and move to the discrete domain. In the continuous domain, time can be infinitely subdivided. If we look at the plot of $y(t) = t$, we see a smooth, continuous forty-five-degree line. On the other hand, if we plot the function $y_k = t_k$, we see a series of points, all lined up on a forty-five-degree line. Both of these functions are shown in Figure 3.9, with the continuous domain on the left and the discrete on the right. The continuous function is valid for every point, $0 < t < 5$, along the line in the left half of the figure. However, the discrete function on the right is only valid for six points, those in the discrete set $t_k \in \{0, 1, 2, 3, 4, 5\}$.

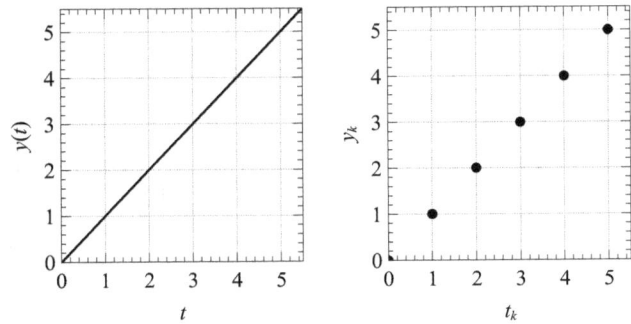

Figure 3.9: A comparison between continuous (left) and discrete (right) functions $y(t) = t$ and $y_k = t_k$, respectively.

Discrete functions are used in many areas of study but are ubiquitous in the design and modeling of digital computers and their programming. In fact, all quantities, including those in simulations, that exist within a digital computer are discretized. Discrete systems must be in some ways handled differently than continuous systems, and since most of the systems discussed in the remainder of this book are discrete, we will make the switch now.

An example of a discrete equation, (3.7), is given below. It iterates the variable y_k and makes use of the constant parameter μ. There are no inputs to this system. It is very straightforward to iterate. For example, assume $\mu = 2.5$ and $y_0 = 0.75$, then $y_1 = (2.5)(0.75)(1 - 0.75) = 0.46875$, then $y_2 = (2.5)(0.46875)(1 - 0.46875) = 0.62256$, and so on.

$$y_{k+1} = \mu\, y_k\, (1 - y_k) \tag{3.7}$$

Mitchell Feigenbaum studied this equation, sometimes called the *logistic equation*, and its propensity for chaotic behavior in 1975 [23]. At low values of μ, this system is well behaved and approaches a constant value when it is iterated, as shown in the top-left plot of Figure 3.10. In this plot, the initial condition $y_0 = 0.5$ is used with the constant $\mu = 2.152$. Iterating the equation, the value of y_k quickly rises and converges to the value 0.53532.

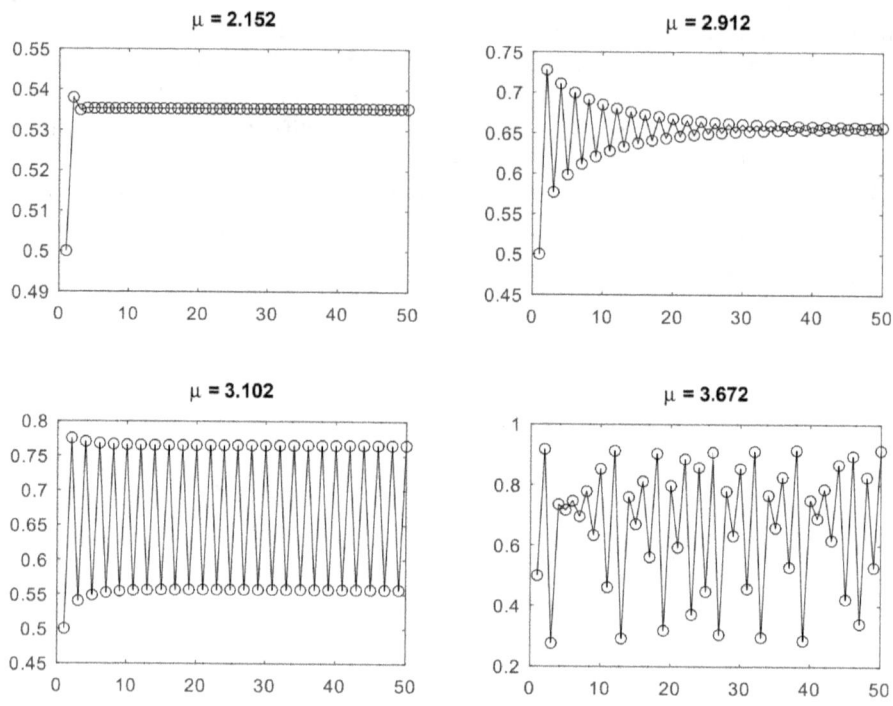

Figure 3.10: Iteration of the discrete logistic equation, all with initial condition $y_0 = 0.5$, and for four different gains $\mu = 2.152$ (top left), $\mu = 2.912$ (top right), $\mu = 3.102$ (bottom left), and $\mu = 3.672$ (bottom right).

The situation gets more interesting as the value of μ approaches three as in the plot in the top right of the figure. The response suddenly *bifurcates* into two envelopes and oscillates between them, eventually converging to about 0.65. Then, the oscillation becomes sustained and flat when μ increases to 3.102 (see the lower-left plot). At first glance, one might think adjusting μ in this system is akin to changing the loop gain in a linear system, and as the gain increases, the system passes from stable to marginally stable to unbounded exponential growth. That, however, does not happen. Instead, something entirely unexpected happens.

As Feigenbaum showed, when μ increases above about 3.6, the system does not become unstable with unbounded growth as linear systems do; rather, it continues to bifurcate until it reaches a region of complete chaos. This is shown in the lower-right plot of Figure 3.10 where the response y_k appears to be random.

3.4. CHAOTIC SYSTEMS

Actually, as μ increases, the response initially converges to one value, then with μ a little larger, a bifurcation to two values takes place where the response oscillates between the two. A further increase in μ will result in another bifurcation into four levels, then eight levels, and so on. As the number of levels grows very large, suddenly, the response becomes completely unpredictable with an infinite number of levels. This is formally known as chaos.

The plot in Figure 3.11 shows this progression graphically. In this plot, the gain μ is increased on the horizontal axis, and the points between which the response of the equation oscillates are plotted on the vertical axis. Below a gain of about three, the system always converges to a single point. However, when the gain increases above three, a bifurcation into two levels takes place. The single split continues until about 3.45, and then a further bifurcation into four levels occurs. Bifurcation continues as μ is increased until chaos ensues around 3.6. Interestingly, at about 3.85, order is restored, and the process repeats (it is hard to see in the plot).

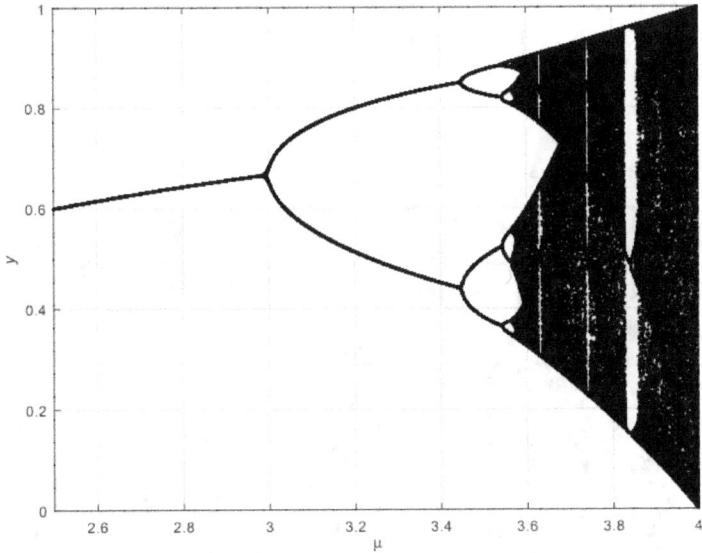

Figure 3.11: The period doubling cascade into chaos exhibited by the logistic equation.

The diagram shows obvious traits of a fractal pattern. Although it is not visible in the plot, self-similar bifurcation continues infinitely as μ is increased, no matter how closely one zooms in on a particular area. There is no method of predicting this behavior without just simulating it. The vital item for our purposes here is that a staggering amount of complexity emerges spontaneously from a seemingly innocuous equation. Before leaving this topic, let's look at one more example that is even more impressive.

In 1980, Benoit Mandelbrot first plotted on an IBM computer what has become since then to be known as the *Mandelbrot set*. The Mandelbrot set is generated from the complex equation $z_{k+1} = z_k^2 + C$. Here I'm using the word complex in the mathematical sense. A complex number is one that contains a real part along with an imaginary part, such as $x = 1 + i2$. The number x contains a real part of 1 and an imaginary part of size 2. Note the use of the imaginary operator $i = \sqrt{-1}$. Just as the square root of 1 is ± 1 because $(+1)^2 = 1$ and $(-1)^2 = 1$, the square root of -1 is $\pm i1$ because $(+i1)^2 = -1$ and $(-i1)^2 = -1$ by definition.

The equation $z_{k+1} = z_k^2 + C$ is to be iterated the same as the logistic equation above. For example, a stable point of the equation is $C = 0 + i0$. (We're always going to start with $z_0 = 0 + i0$.) The first iteration with this constant is $(0 + i0)^2 + (0 + i0) = 0 + i0$. And if we calculate the second step, we get the same thing, $(0 + i0)^2 + (0 + i0) = 0 + i0$. Therefore, the response started at $0 + i0$ and remained at $0 + i0$. Now, let's look at the unstable point $C = -1 - i1$. The first step yields $(0 + i0)^2 + (-1 - i1) = -1 - i1$. The second iteration is $(-1 - i1)^2 + (-1 - i1) = -1 + i1$. The third is $(-1 + i1)^2 + (-1 - i1) = -1 - i3$. The fourth and fifth steps are $(-1 - i3)^2 + (-1 - i1) = -9 + i5$ and $(-9 + i5)^2 + (-1 - i1) = 55 - i91$, respectively. And going once more, the sixth iteration is $(55 - i91)^2 + (-1 - i1) = -5257 - i10011$. It is a simple matter to see that the response is growing very rapidly and, thus, the point $C = -1 - i1$ is outside the region of stability.

Figure 3.12: The Mandelbrot set that comprises the attraction basin of the equation $z_{k+1} = z_k^2 + C$, where C is a complex number. The black regions are stable. The lighter the shade the more unstable a point is.

What Mandelbrot did was perform the calculations for every C within the plane and plot the stable points in different colors than those that were unstable. The Mandelbrot plot is shown in Figure 3.12. Black denotes stable points. Lighter colors correspond to points that diverge faster. The result is a fractal of astounding complexity that has intrigued viewers for decades. As one zooms

3.4. CHAOTIC SYSTEMS

further in, some patterns repeat, and other new patterns continue to emerge. This very intricate fractal defines the region of stability of the corresponding second-order equation.

This section started with a summary of the behavior of linear systems. Linear systems are the easiest to conceptualize and are what people usually think of when they ponder dynamic systems. The reason for our intuitive understanding of linear systems is because they obey two rules that our minds find easy to grasp, scaling and superposition. That is, for a linear system, $f(ax) = af(x)$ and $f(x + y) = f(x) + f(y)$, and so nothing surprising ever really shows up. Although many nonlinear systems can be approximated as linear, leading to ease of simulation and analysis, there exist nonlinear systems that are completely intractable and do not admit to analysis.

In many situations, we can make qualitative, or even quantitative, statements about the long-run behavior of systems. For example, the first differential equation we looked at in this section always converged to the input value after some time. Although possessing a dynamic response, the outcome was predictable and could be easily described. Even the second-order system with a sinusoidal response falls into this category. Its output can be entirely characterized by specifying a frequency and an amplitude. Many systems behave in this way. And, then, there are some systems that do not behave this nicely.

There are some nonlinear systems for which the behavior of their outputs cannot be predicted at all. This seems almost an absurd state of affairs. How could the behavior of a simple second-order equation not be predictable? This is simply the nature of the mathematics. One might as well ask why pi, the ratio of a circle's circumference to its diameter, is an irrational number. As the *ratio* of the circumference to the diameter is a *ratio*, how could the answer, pi, not be a ratio? This speaks to something very deep about mathematics and reality. My point is that very innocuous mathematical statements can provide incredibly complex results.

Some scientists would argue that the physical world, nature, is constructed in the language of mathematics. Pick up a rock and drop it, and its motion due to the force of gravity will be described almost exactly by Newton's second law of motion. The spectral lines observed in starlight definitively specify which elements are present within the star. The behavior of the electric and magnetic fields is governed by the mathematical beauty of Maxwell's equations. But, if nature's rules really are written in the language of mathematics, it should not be surprising at all to see the emergence of complexity in nature. Of course, it is not this statement that surprises people. Rather, it is when the mathematical complexity is translated to biological complexity that people are astonished and even unbelieving of the result. Again, this is a thesis point of this book. If one part is accepted, that the behavior of the natural world is written in the language of mathematics, then the second part must also be accepted, that complexity appearing in the natural world is an inevitable result.

No one really questions it when immense complexity shows up from the differential equation describing the dynamics of atmospheric weather patterns or in the prediction of planetary motion in a three-body problem, but when it appears within biology, the proverbial crap hits the fan. It is as if mathematical complexity can be evinced in the one area but should not be apparent in the other. If one can put away all preconceptions, the admission would have to be made that it would be more surprising if complexity did not show up in biology. There will be much more to say about this subject later.

3.5 Fractals

In 1967, the same Benoit Mandelbrot from the previous section wrote a paper on characterizing the coastline of Great Britain in which he came to some conclusions about geographical curves [24]:

> Geographical curves are so involved in their detail that their lengths are often infinite or, rather, undefinable. However, many are statistically "self-similar," meaning that each portion can be considered a reduced-scale image of the whole. In that case, the degree of complication can be described by a quantity **D** that has many properties of a "dimension," though it is fractional; that is, it exceeds the value unity associated with the ordinary, rectifiable, curves.

Most of us are accustomed to referring to dimension as the number of orthogonal dimensions defining a given system. For instance, a straight line is one-dimensional, corresponding to the fact that there is only one direction in which to move (moving backward means moving negatively in the positive direction). A two-dimensional system has two directions in which to move and so on, and is formed by adding an additional orthogonal axis to the original one-dimensional axis. The same is true when a third orthogonal dimension is added to obtain a three-dimensional system. But what does it mean for a system to have a fractional dimension?

Let's first follow Benoit and define the dimension of a system by how it scales. For example, consider a simple line segment. If we zoom in on one-fourth length sections of the line segment, we end up with four segments placed end-to-end. The scaling factor is four. The number of resulting self-similar pieces (the four segments) is four. We can write this as follows:

$$\text{Number of pieces} = (\text{Scaling factor})^d. \tag{3.8}$$

In the case of the line segment scaled by a factor of four, the resulting equation would be $4 = 4^d$. The value of d that makes this equality true is 1. Benoit termed the exponent in this expression the *fractal dimension*.

Now, consider the squares in Figure 3.13. The left side of the figure shows a single square with sides of length one unit, and the right side shows that same square with each edge subdivided (magnified or scaled) by a factor of four, resulting in 16 sub-squares. Therefore, the scaling is four, and the number of

3.5. FRACTALS

self-similar pieces is 16. We can rewrite the above equality (3.8) as $16 = 4^d$. The fractal dimension of the square can now be found as $d = 2$. This result makes sense as the square is a two-dimensional figure.

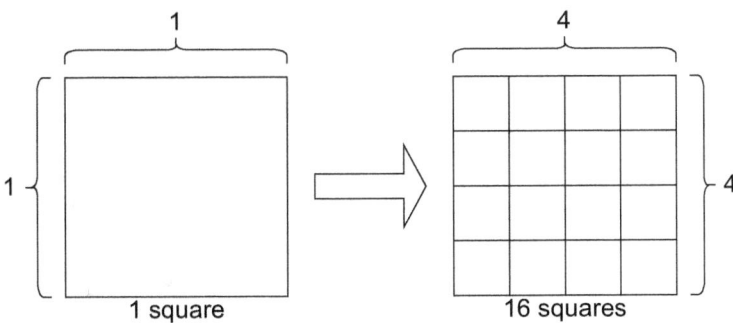

Figure 3.13: As the square on the left is magnified by a factor of four, that is, each axis is split into four pieces, the number of constituent squares increases by a factor of 16.

The same process can be repeated for a cube to arrive at $64 = 4^d$, meaning that the fractal dimension of a cube is three, $d = 3$. So far, we have been writing an equality and simply reading off an exponent to obtain the system dimension. However, the equation is easily solved explicitly for the dimension. Let N represent the total number of self-similar pieces after scaling by s and let d denote the system dimension. The pertinent equation then becomes $N = s^d$. Taking the logarithm of both sides yields $\log N = \log s^d = d \log s$. Dividing through, the final result is $d = (\log N)/(\log s)$. Applying this to the cube gives $d = \log(4^3)/\log(4) = 3\log(4)/\log(4) = 3$. (The base of the logarithm used does not matter.)

This formulation works quite well for solving dimensionality problems to which we already know the answers, but how does it work in general? Several iterations of what is called the Koch curve are plotted in Figure 3.14. It is a fractal that starts as a line segment like that shown in (a) of the figure. The line segment is then split into four equal lengths, with the center section being transformed into an equilateral triangle as in (b). The length of each of the four resulting segments is one-third that of the original line segment, meaning the total length of the resulting line is 4/3 that of the original line segment (the line started with three of these segments, but it contains four after the triangle is added). The same process is repeated in (C), resulting in a length of $(4/3)^2 = 16/9$, and again in subplots (d) through (f), ending in a shape that looks sort of like the edge of a snowflake.

Note that every time the line is split by a factor of three, the total number of segments increases by a factor of four. The scaling is by three, and the resulting number of pieces is four. Therefore, for the kth step, the fractal dimension is $d_k = \log(4^k)/\log(3^k)$ as k goes to infinity. Simplifying, $d_k = (k \log 4)/(k \log 3) =$

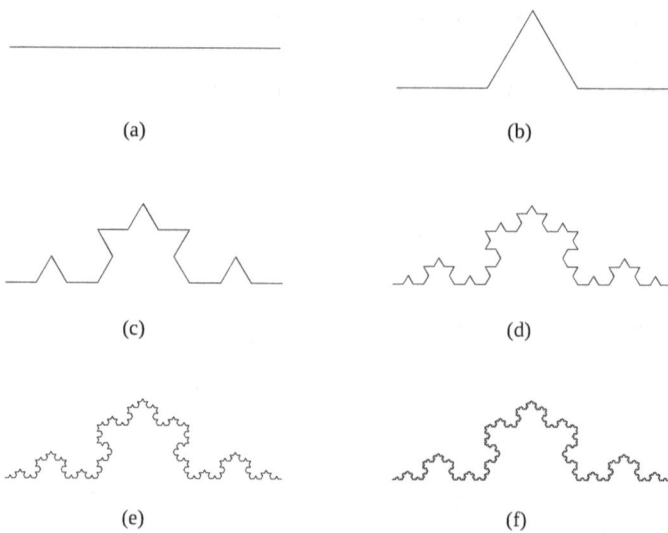

Figure 3.14: The Koch curve fractal starts with a single line in (a) that is subdivided in (b) into four lines that are each 1/3 the length of the original line. An equilateral triangle (without the base) is formed by the two new lines, meaning the length of the line in (b) is 4/3 the length of that in (a). This process is repeated for each of the four line segments in (c) with a total length of 16/9 the original length. The process is again repeated in the remaining plots (d) - (f).

$(\log 4)/(\log 3) = 1.2619$. Due to the simplification, we can drop the k subscript and denote the dimension as simply $d = 1.2619$. This is a strange answer: the dimension of the Koch fractal is between one and two. It is not an integer. Another interesting aspect of this curve is that its length increases toward infinity as the number of iterations increases, but the area within which the curve resides remains finite. An infinite length line is contained in a finite area—this is quite an oddity.

As a second example, the box fractal is shown in Figure 3.15. This fractal pattern replicates itself by replacing each existing square with five smaller squares, as shown in plots (a) through (c) in the figure. In each iteration, the box edges are cut into thirds, meaning the scaling factor is three. The number of self-similar pieces acquired in each iteration is five (five new boxes show up). This means that the fractal dimension is $d = (\log 5)/(\log 3) = 1.465$.

Finally, flip a few pages ahead and take a look at Figure 3.19. The shape in the plot is called a Sierpinski triangle, and it is constructed by first drawing a simple equilateral triangle. Then an upside-down equilateral triangle is drawn in the center, circumscribed by the original triangle. That results in three right-side-up triangles, each formed by an edge of the newly inscribed triangle and the edges of the original triangle. These three new triangles are one-half the size of

3.5. FRACTALS

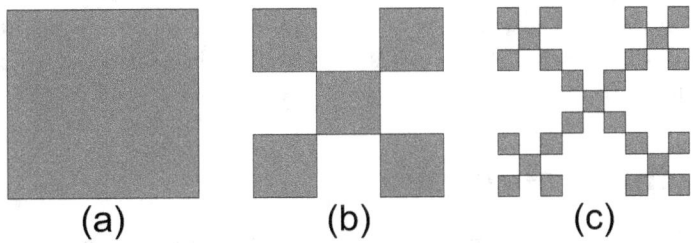

Figure 3.15: The box fractal starts with a simple black square in (a). The original square is then split into five squares, as shown in (b). With another iteration, each of the five squares is then split into five additional squares as in (c).

the original, leading to a scaling factor of two and, since each iteration leads to the creation of three times as many new triangles, the fractal dimension of the Sierpinski triangle is $d = (\log 3)/(\log 2) = 1.585$.

Fractal patterns are important for multiple reasons. The first is that they allow for the design of structures that seem to defy our intuition. For example, in the limit, the Koch curve above has an infinite line length and yet a finite area. This is very important in areas such as the design of wide-bandwidth antennae. The greater the physical length of the antenna, the lower the frequencies it can efficiently receive. The problem is that a longer antenna means a larger antenna. Building compact devices requires designing small antennae with wide bandwidths, which cannot be done using conventional methods. Hence, fractal antennae have become very common over the past few decades.

Another critical application in which fractal designs are employed is in the construction of roadways that reach as many places as possible with the shortest total road. This is similar to the blood vessels in animals, where the circulatory system needs to span the entire body and do it while taking up as little volume as possible. Similarly, the lungs of an animal must provide the largest surface area possible with minimal volume. Fractal patterns show up everywhere as solutions to problems that cannot be optimally solved using conventional methods. The fractal concept as a geometrical solution will be essential later on in explaining complex biological structures.

As we have already seen in the previous section, the stability regions of nonlinear, chaotic systems are fractal in their nature. This has some strong implications. For example, why do the placement and length of the stalk and limbs of a tree form a fractal pattern? The geometrical answer is that this arrangement provides the best means of optimizing the locations of the leaves for the absorption of sunlight and atmospheric carbon dioxide. Moreover, this is a legitimate answer. However, looking at it from another angle, the structure of the trees is not chosen to satisfy any goal or objective. Instead, it is the most stable arrangement for trees needing access to sunlight and atmospheric carbon dioxide. Tree varieties that did not form with a fractal geometry tended to die faster than those that did. Only the trees with the forms we see today were stable enough to survive to the present day. We will get back to the subject of fractals in the coming chapters.

3.6 Cellular automata

John von Neumann first envisioned the discrete computational model known today as the *cellular automaton*, and the concept was later placed in a practical framework by the likes of John Conway and Stephen Wolfram [25, 26]. In its simplest form, it is a distributed network of binary states where the value of any state is determined solely by the previous values of surrounding states. In other words, in its simplest case, the cellular automata process makes use of information localized in time and space.

To see how this works, consider the example in Figure 3.16. The simulation starts in (a) with a single black cell on row 1 of the grid; this is the initial condition. The process then continues by iterating through each cell in row 2. For each cell in row 2, the cells directly above in row 1 are checked to see if they match the rules of the automaton in question, (100) in this case. If there is a match, the cell in row 2 is changed to black, or a 1, and if it is not true, the cell is left white, or 0. This process is continued for each cell in row 2.

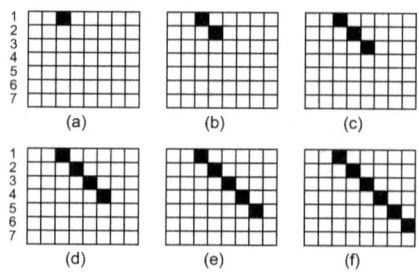

Figure 3.16: Five iterations of the cellular automata rule (100) beginning with the initial condition in (a).

Note that there is only one instance where the rule is satisfied in row 2, as is shown in (b). The process is repeated for row 3, looking for matches to the rule in row 2 this time. The result is shown in (c). As the process is repeated for each row, a diagonal line from left to right emerges. (Note that we have ignored the boundary conditions where the grid ends and assume there is an infinite grid of white squares extending left and right that are not shown in the figure.)

Let the rows in the grid be denoted by $j \in \{0, 1, 2, \ldots, N\}$ and columns by $k \in \{0, 1, 2, \ldots, M\}$. The computation process for evaluating cell $S_{j,k}$ then becomes $S_{j,k} = \prod_i R_i \wedge [S_{j-1,k-1} \ S_{j-1,k} \ S_{j-1,k+1}]$, where $[x \ x \ x]$ indicates a three-bit sequence. R_i denotes a set of rules of the same three-bit form. In our

3.6. CELLULAR AUTOMATA

example, $R_0 = [1\ 0\ 0]$. This expression indicates that each rule is and'd with the three-bit sequence in the row of cells directly above the current cell. It is easy to see why the rule (100) creates a diagonal line from left to right.

There are eight combinations of three-bit rules meaning that there are a total of 256 different rule combinations that a particular automaton can utilize.

Several other combinations of rules are shown in Figure 3.17. In (a), the initial black cell in row 1 is simply repeated in every following row. So, the cell in row 2, directly below the initial black cell, has above it the three-cell combination of white, black, white, which fits the rule (010) printed above the grid. No other cell in row two has this sequence above it. Similarly, sweeping across the cells in row three ends up in the same situation: only the middle cell in row 2 combines white, black, white. This situation repeats forever as the rows are traversed. The example demonstrates no complexity at all and fully resides in the region of order. Many of the rule combinations provide ordered results like this one.

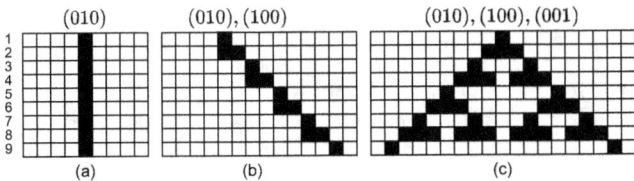

Figure 3.17: The first nine steps of three simple cellular automata rules.

The second example, (b) in the figure, demonstrates an automaton's application of two rules. The cell in row 2 directly below the initial black cell in row 1 is changed to black due to the rule (010). The cell to the right of the one modified in row 2 is also made black because of the other rule (100). It is clear to see that rule (010) generates vertical lines. However, when rule (100) is added to the mix, it causes the line to bend to the right. Once started, this pattern repeats indefinitely in an orderly fashion.

As one last example demonstrating how cellular automata processes behave, (c) in the same figure uses three distinct rules. The additional rule is (001), and the result is continual bifurcation to the left and right. This pattern also continues indefinitely. Although this pattern is more sophisticated than the previous two, it is not what we would call complex.

The next example, Figure 3.18, uses the rules (111), (110), (100), (010) and (001) and is iterated over 200 rows. It is constructed from a simple repeating structure that is uniform throughout the entire grid. There is nothing unusual going on here; it is exactly the kind of order one might expect from applying such simple rules. But things begin to change with the next example.

In Figure 3.19 the first 200 steps of an automaton using rules (111), (100) and (001) are plotted. The interesting thing about this example is that these three

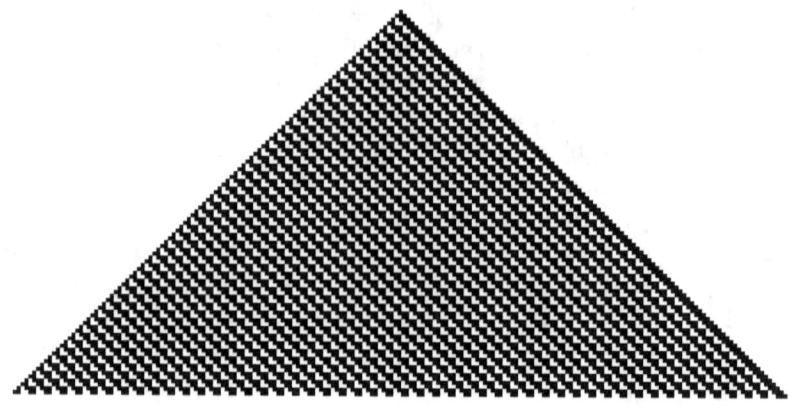

Figure 3.18: Two hundred steps of the cellular automaton using rules (111), (110), (100), (010) and (001).

simple rules compute a Sierpinski triangle. This fractal pattern repeats forever with triangles being circumscribed within other triangles and it is generated by applying three simple rules.

Figure 3.19: Two hundred steps of a cellular automaton using rules (111), (100) and (001).

Now, let's look at what happens when rules (100), (011), (010) and (001) are applied. The first 375 steps are shown in Figure 3.20, and it is quite apparent that something new has shown up. On the left of the plot, there is still some apparent sense of regularity, but the right half of the plot looks virtually random. It is not truly random because a deterministic process generated it, but it certainly is not as regular as the previous plots we have been looking at. As it turns out,

3.6. CELLULAR AUTOMATA

the right-hand side of the plot is not random, but, rather, the computation of the automaton becomes chaotic to produce complexity in the formal sense (see the previous section). To see this more clearly, Figure 3.21 zooms in on a section of the right-hand-side of the plot of Figure 3.20.

Figure 3.22 explicitly shows the chaotic behavior of the automaton of Figure 3.21. The center 16 columns of the figure are used to calculate a number. The cells in each row, across these 16 columns, are considered bits of a binary number, black being a 1 and white being a 0. This number is divided by 2^{16} to normalize the values between zero and one and plotted for 250 rows in the figure. The result is what appears to be completely random noise. Since a deterministic system created the data in the plot, it is not random but rather exhibits a significant degree of complexity.

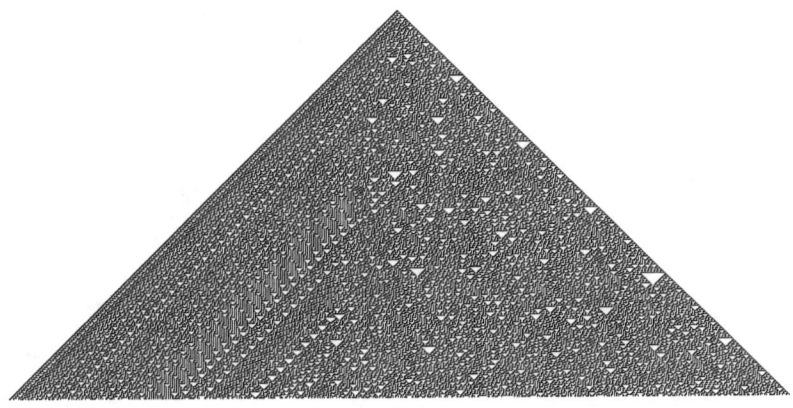

Figure 3.20: Three hundred and seventy-five steps of a cellular automaton using rules (100), (011), (010) and (001).

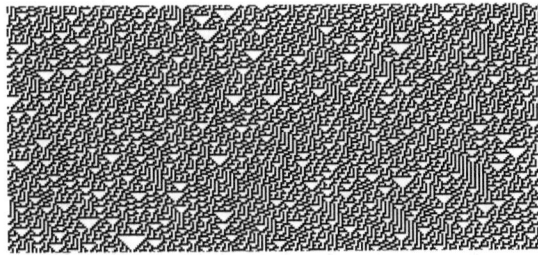

Figure 3.21: A zoom in of a small section on the right of Figure 3.20.

As Wolfram showed, many of these simulations, including more complicated versions, are capable of generating order, chaos, and mixtures of the two, and of exhibiting convergence—arriving at the same outcome despite the initial con-

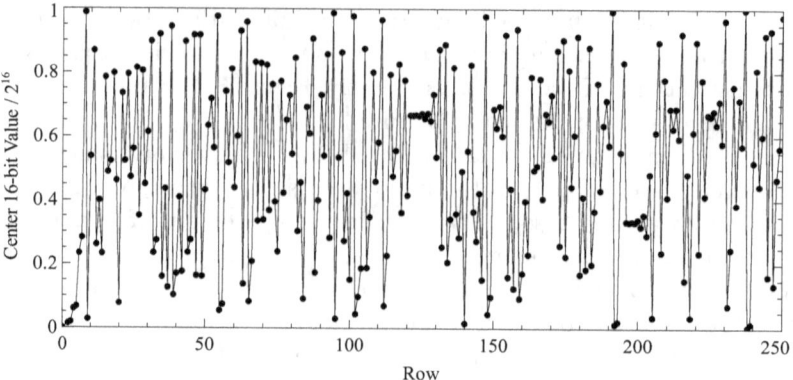

Figure 3.22: The numerical value (divided by 2^{16}) of the 16 center cells of each row in Figure 3.21. Since the process is deterministic, despite appearing random, these numbers represent a pseudo-random sequence.

ditions being very different [25]. The property of convergence is an important one. Convergence means that a specific state will be reached every time by the automaton when starting at any initial state.

Although it can be sequentially simulated (which it was in the cases shown here—that is the only way to do it on a standard computer), the nature of cellular automata computation is actually parallel in that every cell in the present row can be computed simultaneously as they only depend upon the state of the previous row. Using the expression given above, it is a simple matter to convert the simulation into a discrete-time recursive process. Each successive row represents a tick forward in time in the evolution of the automaton.

The behavior of simulated cellular automata has definite implications for understanding the complexity and behavior we observe in biological systems. Although this will be discussed in detail in later chapters, a few defining characteristics of these systems should be mentioned here. Although the systems we looked at in this section were very simple, having small working spaces and a minimal set of eight rules, they are indicative of how all cellular automata behave.

The first thing to note is that the outcome is very sensitive to initial conditions and the iteration rules. Slight changes in the initial state of the system can result in entirely different outcomes. There is no way to determine ahead of time what the outcome will look like given a particular initial condition.

Second, there is no way to know the behavior of a given ruleset *a priori*; it must be simulated or experienced to discover its ultimate behavior. The particular complex pattern the automaton will generate as it is sequenced cannot be determined ahead of time.

Third, despite being based on a small set of elementary rules, cellular automata can display incredible complexity, which is not expected. One of the primary questions asked by this book is how complexity comes about from elementary processes, for example, how the electrostatic forces involved in molecular interactions come to generate the complexity apparent in biological proteins and DNA.

Finally, many initial starting conditions can produce the same outcome for some rulesets. This kind of convergence is observed in biological evolution repeatedly. Many sense organs such as eyesight, locomotive morphologies such as four legs, and the ability to fly have evolved multiple times as convergent processes. As we discussed with chaotic systems above, initial conditions which differ by an infinitesimal amount can result in wholly different outcomes. It turns out that this goes in both directions. Convergent final states that differ by an infinitesimal (or zero) amount can start from entirely different initial states. This is how biological functions can evolve multiple times from different starting points. It is not by complete chance that biological functions "just happen" to evolve multiple times. It is because these morphologies represent stable forms within the evolutionary landscape.

These complex systems provide avenues useful in describing biological processes. They can generate immense complexity from simple computational forms, and the complexity can vary by extraordinary amounts due to minute changes in the initial conditions and rulesets. On the other hand, they can also generate convergent patterns from initial states that are vastly different. All of these features of cellular automaton will be important in understanding biological complexity.

3.7 Randomness and information

The concept of randomness is probably the most important constituent of the thesis of this book in defending natural processes as the progenitor of biological diversity and the emergence of intelligence. At first glance, randomness seems to be an incidental property of nature that, rather than being something useful, is instead a problem that must be dealt with to bring about order. Engineers generally see noise as a problem. It is an issue that all contemporary electronic designs must mitigate. It sets inherent limits on the signal-to-noise ratio of amplifiers, constrains the accuracy of sensors, and means that control systems are always suboptimal. Nevertheless, as we will see, it is essential in generating the novel information that deterministic systems cannot. However, let us not get ahead of ourselves; right now, an examination of the properties of randomness is necessary.

A common definition of randomness is the apparent lack of pattern or predictability. However, that is only an intuitive statement. For example, if a fair die is rolled, with no cheating involved, it is impossible to guess which numbered face will appear when the die stops rolling. Of course, someone can get lucky and guess correctly in any particular instance. But roll the die over and over and continue guessing, and it soon becomes apparent that the outcome cannot be predicted on a perpetual basis.

Imagine ten identical balls numbered zero through nine placed in a jar and selected one at a time with replacement. An individual selects a ball at random, records its value, and returns it to the jar. This process is repeated to form the sequence x_k, where x is the value printed on the ball, and k is the enumeration of the selections. Note that since all the balls are identical and the individual doing the selecting cannot see into the jar, there is no way to tell which ball is chosen during any selection event until it is observed. Also, the likelihood of selecting ball 0 is the same as that of selecting ball 1, which is the same as selecting any other of the balls. As we have discussed previously in this book, we call this the *probability* of selecting any one of the balls, and it is found by dividing the size of the outcome space by the size of the sample space. In this case, the size of the outcome space is one as each number corresponds to only one ball, and the size of the sample space is ten since there are ten balls from which to choose. Since the probability is the same for all outcomes, the selection process is represented by a *uniform* probability distribution, and the probability density function is a horizontal line of height 1/10. The important thing here is that there is *no means of determining which ball will be selected.*

There is something intrinsically subtle and mysterious about genuinely random events. This mysteriousness rests solely on the idea of how the outcome of a random event is decided. It may not be so mysterious when we are talking about selecting balls at random out of a jar as it seems "obvious" that the result will be unknown. But how does this randomness enter the system? Is it an error in the individual's brain doing the selecting that causes the selection of one ball versus another? Is it the interplay of the electric fields in the outer electron shells of the atoms making up the balls that causes one to be shifted into just the proper position to be selected over another? Is it a combination of these and a dozen other factors? It is difficult to sort out in most scenarios like this, but there are examples where things become less complicated and, astonishingly, more mysterious.

The radioactive decay of an unstable element is one such mysterious event. On average, the atoms of a particular element will decay according to the element's radioactive half-life period. The half-life is a statistical mean—half of the atoms in a mass of the element will decay faster than the half-life, and the other half will take longer. However, this is not really important here; what we are interested in is that there is no way to pin down exactly when the decay of a particular atom will occur. This is as close to a supernatural event as we get in this Universe. That statement is somewhat facetious, but not entirely. Nothing in this Universe that science is aware of will allow one to precisely pin down in time when the radioactive decay of a single atom will occur. This fact should be surprising.

Some will argue that a random process is wholly defined by statistical properties, such as the corresponding standard deviation and mean, which are attached to and define the process. And, in a sense, this is true when viewing the decay

3.7. RANDOMNESS AND INFORMATION

of an ensemble of atoms, but it says nothing about the decay of an individual atom. Individual atoms decay after a completely unknown time. The decay has no dependency on anything in physical reality, as will be shown in the next section. If it were dependent, it would be, in theory, be predictable.

Deterministic systems cannot produce random events. Moreover, they cannot produce new information other than the information contained in their construction or definition. This fact is intuitively clear, but let us show it explicitly.

A deterministic system, $f(\cdot)$, can be represented as a set of sequential, programmatic rules, like a computer program. Here, this system will be modeled as a set of programmatic rules that act upon a vector of input data $\mathbf{d} = [a_0, a_1, \cdots, a_N]^T$. The program can perform as many operations as it wishes upon this sequence to produce an output. The only requirement is that all of these operations are performed by fixed instructions. For example, if the first bit of data is acted upon by the function $f_0()$ to produce the output x_0, then it is always operated upon in this fashion and produces this output. The same is true for every instruction in the program, producing an output vector $\mathbf{x}$ from the input vector. We can write this program in a succinct form as $\mathbf{x} = f(\mathbf{d})$, where $\mathbf{d}$ is an input vector, $f(\cdot)$ is a function modeling the program, and $\mathbf{x}$ is an output vector. This expression states that the fixed function $f(\cdot)$ acts upon the input vector $\mathbf{d}$ to produce the output $\mathbf{x}$. The point is that no matter what input combination is chosen, the mapping between the input and output is always determined by the same function or program $f(\cdot)$. Nothing ever changes. Now, if we assume that the function has just been formed and this is the first time it has been applied, then it would seem that some new information has been gained from this function (it depends on just how $f(\cdot)$ was created), but no new information will be created after the first usage.

Assume that the information contained within the construction of $f(\cdot)$ is taken into account, and this function does not change with time. (Of course, no generality is lost here. If the function does change deterministically with time, we can roll these changes into a more complicated function.) The question at hand is whether or not such a function can generate new information.

Another way of looking at it is that a deterministic system's functioning and, hence, its information content, is known *a priori*, and that does not change during the time the system operates. In other words, the operation of the programmatic rules is precisely known from the outset. The output from a program is simply a fixed set of rules applied to input data. It does not allow any changes to the way it processes data. Therefore, a program's output simply repeats for a specific input, although it may well have an extended period. Pseudo-random number generators are designed in this way. They are fixed, discrete systems that produce outputs that appear random over long periods but are not random. Given enough time, they repeat.

Any deterministic function or program can be written as above. Again, if we let this function act on an input data vector, we have $\mathbf{x} = f(\mathbf{d})$. The function produces a specific output vector from a specific input vector. There can be more information contained within the output vector than in the input vector that derives from the function $f(\cdot)$. However, the total information content in the output cannot exceed the sum of information contained in the input data and the function $f(\cdot)$. Since $f(\cdot)$ does not change, the total system information is fixed from this perspective. The only place new information can enter the system is through the input data.

If the function $f(\cdot)$ happens to be invertible, that is $\mathbf{d} = f^{-1}(\mathbf{x})$, then we can say that the system conserves information. If this function is not invertible, then there is no one-to-one correspondence from input to output, and some information within the input must be lost at the output.

I have taken a heuristic approach to demonstrate that deterministic systems cannot generate new information, but I think the result is obvious enough. Deterministic systems cannot produce new information because their complete behavior is known beforehand and it does not change.

Let's now switch back to the subject of randomness. Switching between these two topics will make sense shortly.

One measure of randomness is called *Kolmogorov randomness* and was developed by Andrey Kolmogorov. It says that a string of bits is random if it is impossible to reproduce the string from a computer program whose bit string is *shorter* than the bit string in question. This statement has strong implications for the definition of information. In fact, we will see as we go on that the degree of randomness expressed by a particular sequence of symbols is directly related to its information content. Randomness is not just simple noise with no relation to information. The same Kolmogorov measure is used to measure information. In other words, the information content within a string of bits is measured by the length in bits of the shortest computer program that can recreate that string. In either case, if the length of the program required to reproduce the random or informative string is not shorter than the string, then the random string is indeed random, and the informative string contains information.

Information compression is an example of representing a sequence of data by a computer program. When a software program is used to compress the data contained in a file, the size of the compressed file compared to the size of the original data is a measure of the information content of the original data file. The more the data can be compressed, the less information was contained in the original file.

Interestingly, the same holds true for random sequences. As mentioned above, a pseudorandom number generator is a program that generates a sequence of numbers that appear to be random but actually repeat with an extremely long period. The output sequence may not repeat until after 10^9 or more numbers have been generated, but it does repeat. The size of the program required to produce the output sequence is a measure of the randomness contained in the sequence. True random number generators do not repeat, and the length of the program required to produce the random output has infinite length.

The amount of information or randomness that a sequence contains is ex-

3.7. RANDOMNESS AND INFORMATION

pressed in the difficulty of reproducing the sequence. Consider the two sequences: $A = 1011101001$ and $B = 1010101010$, each of which is ten-bits long. If they were formed by simply flipping a coin and assigning 1 to tails and 0 to heads, the probability of *both* of them occurring would be $(1/2)^{10} = 1/1,024$. However, the amount of information or randomness contained in them in the Kolmogorov sense is not signified by this probability number. Obviously, the first number cannot be easily represented by a sequence smaller than itself. However, sequence B can be represented by a shorter set of bits, for example, 101 : 10 where the first three bits from the left denote the number of times, $101b = 5$, to repeat the last two bits, 10. In this sense, there is more information contained within A than in B. However, the same holds true for the measure of randomness. The measure of randomness in sequence A is larger than that in B simply because A cannot be represented by a shorter sequence of bits.

We have already established that the source of randomness cannot be found; it is a mystery to us. This point has even bothered some of the giants in physics, theology, and mathematics. Einstein did not enjoy the idea that the Universe was not ultimately deterministic. Randomness shows up as noise in physical systems, provides justification for accidents, and places some physical predictions out of reach of science. This book intends to take the notion of uncertainty one step further by including information in the same category as randomness. This may sound counterintuitive the first time it is heard, but it will become more apparent as we go along. The same alchemy that produces randomness also generates new information.

It is important to note that when the term information is used in this book, we will be referring to new or novel information. Of course, a snapshot of a picturesque landscape is information, but it is not original information. It is a replica of a portion of the information already present in reality. Many ID proponents are correct when they make the claim that specified complexity only comes from an intelligence. However, the assessment can be made more explicit. As we will see later, randomness is one of the primary drivers of intelligence. It is the source of originality. In that sense, specified complexity and novel information are spawned directly from uncertainty as it passes through physical processes.

Systems that have a fixed structure and are deterministic cannot produce randomness. Their output is entirely determined from the input and the system model at any point in time. Systems that produce a random output violate this constraint; their outputs cannot be determined solely from their inputs and a system model. Their outputs are completely unpredictable except in the average.

The evolution of a deterministic system can be written as

$$\mathbf{x}(t) = \int_t f(\mathbf{x})dt + \mathbf{x}(0), \qquad (3.9)$$

where $\mathbf{x}$ is a state vector representation of the system and $\mathbf{x}(0)$ is the initial condition vector. This is Newton's clockwork Universe. It can be integrated both forward and backward, into the future or the past, and the deterministic trajectories never change. Everything is predetermined or destined, and there is not a thing that anyone can do to change it. Of course, this framework caused many problems for theologians, for how can anyone, not truly in charge of his or her destiny, be justifiably culpable for their actions? The way around this problem was to introduce (or reiterate upon) a non-physical soul that could, through some mysterious power, bring about non-contingent events or actions. Descartes was one of the first to attempt this synthesis of mind versus matter. Intelligent design advocates are attempting a contemporary version of the same thing by claiming that intelligence or agency lies within the supernatural.

Rewrite the above equation as

$$\mathbf{x}(t) = \int_t f(\mathbf{x}) + \delta(t) dt + \mathbf{x}(0). \tag{3.10}$$

Now things have changed substantially. A random function $\delta(t)$ has been added to the integrand, and Newton's clockwork Universe is broken. As this new system is evolved through time, its output will be different than before and unpredictable. If we look at the state vector at any particular instant, we will likely see that new information has been added to the system. For instance, if this is the model for the Universe, in the deterministic case, the output would be entirely predictable and would not change if one traveled backward in time. No new information would ever enter the system. However, with the second system, new things start happening immediately. If one looks closely, it will be seen that the positions of the stars have deviated ever so slightly from their deterministically-predicted positions. Possibility has been breathed into the Universe.

With the discoveries that led to the development of quantum mechanics in the early 20th century, Newton's deterministic view was dethroned, and uncertainty reentered into reality, and the world became alive again. No longer were we stuck in a predestined timeline; fate and our choices once again played an essential part in our lives. But we are getting ahead of ourselves here. We will leave the philosophical and theological implications for a later chapter. Nonetheless, the description of randomness and information needs to be finished.

The creation of novel information is intimately related to the functioning of intelligence. Merriam-Webster defines *intelligence* as *the ability to learn or understand or to deal with new or trying situations; to reason.* This definition appears to leave out something essential. The fresh insight of the poet is not captured. The unbounded genius of an Einstein seems to not fall under this definition either. There is something new and difficult to grasp that arises in both of these cases. It was not there beforehand but is there when the work is done. We will talk more about this later, but there is a "newness" that emerges from intelligence that can not be qualified as simple understanding or smartness.

The randomness associated with quantum-mechanical events such as the decay of atomic nuclei is completely unpredictable. It is unprecedented, the uncaused cause. The intelligent creation of new information is also completely

3.7. RANDOMNESS AND INFORMATION

unpredictable. Because we have become accustomed to thinking we know what intelligence is—we associate it with being able to modify the world around us, coming up with new ideas, understanding mathematics, etc.—we forget that we may not be able to understand it fully. What makes intelligence so unique? In other words, when presented with a problem that needs a solution, such as the Tower of Hanoi puzzle [27], do we really believe that the solution simply appears out of thin air for an intelligent person?

The Tower of Hanoi puzzle is a game that has three colinear posts and five disks of different sizes with holes in their centers. It starts with all five disks stacked up on the leftmost post, largest to smallest, from the bottom to the top. The goal of the game is to move the disks from the leftmost to the rightmost post subject to three rules: (1) only one disk may be moved at a time; (2) disks may be moved by placing them on top of another stack or post; (3) no disk may be placed on a smaller disk. I remember solving this problem in my head on a boring trip back from Kansas City, MO, sometime near the year 2005. The solution involves finding a perfunctory pattern to be followed. How is this pattern found? It is not by deductive or inductive reasoning—those only follow afterward. The solution process starts with a random thought of how to proceed, and that solution is then tested virtually in the mind.

Our thinking process is strikingly random. Any reader who has ever solved an algebra problem will be familiar with the uncertainty involved in this process. Unless one's proficiency in mathematics has gotten to the point where many solutions are simply functions of rote memory, which is certainly possible, one must come up with a unique solution to a problem that presents itself as being unique (as there is no reference unless one has seen the form of the problem previously). Try it right now on a new problem. Watch close, and it will become immediately apparent that the solution does not just appear out of the ethereal depths of the mind. Instead, the mind struggles and presents seemingly random things to try, which are subsequently validated in the virtual reality of consciousness. Only after a particular solution is validated to a satisfactory extent is it selected, and all others are promptly forgotten. It is an astonishing discovery that intelligence stems from random thoughts over which we have no conscious control. This is the only way it could be if an entirely new concept or process is to be invented. This new concept cannot come from the knowledge already present in the memory. It must come from somewhere outside of our fixed knowledge base. (Of course, I am not saying that a contextual grounding in memory is not needed. Instead, I am claiming that novel ideas do not originate solely from the deterministic usage of memory.) We will pick this topic up again later.

It is not that random behavior is the sole component or the most significant part of intelligence; it is not. Nevertheless, it is indeed the proverbial wellspring of new information and novel ideas. The state of a random variable is seemingly determined exogenously to reality. The justification for this claim is that its precise value cannot be determined from *a priori* information from this reality. The claim is not that random events are caused by supernatural means or magic but that they do happen in this reality irrespective of the cause. Furthermore, by definition, they introduce new information into the universe *as they cannot be predicted from previous state information*. If such information were present, possibly encoded in the laws of the universe, we would be able to predict random events, which we are not capable of doing.

Random events do not cause things to happen. Randomness is a description of what happens in reality; it is not a cause. One cannot say that a flipped coin lands on heads because randomness made it do so. Instead, the behavior of the coin expresses randomness. The distinction is significant. Saying that disorder is introduced into the Universe is incorrect. It is correct to simply state that the Universe evinces disorder. Whether or not anyone likes it, random behavior is a component of reality. Furthermore, it is not some side-show effect. It is an integral component that allows the Universe to modify itself in ways that would otherwise be barred by determinism. As we will see in later chapters, randomness is not simply the malady of a forgetful mind, and it is not the random genetic mutation that leads to disease. As counterintuitive as it might seem, randomness is the source of creation.

3.8 Bell's theorem and indeterminism

In 1935 three physicists—Einstein, Podolsky, and Rosen—submitted a paper to Physical Review entitled *Can quantum-mechanical description of physical reality be considered complete?* [28,29]. The purpose of this paper was to address what Einstein perceived as a blatant fault within the still burgeoning field of quantum mechanics. As we have already talked about in Chapter 2, the Copenhagen interpretation of quantum mechanics states that the properties, such as position, momentum, etc., of quantum objects are represented by probabilistic wave functions. Each property of a particle has a probability wave defined by the solution of the Schrödinger equation associated with it, the square of which relates the probability of the property taking a particular value.

This indeterministic character of quantum mechanics leads to a set of uncertainty relations developed by Werner Heisenberg and given by $\Delta x \Delta p \geq \hbar$ and $\Delta t \Delta E \geq \hbar$, where we have defined $\hbar$ previously. The variables Δx, Δp, Δt, and ΔE represent the inherent uncertainty in position, momentum, time, and energy, respectively. The first relationship says that the uncertainty in a particle's position multiplied by its momentum uncertainty can never be less than $\hbar$, meaning that if the position is known precisely, then the momentum is entirely unknown and vice-versa. By the second relation, the same is true for the uncertainty in a particle's energy and the time its energy measurement is made.

Einstein's intent with the 1935 paper was to demonstrate that these relations were incorrect by using an ingenious argument in a thought experiment. The wave function does not just describe the properties of individual particles. If two particles are brought together so they interact with one another and then are separated, they subsequently share a composite wave function describing the properties of the particles as a whole. Although the term was not used at the time, this process would come to be known as *entanglement*. In other words, the wave functions of two particles become "tangled" up with one another to form an overall wave function. From that point on, the two are linked; interacting with one affects the other instantaneously no matter how far apart they are. I want to forego repeating the hackneyed quote of Einstein's sentiment on this topic, but I had better relay it for completeness. Einstein called this behavior "spooky action at a distance" and believed it indicated that the theory of quantum mechanics was incomplete because instantaneous interaction shouldn't be allowed.

Quantum mechanical theory said that measuring, for example, the position of the first particle not only precisely determined its position but also the position of the second entangled particle. Moreover, from the Heisenberg relations, knowledge of the first particle's momentum became undefined with the measurement of its position, and so did the momentum of the second particle. The part that bothered Einstein was how the second particle knew instantaneously that its position had been determined by measurement of the first particle.

The three authors of the paper suggested the following thought experiment. Suppose two particles are brought into contact with one another such that their wave functions become entangled, resulting in the composite wave function $\Psi(\alpha_1, \alpha_2)$, where the numerical subscripts indicate particles one and two, and then are allowed to separate. We are left with the two particles separated in space but sharing the same wave function.

Now, assume we measure the position of particle 1, causing a collapse of the wave function to give a new total wave function Ψ_α. Since the original wave function described the complete system, the subsequent function does as well and ultimately defines the properties of particle 2. The authors then assumed that instead of measuring the position of particle 1, they had measured its momentum. Since this measurement will cause the wave function to collapse differently, the resulting wave function is denoted Ψ_β in this case. This wave function also defines the properties of particle 2. And this is the EPR paradox: despite no interaction between the two particles (they assumed there was no instantaneous communication between the particles), for different measurements made of two different properties of particle 1, there are *two different realities presented for particle 2*, Ψ_α and Ψ_β. Based on this result, Einstein, Podolsky, and Rosen concluded that the quantum mechanical description of reality was incomplete.

So, starting with the proposition of quantum mechanics being true, the EPR experiment resulted in what appeared to be a contradiction. The authors of the paper bet their money on the contradiction being a result of a problem with the quantum-mechanical theory. However, as we shall see, the problem was not with quantum mechanics but rather was due to an incorrect assumption about the determinism and local behavior of reality itself.

One possible way out of the dilemma of the EPR paradox was to dispense with the assumption that reality is local, that is, that information cannot be transferred from one location to another instantaneously. After all, Einstein had been the one to show by his special relativity that nothing, including information, could travel faster than the speed of light, and certainly not instantaneously, and he was far more predisposed to conclude there was a problem with quantum-mechanical theory than he was to admit that the Universe can be interconnected at a distance.

If quantum mechanics were true and entanglement does lead to an encompassing wave function defining the behavior of the two particles, then when particle 1 is measured, the overall wave function is modified, and this takes effect *immediately* for particle 2 as well. As we will show in a moment, this is the correct explanation for the EPR paradox. Interaction with one of the particles is equivalent to an interaction with the other. Reality is nonlocal.

It is important to understand that nonlocality does not imply that there is some amount of information physically traveling from one particle to the other, the purpose of which is to inform particle 2 of the state of particle 1. Instantaneous interaction is another way of saying that two things are intimately connected. Sharing a common wave function says that the two particles are essentially one system despite being separated by a distance—one changes when the other changes because they are parts of the same system. In fact, we could go further and say that they are essentially the same particle.

In today's world, the EPR result does not elicit a great deal of consternation, but it certainly did when it was presented in 1935. Many physicists tried to find a way around the perceived contradiction. One of the prevailing theories at the time was called *hidden variables* and was the notion that the particles carried state information with them from their point of interaction. In other words, it is not that information is transferred instantaneously from particle 1 to 2 but, rather, the information was already there in particle 2. When the two systems interacted and became entangled, information was imparted to both, analogous to data written to their internal hard drives instructing them how to behave when later measured.

Niels Bohr responded to the EPR paradox paper and pointed out that the physical sciences only demonstrates *how* nature behaves, not *why*, and that the claims being made about hidden variables were untestable and therefore outside the purview of scientific inquiry. It was left up to an Irish physicist named John Bell to take up this subject in a 1964 paper entitled *On the Einstein Podolsky Rosen paradox* [30] and demonstrate that it was indeed possible to test the notion of hidden variables. The result was one of the most ingenious experiments ever conceived.

Consider the diagram depicted in Figure 3.23 where an emitter emits an entangled electron-positron pair, the electron traveling to the left and the positron to the right. Conservation of momentum dictates that one of these particles has spin up and the other spin down. Every time two particles are emitted, one is always measured spin up and the other down. This experiment has been performed so many times that we are certain that the particles always have opposite spins.

3.8. BELL'S THEOREM AND INDETERMINISM

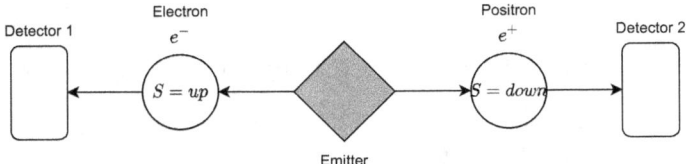

Figure 3.23: An emitter of entangled electron/positron pairs ejects these particles in opposite directions toward two detectors. The spin of one particle is up and the other down.

Spin has a connotation of angular momentum in quantum mechanics. I use the term connotation because spin is only akin to momentum; it is not precisely momentum. However, just like momentum, the spin of a particle can be measured along any axis, for example, the x, y, and z axes in a Cartesian system. One striking feature is that the spin of an electron or positron is always up or down when measured along any axis. That is, the angular momentum is quantized; it is not continuous.

In order to show what Bell proposed, I am going to follow an approach taken by Brian Greene and Bob Eagle [31, 32]. Let's set up our experiment like that given Figure 3.23. Then assume that we are going to measure the spin along the axes shown in Figure 3.24. Each of these numbered axes is separated by 120°. Many pairs of particles are fired from the emitter and the spin of each is measured along one of the three axes. The particular axis is chosen at random. Remember, we cannot measure all three axes for a given particle because any one of the measurements leaves the quantum wave function of the particle in a different state. Therefore, we have to make large numbers of single-axis measurements and average them to obtain the total system response.

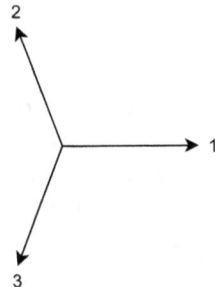

Figure 3.24: The three possible measurement directions of the spin of the electrons and positrons. (The vectors are 120° apart.)

So, the process goes like this:

1. Fire an entangled electron-positron pair from the emitter.

2. Detector 1 randomly selects an axis along which to measure and makes the measurement.

3. Detector 2 randomly selects an axis along which to measure and makes the measurement.

This process continues until we have a large amount of data that can be averaged to work out the overall results. However, if nature is really using hidden variables that are defined during the interaction of the particles within the emitter to record the spin values in each axis, then there can only be a finite number of measurement outcomes, and we can delineate these.

There are eight combinations for the spin measurements for any pair of particles in this experiment. This value comes from the fact that there are three axes along which the spin can be measured, and the spin can only take on two values, up or down. Therefore, there are $2^3 = 8$ combinations. (Each particle has three possible spin measurements, but one determines the other as they must be opposites along any axis. This means there are only 2^3 combinations and not 2^6.) Four of these combinations are shown in Figure 3.25. The electron and positron spin configurations are given at the top of each sub-figure, with the first arrow representing the spin along axis 1 of Figure 3.24, the second arrow corresponding to axis 2, and the third to 3. Note that the spin for any of the three locations must be opposites for the two particles.

The numbered pairs in the matrix below the particle descriptions shows the possible outcomes for any one measurement. The number pair at the top of each element of the matrix denotes the axis along which each of the detectors is measured. For example, an entry of 1, 2 means that the first detector measured along axis 1 and the second detector along axis 2. The arrows below the indices of the axes show the spin values for each of the measurements. The top-left entry of Case A shows that the first detector measured along axis 1 and obtained spin down while the second detector also measured along axis 1 and found spin up. The top-right entry of Case A shows that the first detector measured along axis 1 and measured spin down while the second detector measured along axis 3 and measured spin up. There are nine combinations for each electron-positron measurement combination, hence the three-by-three matrix of outcomes. (Don't get this confused with the possible axes' combinations from above.)

Case A showed the outcomes for an electron spin configuration of {*down, down, down*}. Case B gives the results for an electron configuration of {*down, up, down*}. Cases C and D display two other possible configurations. Again, there are eight in total.

And here is the point of Bell's assertion: we can set a bound on the probability of making measurements of opposite spin, that is, {*up, down*} or {*down, up*}, across all possibilities. For Case A in the figure, the probability of measuring opposite spins is obviously unity as all possible measurements render opposite

3.8. BELL'S THEOREM AND INDETERMINISM

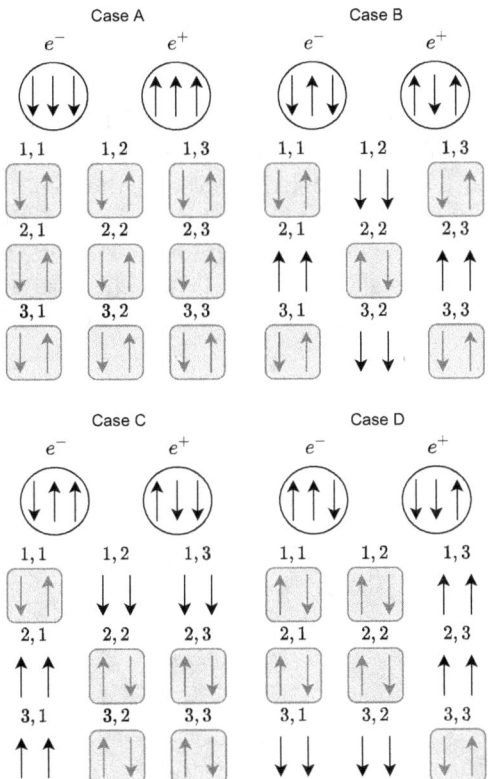

Figure 3.25: Under the assumption of hidden variables, the possible electron and positron spin values in the three axes are shown above each of the four figures. Below the particles are shown the possible measurements by the two detectors given the selected axes.

spins. However, for the remaining cases in the figure, the probability is 5/9 because five out of the nine possible measurement scenarios give opposite spins (they are emphasized with the gray backgrounds). If we were to list the remaining four cases of electron spin configuration, we would again obtain 5/9. Therefore, the ultimate conclusion is, if there is hidden information carried by the particles from the emitter, the probability of measuring opposite spins by the detectors, given a random selection of measurement axes, *must be greater than* 5/9 (this is a greater-than relation because of the two cases with probability equal to one). This result is often called Bell's inequality and can be written as:

$$\text{Probability of opposite spins} \geq \frac{5}{9} \qquad (3.11)$$

If we actually perform this experiment, which scientists have done many times [33, 34], the inequality above is violated. Quantum mechanical theory, on the other hand, predicts exactly the measured value. The EPR paradox resulted not from quantum mechanics being incomplete in its description of reality but rather because of the author's assumptions about reality. Contrary to their assumptions, reality really is nondeterministic and nonlocal.

In order to see this discrepancy a little clearer, consider the same experiment as above only with both detectors fixed to measure the spin in only axis 1. In that case, they are guaranteed to measure opposite spins. Then the angle between the sensors' axes of measurement is adjusted from 0° to 360° and the measurement correlation between the sensors recorded [35]. The result is shown in Figure 3.26 where the classical (hidden variable) and quantum-mechanical responses are plotted.

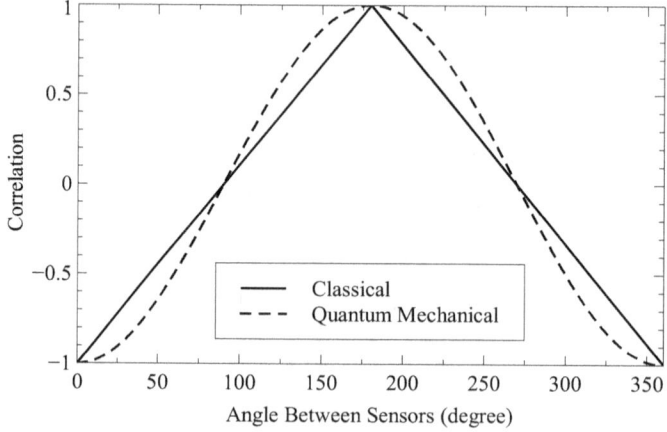

Figure 3.26: The correlation of spin measurements of two entangled particles as the angle of the detectors is varied. Both the expected classical (solid) and the quantum mechanical (dashed) results are shown in the plot.

Note the sharp responses at 0°, 180° and 360° of the classical response vs. the sinusoidal response of the quantum-mechanical result. This difference has been enough to violate Bell's inequality many times in experiments. The implications are manifold. The first and potentially weirdest implication is that the Universe is nonlocal. That is, different regions of the Universe are connected to other regions as if they *were those regions*. For example, in our experiment above, if the two detectors are separated by the width of our galaxy, one-hundred thousand light-years, when the spin of one particle is measured to be up, the spin of the other is immediately selected to be down. Astonishingly, Bell's inequality shows that the results obtained by measuring these two particles were not predetermined. Therefore, these two particles must be connected across an incredible distance.

The second implication, and the most pertinent in this book, is that the spins of the two particles are not determined before they are measured. *Reality is not deterministic.* This is an astounding revelation and opens up all of reality to profound possibility.

Using Heisenberg's uncertainty relationships above, if a measurement of the momentum of an electron is made, what do you think the possibilities are for subsequent measurement of its location in space? Let's rearrange the uncertainty relation to calculate its bound: $\Delta x = \lim_{\Delta p \to 0} \hbar/\Delta p \to \infty$. As the uncertainty in the electron's momentum goes to zero due to measurement, the uncertainty in its position goes to infinity, meaning it can be anywhere. This is real magic, the wellspring of real miracles, and opens up the potential for all things.

In a deterministic universe, the future is fixed. In theory, it would be possible to simulate such a universe and predict the future, even the actions of humans, as there would be no free will. We can never be responsible for our actions in such a universe as we do not author them. Creativity is only a facade. Anything we do is what we have been made do by determinism. There is no will to do anything else as all things must play out as the initial conditions of the Universe destined them.

It would not be fun to live in a deterministic universe unless one did not realize it was the case. The ID solution to this problem is the same as the religious solution: to bring a god or designer into the picture who is capable of spontaneous, or dare we say, random, actions. Randomness is banished from the natural world and reintroduced via the supernatural. In doing so, free will and spontaneity are added back into reality in a contrived fashion. It is a superfluous addition.

But don't come to any fixed conclusions about free will just yet. It may turn out before we're finished that free will isn't exactly what you might think.

Of course, there is no way to prove that the ID vision of reality is incorrect. Indeterminism is indeterminism, no matter if it is a part of the natural world or simulated by a deity; we cannot tell the difference from our vantage point. We simply see random events happening. However, Laplace's answer to Napoleon when he asked why Laplace's theory of the Universe contained no mention of God seems appropriate here as well. Laplace simply stated, "I had no need of that hypothesis." And neither is it needed here.

3.9 The end of reductionism

Reductionism, at least by one common definition [36], is "[...] any of several related philosophical ideas regarding the associations between phenomena, which can be described in terms of other simpler or more fundamental phenomena." In other words, reductionism, as the root "to reduce" implies, comprises replacing complex concepts with simpler ones. This is a fundamental method by which science makes complex phenomena understandable and expressible.

An example is the set of Newton's laws, which derive from the relativistic formulation when speeds much less than that of light are considered. The Lorentz scaling $\sqrt{1 - v^2/c^2}$ at speeds much less than light, $v << c$, is approximated by a value of unity. It is not until v becomes a significant fraction of c that the value starts falling appreciably below one. Another example would be the ideal gas law that, for example, reduces the momentum of a large number of gas particles over a specified surface area to a pressure. Although there are billions of particles in motion, the average net effect of all of them can be expressed as a force per unit area or pressure.

The expression for momentum for a single object is $p = mv$ where p is momentum, m is the object's mass, and v is its velocity. If we take a time-derivative of this expression, we get $dp/dt = m\,dv/dt$, and since acceleration is $a = dv/dt$, this expression reduces to Newton's second law with force equal to dp/dt. Now, let's expand this to an ensemble of particles as shown in Figure 3.27.

Figure 3.27 shows an ensemble of gas particles moving in random directions impacting a flat surface. We do not need to go through the math here, but each of these particles is moving at a particular speed and has a corresponding momentum. Using the expression for force we just derived, as these particles hit the surface, their momentum translates to a force on the face of the surface. Each particle is small, but a large group of particles can create a large force per unit area, or pressure, on the surface. Pressure is the averaged (actually root-mean-square) force per-unit-area the particles impart to the surface. In other words, pressure is a quantity found through the reduction of the behavior of an ensemble of particles; it is an emergent property, and it is also an approximation of reality. The expression for pressure only becomes exact in the limit as the number of gas particles per unit volume approaches infinity, which is never really the case.

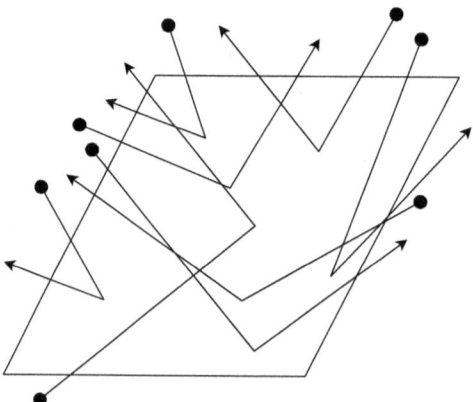

Figure 3.27: Particles of a gas impacting a surface. The result is a pressure upon that surface.

Of course, as the measurement precision increases, no system admits reduction. The indeterminism inherent in quantum mechanics evinces the failure of reductionism particularly well. Macroscopic physical laws appear to describe the behavior of large objects quite well. However, when one looks at the microscopic level, it becomes clear that the macroscopic laws are only an approximation.

3.9. THE END OF REDUCTIONISM

Reductionism also fails for chaotic and complex systems. Although the behavior of these systems is deterministic, their sensitivity to initial conditions renders them unpredictable in a world that admits quantum randomness. The physical laws that govern the dynamics of the atmosphere can be expressed to a high degree of accuracy. Despite this fact, from these laws chaotic behavior emerges, not allowing long-term prediction of the system states no matter how accurately the initial conditions are estimated. Complex systems are notoriously unyielding to reductionism. In general, the behavior of cellular automata cannot be encapsulated in a simple expression. These are simulate and see systems. The only way to ascertain the final state of a complex system is to actually evolve it with respect to time and see what comes out.

As we will see later, the refusal of these systems to admit to reductionism is intimately related to what Michael Behe would term as irreducible complexity [37]. These systems are emergent in nature—they produce behavior that is seemingly more than the sum of their parts. In his book *A World Beyond Physics* [38], Stuart Kauffman refers to the notion that we cannot predict what type of biological complexity will emerge. Although I do not agree with his assessment that these systems are not describable by physics, I do agree that there is no way to predict their outcomes without actually letting them evolve. This is a consequence of the mathematical nature of complex systems. It is still physics as physics is simply observation and description of how the natural world behaves. The fact that mathematics and physics bar us from writing a reductionist framework that describes and predicts biological complexity is a wonderful thing. Nature will not readily submit to our attempts to control it intellectually. However, physics is still intact.

As it turns out, reality does not admit to reductionism without error. This is a direct conclusion of the result of the previous section. Reality is not deterministic, while mathematics is, and this causes an intractable breach between the two. Although there is a way to write an expression that includes the half-life of a radioactive material which describes exactly how the complete ensemble of atoms in a sample will decay, there is no way to write down an expression for the decay of a single atom in the sample. Even if the experimenter knows every parameter describing the atom, there will still be no means of determining when the decay event will occur.

Have you ever heard or read about a scientist stating that the more she learns about the Universe, the more complicated it becomes? That discovering the answer to one question opens up ten other unanswered questions? Reductionism is the reason why. It appears this way because the first question was never answered—it was a reductionist approximation—and the emerging ten questions represent the error in the reduction. For example, Newton's laws of motion were the answer to the first question. Einstein's relativity was the answer to the question elicited by the error in Newton's laws. A complete theory of quantum gravity will be the answer to the question brought to light by relativity. And so it goes, level after level.

It is very tempting for engineers and physicists to believe that there is a bottom to this well. That is, there is a final answer to the string of questions. And, in a sense, there is, but it will not be a simple law. And I cannot prove this, so consider the remainder of this section speculation. No one has an answer to this question, but let's see what we can intuit about it through discussion.

Recall the cellular automata of section 3.6. Although these systems can be described by a very simple equation that tells one how to generate the next row from the previous, there is no means by which to write a function that outputs a given row from only the initial state. That is, the function $\mathbf{x}_j = f(\mathbf{x}_0, j)$, that takes as inputs the initial row $\mathbf{x}_0$ and the desired row j, and returns the row $\mathbf{x}_j$, does not exist.

The same is true for chaotic systems. Although the Lorenz equations are elementary to write down and simulate, it is, at this point, impossible to come up with a general expression that provides the state of the system given an initial condition and a duration of time. In other words, no closed-form solution exists for the Lorenz system. The only way to calculate the state of the system at time t given an initial state $\mathbf{x}_0$ is to simulate it and see.

It is interesting to note that with complex and chaotic systems, the underlying dynamics of a system can be precisely expressed mathematically, and still, there is no way to predict its behavior. This, in and of itself, is enough to say something fundamental about reality, including abstract, conceptual reality. Complex systems are the first Achilles' heel of reductionism.

The second problem is even more subtle and insidious. I will not go through the details here, but the mathematician Kurt Gödel proved that mathematics as a whole cannot be simultaneously complete and consistent. What this means practically is that there are mathematical statements that cannot be proven true or false. Now, how is this possible? For anyone who deals with mathematics on a daily basis, this claim seems absurd. However, it is true. To be fair, Gödel's proof never showed how this was possible since it was proof by contradiction. Nevertheless, it is quite unsettling to be working within a framework that we know will fail at some point. This doesn't mean that reality is inconsistent. Nature will go right on. But it is possible our analysis can be stymied.

Science sets out to demonstrate that aspects of reality are predictable. In order to do this, at least initially, it chooses low bars. Simplified expressions that only approximately model the behaviors of collections of components of reality are natural starting points. The containers used to express these collective behaviors are norms, such as the 1-norm (average) or 2-norm (root-mean-square). They define precisely only the average behavior of groups and say absolutely nothing about individual behavior. Is this important? It most assuredly is.

There is no real way to model what cannot be modeled. Uncertainty can be characterized, but it cannot be predicted. When randomness is coupled with the large sensitivities of complex systems, the time-evolution of these systems potentially becomes completely unpredictable. One minute change induced through random noise can completely alter the trajectory of the system. It is certainly possible to mathematically parameterize the possible evolutionary paths of the system, but it is impossible to determine in advance which path is taken. Since nature exhibits chaos and complexity, randomness becomes the final nail in the coffin of reductionism.

Even if one had a god-like computer that could simulate the assumed 10^{80} particles in the Universe, how could the Universe be accurately simulated? We just demonstrated in the previous section that reality is nondeterministic. This means that the states of these particles cannot be accurately calculated as time advances. Generally, a simulation is designed to integrate state equations from one time step to the next. In this case, there are no inputs, and this simulation is a self-contained set of differential (or difference) equations. However, the bugaboo is caused by the states being indeterministic. Although it is possible to constrain the uncertainty, there is no way to calculate the equations from one time step to the next, and this is severely compounded when chaotic systems are involved.

It is not the intent of this section to claim that reductionism is not useful—it certainly is useful. Reductionism has been incredibly useful up to this point. However, with the emergence of chaotic and complex systems, we are entering a new territory where the power of reductionism is greatly reduced. There is no way to reduce a system that cannot be reduced. At its best, reductionism is an approximation, and, at its worst, it is not even applicable. We are entering the worst-case territory. Although reductionism will remain a constant companion as we move forward, it is likely to fail in its attack on biological complexity. As we'll see later, this is one of the mistakes the ID community is making—assuming that something more exists, and yet still relying on reductionism.

3.10 An uncaused cause

A general talking point in theological philosophy and scientific debates is the question concerning a prime mover or uncaused cause. In other words, the argument is that for anything that begins, there must be a cause for its beginning. For example, the Kalam cosmological argument is a very well-known syllogism that deductively proves there is a first cause for the Universe [39]. The argument goes like this:

1. Everything that begins to exist has a cause.

2. The Universe began to exist.

3. Therefore, the Universe had a cause.

Philosopher William Lane Craig further expands this syllogism by adding two more premises and a second conclusion:

1. The Universe has a cause.

2. If the Universe has a cause, then an uncaused, personal Creator of the Universe exists who *sans* (without) the Universe is beginningless, changeless, immaterial, timeless, spaceless and enormously powerful.

3. Therefore, an uncaused, personal Creator of the Universe exists, who *sans* the Universe is beginningless, changeless, immaterial, timeless, spaceless and enormously powerful.

The original form of the argument is believable if we buy premise one that everything that begins to exist has a cause. However, the other syllogism by Craig I add here only for the reader's amusement. The second premise is the granddaddy of all presumptuous statements.

I do not want to draw this discussion out too far because it is not directly related to the thesis of this book; however, I want to demonstrate a mathematical violation of the first premise of the cosmological argument above. It is only relevant in that it again demonstrates how the religious community tends to make blanket statements that cannot be proven.

We've already discussed that quantum randomness is essentially an uncaused event. However, in this short anecdotal section, I want to demonstrate how an event without a cause comes about in a deterministic environment.

Consider the existence of an infinite number of sinusoids extending in time from negative infinity to positive infinity. This means that these waveforms have no beginning or end; they exist for all time, having no beginning and no end. (Of course, for this simple demonstration, I am ignoring the fact that there are situations, like the big bang, before which time probably did not exist. The time we experience today need not be the independent variable.) A small sampling of such a set of sinusoids is shown in Figure 3.28 below. Naturally, there are only a small number of sinusoids in this figure for clarity. It is only intended to provide a flavor of the situation we are looking at here. Although the horizontal axis has been clipped in the plot in the figure, in our example, the sinusoids extend infinitely in both directions.

The question is, what happens when we add up this infinite number of sinusoids? We cannot do this in a computer simulation—the number of sinusoids and their extents must be finite. (We could work this problem out mathematically using a Fourier transform, but that would not be as illuminating.) Figure 3.29 shows the summation of five hundred sinusoids whose frequencies range from -25 to $+25$ rad/s in 0.1 rad/s steps. The result of the summation is astonishing. From negative infinity to zero, the cumulative signal is zero. The same is true from zero to positive infinity. However, right at time equal to zero, a delta function emerges. Formally, a delta function is a short-lived spike of zero width and infinite height. (The width is not zero and the height is not infinity in this approximation. I have scaled the height here to fit in the plot.)

It is important to understand that there is nothing to the left of the origin. If the horizontal axis in the plot represents time and the vertical axis events, then nothing is happening for all past eternity before time is equal to zero. Moreover,

3.10. AN UNCAUSED CAUSE

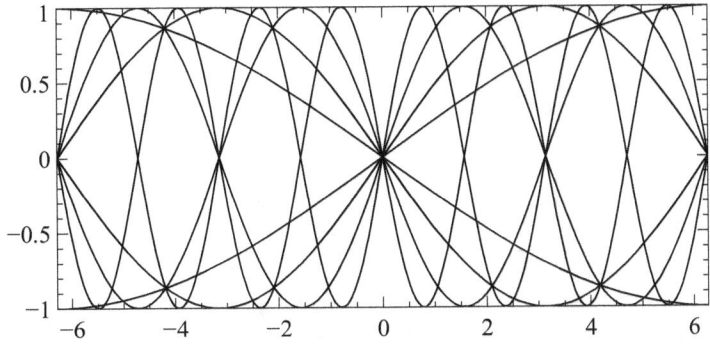

Figure 3.28: A series of sinusoids of different frequencies.

at that point, suddenly, there transpires an event that has no cause. Keep in mind that it is not '"as if" nothing transpires before $t = 0$, *it is exactly that nothing happens before $t = 0$*. Before the time origin, the infinitude of sinusoids adds to zero. Nothing is the actual reality outside of the delta function at the origin.

Some readers may observe that the function in the plot is not exactly zero outside of $t = 0$, resulting in the possible question of whether or not the impulse at the origin stands alone along the time axis. Like I said above, the waveform in the figure is an approximation that is practical to simulate on a computer. However, we can solve this problem mathematically from the definition of a delta function (I'm not going to derive it):

$$\delta(t) = \frac{1}{2\pi} \int_{-\infty}^{\infty} e^{j\omega t} d\omega. \tag{3.12}$$

Using Euler's identity, the integrand in the equation is $e^{j\omega t} = \cos(\omega t) + j\sin(\omega t)$. The integral sums over frequencies ω from negative infinity to positive infinity. The sum of sinusoids of all frequencies is a delta function. Ideally, it is the function plotted in Figure 3.30.

The intent of this short section was simply to provide another example that reality is far more sophisticated than simple assertions about what it should be. Some events are uncaused, at least in mathematics, so long as the signals included are allowed to be infinite in duration. Whether or not this same condition holds in physical reality is a separate question. However, considering its validity is certainly no more of a stretch than just asserting that a god is the one uncaused cause.

It was demonstrated here that such topics as an uncaused cause are approachable by science and do not need to end in a silly infinite regress where the religious side of the debate ends the regress by the simple, unjustified claim that their god

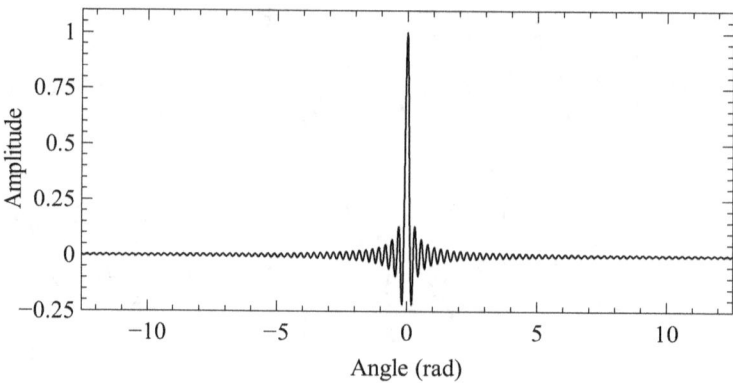

Figure 3.29: The summation of five hundred sinusoids whose frequencies range from -25 to $+25$ rad/s in 0.1 rad/s steps.

does not need a cause. We will not approach this subject again as it is not an integral part of the thesis of this book, but the spirit of the question is in line with where we are headed.

3.11 You weren't there

There is this notion by some creationists and ID advocates that there is no way to investigate events in the past. For example, the evolution of life occurred over billions of years—how could what happened over that span of time be examined today? The subject of evolution in this context will be looked at in later chapters. Here, a couple of simpler examples will be considered to give an idea of how analysis like this is done.

The dwarf planet Pluto was discovered on February 18, 1930, by astronomer Clyde W. Tombaugh at the Lowell Observatory in Flagstaff, Arizona. It has an orbital period around the Sun of approximately 248 years. How can this possibly be known when it was only discovered 91 years ago and has not had time to complete a full revolution? The answer is that it follows Newtonian mechanics as it orbits the Sun, and we can use these mathematical rules to essentially rewind (or fast forward) time and plot a complete orbit.

In the simplest case, if we have almost a century of data in the form of position versus time for the planet, we can fit this data to an elliptical orbit and obtain a mathematical expression from which we can extrapolate the planet's position forward or backward in time. There are many ways of doing this, but one of the most often used means in the sciences is a Least Squares fit. Given a data set $\{x_i, y_i\}$ and a function $y = f(x) = a_1 g_1(x) + a_2 g_2(x) + \cdots + a_n g_n(x)$ we can use a little bit of calculus find an optimal set of coefficients a_i that makes the function $f(x)$ fit the data. An example of a least-squares fit to data was given previously

3.11. YOU WEREN'T THERE

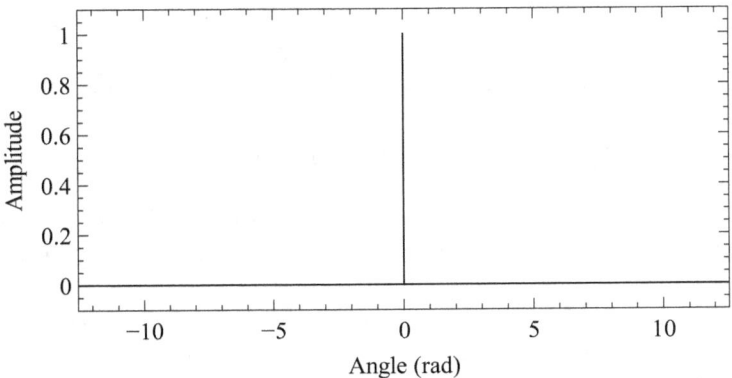

Figure 3.30: A real delta function when an infinite number of sinusoids at an infinite number of frequencies are added together.

in Figure 3.2.

Once we have this function, it is a simple matter to plot it on a computer and find the revolution time of the planet. Of course, some assumptions are being made that we can't get around. The most important of these is that the universal gravitation constant doesn't change with time; that is, the strength of gravity does not change over time. In other words, we assume that a ten-pound object weighs the same today as it did at any other time in the past. Without this assumption, and without knowing how the strength of gravity changed over time, if it did, there is no means of calculating the actual period of Pluto's orbit. The best we can say is that if all things remain constant, it will be such and such.

However, scientific assumptions are rarely just assumptions, and conclusions are hardly ever based on only one line of reasoning. The force of gravity has been measured on a regular basis since 1798 when Henry Cavendish first made a measurement using a torsion balance. Although there is some discrepancy in the measurements using laboratory-based setups, the error is below a few hundred parts per million.

A more interesting measurement was made over the course of 21 years on pulsar binary J1713 + 0747 [40]. The pulsar revolves in a binary system with a white dwarf companion. The regularity of the pulsar's spin frequency has been well characterized over the years and is approximately 218.8 Hz. Using the regularity of the pulsar emission, the researchers measured the change in the system's orbital period to be -0.2 ± 0.17 picoseconds per second. This is far less than one part in a trillion. If the gravitation constant had changed, the orbital period of this pair would have changed with it, which it did not, implying that gravity has remained constant for the 21-year period over which the system has been observed.

As a final piece of evidence, the orbits of the planets and moons in our solar system have been measured for a couple of hundred years. If gravity had changed, the result would have been an observable change in their orbital paths. No such change has ever been measured.

All of these pieces of corroborating evidence (and these are just a few of many more) come together to reinforce the assumption that the gravitational constant does not change, which in turn provides validity to our orbital model for Pluto, allowing it to be predicted with confidence even though we have never actually observed a full revolution.

As a final example, let's look at some evidence for the Big Bang. In the 1920's, Edwin Hubble was an astronomer at the Mount Wilson Observatory in California. He was the first to discover with this large 100-inch telescope that the so-called nebulae that astronomers saw in the night sky were actually galaxies unto themselves, very similar to our own Milky Way. Over the course of about a decade, Hubble calculated the distances of many of these galaxies from Earth based on their luminosities. He also calculated their radial speed away from Earth based on their *red shift*.

The redshift of light literally means that an object's spectrum is shifted further toward the low-frequency, red end the faster an object is moving away from an observer. This is a consequence of the *Doppler effect*. Everyone has probably, at one time or another, been introduced to the Doppler effect as a train passes by, a jet flies overhead, or a semi-truck blows its horn as it goes by. If you are standing by the road as the truck approaches, the wavelength of the sound from its horn is compressed, producing a shorter wavelength and making the horn sound higher in pitch. As the truck passes by and is moving away, the opposite occurs, the wavelength of the sound is stretched, and the horn drops in pitch.

The same thing happens with light. Light is a wave made up of electric and magnetic fields. If it is emitted by an object moving away from an observer, its wavelength appears to lengthen, as shown in Figure 3.31. Humans see the light in the visible spectrum with low-frequency red at one end and high-frequency violet at the other end. If a visible star or galaxy is moving away from Earth, its color will appear in a telescope to shift toward red. This is exactly what Hubble observed.

Hubble observed two notable characteristics of the light from distant galaxies: the light of every galaxy was redshifted, and the further a galaxy was away, the more redshifted it was [41]. Hubble correctly interpreted these two observations to mean that all galaxies are moving away from the Earth, and the further away a galaxy is, the faster it is receding from Earth. In fact, the implications were even more profound. As it would later come to be understood, the implication is that space itself is expanding. All points of space are moving away from all other points.

Figure 3.32 illustrates this effect from the vantage point of an observer at two different locations. The two plots in the figures are two-dimensional representations of space, and the circles represent stars spread evenly throughout space

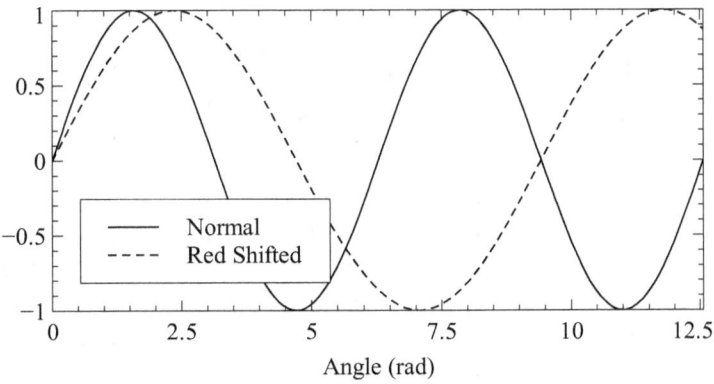

Figure 3.31: The dashed-line sine wave is a red-shifted version of the solid-line sine wave. As a luminous object moves away from an observer, due to the Doppler effect, the observer sees a red-shifted version of the light emanating from the object.

to act as markers so we can see how the space behaves. If the observer is at location $\{x = 30, y = 30\}$ as in the left plot, it appears as if all the rest of the Universe is speeding away from this point. And, if the observer moves to point $\{x = 50, y = 50\}$, it once again appears as if everything is speeding away from this point. Both cases are true.

The Big Bang is not an explosion like a stick of dynamite on flat land. When the dynamite explodes, gas and debris rush away from the focal point located at the stick's position. The Big Bang is something entirely different. Space itself actually exploded out of the Big Bang in all directions. It was not as if there was already space, and star-stuff was flung from the Big Bang's explosion. Space was created at the initial event and began expanding uniformly, and it still is to this day. All the stars and galaxies are moving away from one another not because they have a velocity relative to one another (which they certainly may have as well), but because the space between them is expanding. This was Hubble's discovery, and it had one inevitable conclusion.

If space is expanding everywhere today, in every direction, it must have started from a distinct point somewhere in the past. In other words, walking the expansion backward in time leads back to a single point. This is the fundamental rationale behind the hypothesis that the Universe had a beginning. In the time since then, the evidence for the Big Bang has continued to pile up.

In 1964, two employees of Bell Labs, Arno Penzias and Robert A. Wilson, were experimenting with an extremely sensitive horn antenna designed to detect signals bounced off of balloon satellites. The signals they were trying to detect were so minute that they had to go to great lengths to remove all extraneous noise. Despite their efforts, they found a low-level noise signal that they couldn't

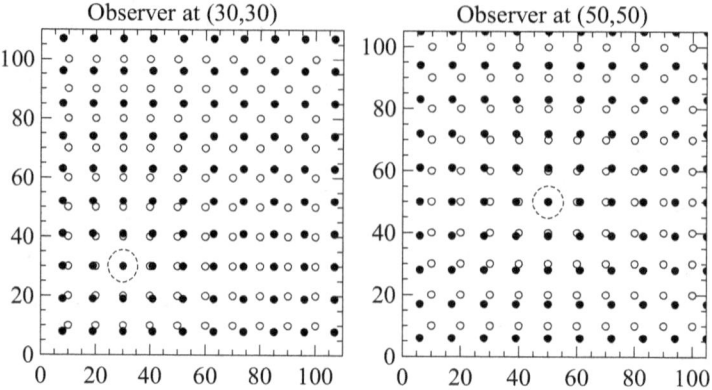

Figure 3.32: Expansion of two-dimensional space from two different vantage points, one on the left at $(30, 30)$, and one on the right at $(50, 50)$. No matter where the observation takes place, it appears to be the center of the expansion.

get rid of no matter which direction the horn was pointed in the sky. They eventually concluded that the signal was not originating in our galaxy.

It turned out that this annoying signal was the residual radiation left over from the Big Bang and has come to be known as the *cosmic microwave background radiation* (CMBR). It is the afterglow of the Big Bang. Since space has been expanding in all directions since the Big Bang, the CMBR can be seen from all directions. Again, there is no centralized location for the Big Bang explosion. It is everywhere.

As one more piece of evidence, consider looking up at the night sky in a rural environment with no street lights. In some places, you will see stars, such as in the arm of the Milky Way, but in other places, you see mostly nothing but darkness. If the Universe were an infinite expanse, there would be an infinite number of stars in any direction in the night sky you chose to view, all stacked up as the distance became progressively greater. And, if the Universe did not have a beginning but rather had been around forever, then there would have been sufficient time for all of that light to make it to your eyes. The night sky should be blindingly bright, but it isn't.

This oddity is called Olbers's paradox, named after physicist Heinrich Olbers who was one of the first to note it. The way out of the contradiction? The Universe cannot be of infinite age and infinite expanse.

Again, all of these lines of evidence (and there are many more) come together to bolster confidence that the Big Bang did happen. We did not have to be there to see it happen; we can look back through the past using many forms of evidence.

The moral of the story is that scientists do not necessarily need to repeat experiments in their labs to confirm a hypothesis. It is certainly possible to model reality and then peer through the lens of those models to view the distant past.

3.12 Conclusion

Science is a rational means of understanding reality. The term rational as used here does not mean "common sense" or "consensus opinion" as it is sometimes used in the common vernacular. Here, rationality refers to developing conclusions about data in a logical, mathematical manner. Science makes observations of reality and then draws logical inferences from those observations. Whereas religion admits and even invites interpretation, science places all ideas on a fair, common ground. Although hypotheses can be subjective, theories cannot be. Science, like mathematics, is, by definition, objective. This is not to say that mistakes cannot be made. Anyone can write a mathematical expression that is in error, and anyone can propose a scientific hypothesis that is incorrect, but these errors are not propagated as there is a built-in correction mechanism in the form of general, rational consensus. Let me give an example of this and contrast it with a similar example in religion.

Assume that Newton had dropped a mass from a set of different heights and timed their fall to the ground. Further assume that he then did a least-squares calculation and arrived at the following expression relating distance traveled and the time it took: $x = (1/2)kt^3$. It would not take long for others worldwide to perform the same experiment and come to the correct solution $x = (1/2)kt^2$. Maybe Newton's watch was broken when he performed the experiment. No one person gets to claim a physical truth as theirs. Observations of nature dictate what is true, not personal opinion. The measurements must be correct, and the mathematics (logic) used to arrive at a conclusion must be performed correctly. If mistakes are made, others will soon find it out. This is something that needs to be understood by non-scientific individuals: scientists *do not* simply believe what they are taught in college as though it is a dogma. The evidence dictates what is and what is not to be believed.

On the other hand, let's look at Christianity as prescribed by the Gospels. These are first-person documents with no corroboration outside themselves and with no independent means of discovering their truth. Furthermore, they are not written in a precise language such as mathematics, and therefore are open to interpretation. Since there are virtually no facts whatsoever that validate the base documents of religions, they also admit speculation as valid truth. An interpretation leads to speculation, which leads to further speculation, which leads to an interpretation of the speculation, and so on. It will be shown in a later chapter how this process leads to the evolution of religious beliefs and the building of complexity as it does so.

Science and religion are typically doing two different things. Science is attempting to reverse engineer the functioning of the natural world, and religion seemingly is attempting to fabricate its own version of reality. The two notions are usually in direct conflict with one another.

Intelligent design falls into this category as well. It is an attempt to use the methods of science, mathematics, inference, and observation to prove a preconceived notion, but we will leave this topic for a later chapter.

In the following chapters, we will take the tools of science and mathematics introduced here and in the previous chapter and bring them to bear on ID and the actual creative forces in the Universe. Up to now, the material has been a bit dry and perfunctory, but things get a bit more exciting from this point on.

Chapter 4

Optimization

> "Nothing is accomplished all at once, and it is one of my great maxims, and one of the most completely verified, that Nature makes no leaps: a maxim which I have called the law of continuity.", Gottfried Wilhelm Leibniz.

The evolution of organisms on Earth can be framed in one sense as a matching of boundary conditions. That is, evolution is the process by which organisms change over time so as to better mesh with the environment in which they find themselves. At its heart, this is an optimization problem.

This does not mean that nature, through evolution, seeks, for lack of a better term, the ultimate forms of life. The idea of superiority of certain forms is a human idea and has no meaning in the biological world. For example, we could think of lions with titanium teeth, eyes that can see in the infrared region of the spectrum, and that can run 100 miles-per-hour. Now, that would be one cool, bad ass lion. If evolution generates superior forms, this appears to be a good candidate. But evolution doesn't generate superior forms; it generates compatible forms. There is no use for a lion to become such a good predator that it drives extinct the very prey upon which it is dependent for food.

That said, evolution can lead to faster predators with better eyesight. And, on the other hand, it can simultaneously generate faster prey with better camouflage. And, along the way, the complexity of these animals grows through an unintelligent optimization process that has no desire to make either superior.

Creationists and ID proponents make the claim that it is impossible for any natural process to generate the biological complexity we see on Earth, for how could optimality be sought by a blind, unintelligent means? They claim that such a thing requires the action of an intelligent agent.

The main thesis of this book is to demonstrate that the complexity we see in the world around us is generated by unintelligent processes and does not need a magical, divine designer. This chapter introduces the concept of system optimization in a generic sense in preparation for understanding biological optimization later. Its sole intent is to serve as a primer for those unfamiliar with what optimization is and to show that system optimization does not require intelligence.

4.1 What is optimization?

The case often arises in engineering, physics, or mathematics to find the point at which a function is minimized or maximized. For example, consider the function $y = f(x)$. As we change the value of x, the resulting value of y changes. The question at hand is, at what value of x is y maximum? It may turn out that $f(x)$ has no maximum—a horizontal line like $y = 5$ has no maximum (or, equivalently, all points along the curve are maximal). Or, $f(x)$ may have multiple maxima. The point is, given that $f(x)$ does have a maximum, how do we go about finding it?

It may seem strange to some readers that we even need to ask such a question. For instance, if I am the author of the equation $y = 27 + 10x - x^2$ (see Figure 4.1), shouldn't I just know that it has a maximum at $x = 5$? Oddly enough, no. Although the maximum is easily found by taking a derivative for this particular case, it is not so easy to determine in many other cases. However, before going any further, it would be profitable to understand why engineers or nature are even concerned about optimization.

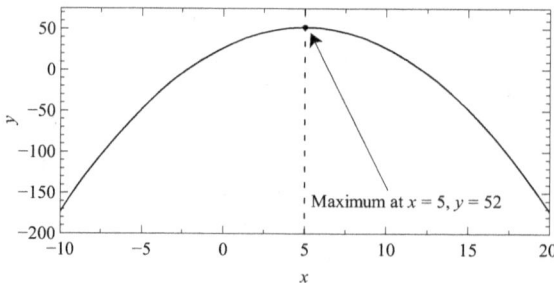

Figure 4.1: A plot of the function $y = 27 + 10x - x^2$, which has a maximum at $x = 5$.

Optimization is usually thought of as the maximization or minimization of a function. An example optimization problem is determining how one should play the cards in Blackjack to win the most money at the local casino. Alternatively, how the ignition timing for a car engine should be adjusted to maximize mileage is also an optimization problem. The optimization process continually adjusts the input to a system (the independent variable) to maximize or minimize the system output (the dependent variable).

4.1. WHAT IS OPTIMIZATION?

Sometimes optimization processes operate all on their own. For example, if you hold on to the spaced-apart ends of a rope and let the rope hang in between, the rope will automatically take the shape of a catenary curve. It does this because the catenary represents a minimal energy state for the rope. In other words, the gravity acting on the rope minimizes the rope's energy. (We'll see later why this is really the case when we look at entropy.) Or, take the example we considered earlier of someone pushing on a car. Newton's third law says that for every action, there is an opposite and equal reaction. Reality automatically optimizes the force the car feels so it just equals the force felt by the person's hands.

It is essential to understand that optimization problems are not always goal-oriented. The catenary curve and the force on the car above are not goal-oriented. They are simply examples of matching boundary conditions. But before we get to the more subtle issues of optimization, let's look at some practical examples that are very relevant to biological evolution.

Consider maximizing the function $f(x) = \sin(x)$. Remember, the sine function is zero at an angle of zero, increases to a value of one at an angle of $\pi/2$ radians, and then again goes to zero at π radians (Figure 4.2 shows $\sin(x)$ from -1 to 4 radians). The point of the optimization, in this case, is to find the peak that occurs at the angle of $\pi/2$, presuming we did not know it was the maximum *a priori*.

Perhaps the first method people think of to attack this problem is to just guess, which is certainly a valid means of finding the maximum. One simply guesses an angle, computes the sine of the angle, and checks if it produces the maximum of one. (In practice, it would be necessary to determine if the answer found was indeed the maximum. One way to do this is by measuring the function's slope a small distance away from each point until it is increasing on the left and decreasing on the right—we will see why this is true later.) Another less reliable way is to randomly sample N points across the function and take the largest one, although this does not guarantee the maximum has been found. The random search method is shown in Figure 4.2, where the starting point is at $(-0.5, \sin(-0.5))$ and random values are selected across the range of angles from -1 to 4 radians until the finishing value is within 0.1 of the peak. (Remember, a radian is an angular measure, and there are 2π radians in 360 degrees.)

In the example in the figure, it takes twenty-five iterations to find the maximum randomly. This is not surprising as the entire domain is five radians wide (-1 to 4 radians), and the target span is ± 0.1 radians, so the probability of landing in this target span is $0.2/5 = 1/25$.

The random search method of optimizing a function like this certainly works. However, it is not a very efficient means of optimization because it necessarily must search the entire space over which the function is to be optimized. I bring it up here because it will be important in later discussions, but let's look at a more intelligent optimization method before we get there.

The random search method suffers from not taking advantage of any infor-

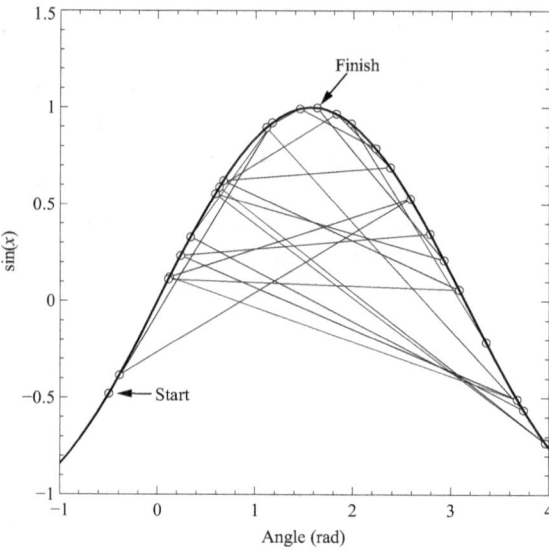

Figure 4.2: Twenty-five steps of a random search of the function sin(x) to get within ±0.1 of the maximum at $\pi/2$ radians.

mation about the function which it is optimizing; all it does is continue to guess what the maximum is until it finds it (or gets arbitrarily close to it). The optimization method can be improved if it uses information about the slope of the function. If we consider the maximum, or peak, of a function, the slope on the left must be positive, as the function is increasing, and the slope on the right of the peak must be negative, as the function is decreasing. These conditions necessarily describe a maximum: around the maximum, the slope is positive on the left and negative on the right. Now, how do we make use of this information?

If we take the same starting point as above, at -0.5 radians we calculate a slope of $\cos(-0.5) = 0.87758$ (I used the fact that the slope of $\sin(x)$ is equal to $\cos(x)$). This slope is positive, and hence the method needs to move toward the right, with the angle increasing, to approach the maximum. The next question concerns how far to move in the positive direction. This is a deep question that involves the shape of the function being optimized. Here, for simplicity, we are simply going to move by a value of the calculated slope (a scaling of one in conventional gradient ascent). Therefore, the algorithm takes a step from -0.5 radians to $-0.5 + 0.87758 = 0.37758$ radians as shown in Figure 4.3. This single step takes the solution over halfway to the peak of the sine function.

This process repeats as the slope is calculated again at 0.37758 radians to be 0.92956. And, again, we add this value to the current angle to get 1.3071, which takes the solution very close to the maximum. With one more step, the solution is at $\pi/2 - 1.5677 = 0.0031$ away from the maximum. Using slope information,

4.1. WHAT IS OPTIMIZATION?

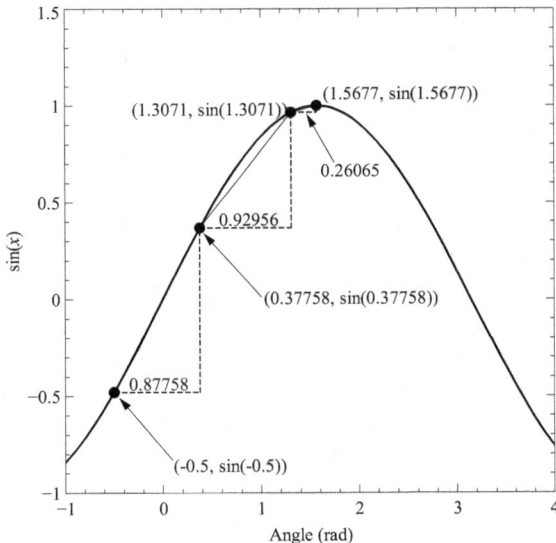

Figure 4.3: Using slope information, the maximum of the function sin(x) is found very quickly.

the optimization takes three steps to get much closer to the maximum than the random search above was able to in twenty-five steps.

An intuitive way of understanding how this algorithm works is to consider climbing to a hilltop in a dense fog where it is only possible to see a few feet in front of you. The direction in which the top of the hill lies is entirely invisible, and all that can be seen lies within a circle of a three-foot radius around you. However, the local slope of the hill can be found by sticking your right foot out, one direction at a time, and choosing to walk in the direction for which your right foot was highest above your left foot—that is, walking in the direction of the largest positive slope. In this way, you continue walking uphill until the top is reached. Of course, here we ignore the possibility of local optima such as an outcropping or a depression on the side of the hill. Mathematical optimization of functions also can run into this problem and get stuck at a local optimum.

The third way to solve this problem is by a straightforward application of calculus. The derivative is the slope of a function at a given point. If this slope is positive on the left of a maximum and negative on the right, then logically, it goes to zero at the maximum. The proof of this is a little bit more involved, but it is the case. As we approach the maximum, the slope is positive, and it falls off until, at the peak, the slope is zero. The slope starts to turn downhill and becomes negative as the peak is passed over. However, right at the peak, it is neither positive nor negative. Therefore, since a derivative is the slope of a function at a given point, finding an optimum involves finding the derivative of the function of interest and solving for when it is equal to zero.

The derivative of $f(x) = \sin(x)$ is $\frac{d}{dx}(\sin(x)) = \cos(x)$. Setting the derivative equal to zero yields $\cos(x) = 0$, the solution of which is $x = n\pi/2$ for n an integer. The case we have been considering is $n = 1$, or $\pi/2 = 1.5708$, which, ignoring rounding errors, is the answer we obtained above with the gradient ascent method.

The random search method is what most people think of when speaking about biological evolution. This is not the case in nature, however. Blind searching would be far too inefficient to accomplish what is observed in nature. Nature does not make use of sophisticated methods like gradient ascent either.

In the following sections, we will examine in detail methods that more closely resemble those employed in biological processes. They all have in common the requirement of persistent excitation—the continual perturbation of system inputs or parameters. This excitation can be random or deterministic so long as it provides variation over the span within which the optimum exists. In other words, in order for the optimum to be found, the excitation must cross over it. Let's see how this works.

4.2 Optimization by random perturbation

An optimization method or algorithm is a process that moves the operating point of a system toward a position that is defined in some way to be the best or optimal. For example, a career person generally strives to make more money over time by improving their skill set, progressing up a particular corporate structure, and wisely using and investing money. These actions continually lead to a more effective means of increasing one's income and, thus, optimize this individual's ability to earn money.

A second example would be an archer trying to hit the bullseye of a target. In the beginning, most people are relatively poor shots and are lucky to hit the target at all. However, after repeated attempts and visual feedback of where each attempt hit the target, the archer's aim begins improving as she learns to adjust the angle of the bow and how to release an arrow smoothly. The arrows hit the target progressively closer to the bullseye with continued practice until they arrive there every time. In this example, the archer's mind is optimizing its control over the body to minimize the distance between arrow impact locations and the bullseye.

Technological advancement also represents an optimization. As we humans learn more about the physical world and the design and application of technology, we get better at it. Computer microprocessors get faster and denser as we continually improve our ability to design and produce silicon (and other material-based) integrated circuits. Our understanding of physics and engineering techniques grows exponentially and, in tandem, our understanding of the application of these technologies grows.

It is interesting to note that the archer example above has a well-defined point of optimality of hitting the bullseye, whereas the other two examples do not. A person can always make more money, and improvements in technology appear to have no end. Regardless, they all make use of methods to move closer to an optimum. However, there is another distinction between the first two and the last. Making money may not have an end target amount of money that indicates optimality as archery has with hitting the bullseye, but they both have at least a well-defined intent: making more money and coming closer to the center of the target. The advancement of technology is different in that as new and more complicated devices are developed, people's views of that technology change as they become acclimated and they want more or different things. Many of the modern innovations we take for granted today were not even envisioned several decades ago. Discontinuous jumps in our understanding of technology can drive entirely new avenues of research and development of devices we never knew we wanted. In this way, society and technology act symbiotically to determine what optimal means. In other words, the technological cost function is dynamic. We'll see later that the same is true for the evolutionary fitness landscape.

Optimization problems routinely show up in the fields of engineering, physics, and mathematics. As an example, think about the efficiency (or mileage) of a car as a function of its speed. In other words, how does the mileage of the average car differ between slow and fast speeds? Let's take a very simplistic view of the operation of the car and separate the energy usage into two components, the first being the idling power the car requires, such as supplying the engine controller, the interior heater, the headlights, etc., and the second being the power required for rolling the tires on the road. The idling power is constant and does not vary with the vehicle's speed, and the power to move the car varies as a square with the vehicle speed. Both of these are shown in the top plot of Figure 4.4 where the solid line is the constant component, and the dashed line is the power needed to move the vehicle.

The efficiency of the vehicle during operation is defined as the distance traveled divided by the energy required to travel that distance or, equivalently, the rate of change of distance, or speed, divided by the rate of change of energy expended, or power.

When stopped and idling, the vehicle is still expending energy to keep the engine crankshaft turning. Therefore the efficiency is zero because no power is being supplied to the wheels to move the car. As the vehicle starts moving, the efficiency starts increasing as displayed in the second plot of Figure 4.4. The maximum efficiency occurs when the power required to move the car equals the idling power of the car (a little calculus and algebra can confirm this). Then, as the speed increases further, the efficiency decreases due to the increased power required to propel the car.

On the other hand, the slower you drive to your destination, the less power is required but, the longer the vehicle's engine must run. In fact, as the speed slows to ridiculous rates, the energy required to just keep the engine spinning outpaces the energy required to move the car. (Think about the car just sitting

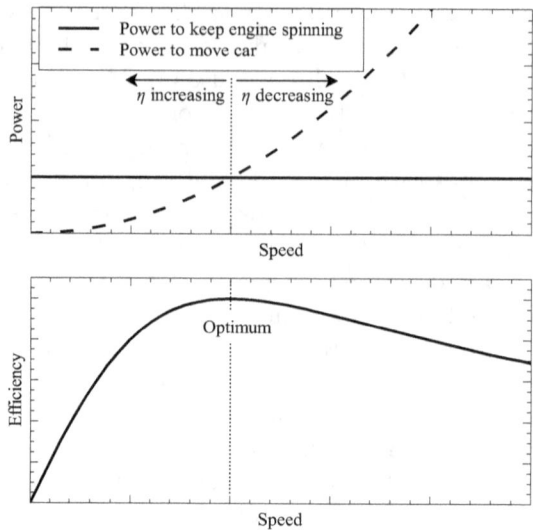

Figure 4.4: The top plot shows the traction (dashed line) and idling (solid line) power required to move a car as a function of the speed of the car. The bottom plot shows the efficiency (mileage) of the vehicle versus speed.

still and idling; the efficiency, in this case, is zero as no useable energy is being output to move the car.)

The point of this example is that at either extreme, the car's efficiency approaches zero, but it is finite between the two endpoints. Hence, there is a speed that is optimal in terms of efficiency. And, if we wished, we could certainly find this optimal speed using any one of several methods.

Again, we have to be careful here. It is easy to conclude that the optimum is to be looked at as a goal or that it carries some meaning, and that would be false. The fact that these systems possess what we have defined as optimal points does not imply that these points are any different from any other. They are simply useful to us. We imbue them with significance.

The efficiency example is typical of the most straightforward class of optimization problems encountered in engineering. In this case, the expression of efficiency would be called the *cost function* of the optimization problem. The cost function, which is generally a mathematical expression, is the statement to be maximized or minimized. In this example, the cost function is written as the vehicle's speed divided by the power it expends.

In addition to a function that needs to be optimized, constraints are generally imposed upon optimization problems. For example, in the money-making example above, one cannot just rob a bank to acquire money and, if employed, must adhere to the rules the employer prescribes. Similarly, technological ad-

4.2. OPTIMIZATION BY RANDOM PERTURBATION

vancement is constrained by the rules of the physical world. The engineer cannot simply violate the speed of light or assume that wire is super-conducting. In fact, often, it can be difficult to distinguish between the constraints of a problem and the original problem one set out to optimize. They are really the same.

In a sense, the constraints are what make a problem solution optimal. For example, consider the case of stockcar racing. The intent is to optimize the track time, or the average track time, during a race of many laps. Obviously, the process involves the team's mechanics continually modifying the structure and operation of the engine to increase its longevity and power output. The driver increasing his skill at driving is also part of the process. However, just as important is the rule that the driver must stay on the track during the race, which would normally be considered a constraint on the system. But this constraint is an integral part of the overall optimization process. The mechanic and driver will undoubtedly alter their strategies to take into account the shape and length of the track.

Let's look at a few mathematical examples to clarify these points.

Consider finding the maximum of the function $f(x) = 1 - x^2$ (see Figure 4.5). In other words, what value of x maximizes this function? The function increases at the left and decreases at the right, meaning that there is at least one optimal value in between. I know it is a simple problem, but let's pretend we do not know the answer. First, let's guess at what value of x maximizes this function. For example, start with $x = 1$. Plugging $x = 1$ into the expression we get $1 - (1)^2 = 0$. Now, we have no idea if this is a maximum since it was arrived at by a pure guess. So, let's guess again and move just a little to the left by setting $x = 0.9$ to get a solution of $1 - (0.9)^2 = 0.19$. This is better since it is greater than zero, so let's move a little further in that direction by letting $x = 0.8$ to arrive at $1 - (0.8)^2 = 0.36$. This is better yet.

If this method is iterated, the solution will approach $x = 0$, which corresponds to the maximum $1 - (0)^2 = 1$. If a step is taken past $x = 0$, such as $x = -0.1$, it will be seen that the value of the function starts falling, meaning that the maximum has been crossed. This process is shown in 4.6 and it is clear that the maximum is indeed $x = 0$.

Note that no intelligence was needed to solve this problem once the algorithm was implemented. The algorithm, if repeated indefinitely, will continue to be attracted to the maximum. In mathematical terms, the maximum is called a *stationary point* and has the special distinction that the derivative, or slope, of the function at that point, is zero. If one thinks about it for a moment, it will be apparent that the algorithm above uses slope information to hone in on the maximum. That is, if the next point computed is larger than the previous, then the slope is positive, meaning that the value of the function is increasing. As long as the slope continues to be positive, the algorithm continues. At the maximum, however, the slope becomes zero, meaning that any further step away from the maximum will provide no increase in the function but rather will result in a decrease. The optimum has been found.

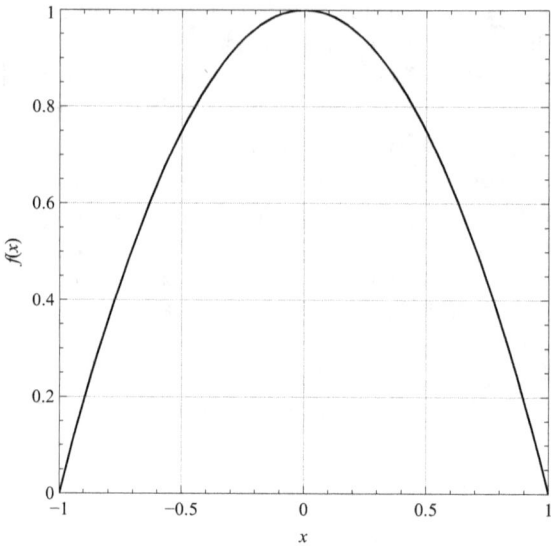

Figure 4.5: Plot of the function $f(x) = 1 - x^2$.

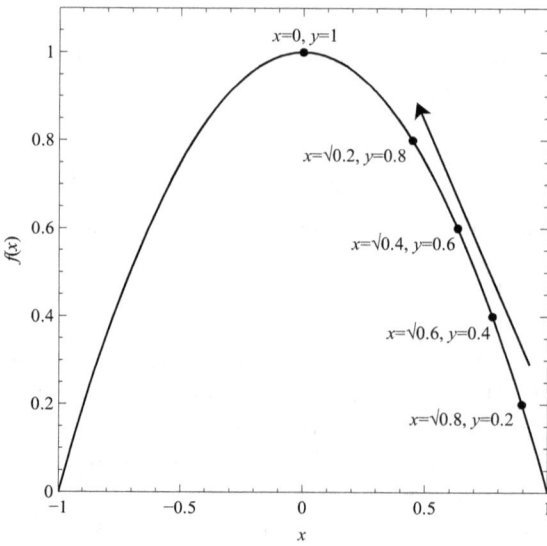

Figure 4.6: Finding the maximum of $f(x) = 1 - x^2$ using a fixed step size.

4.2. OPTIMIZATION BY RANDOM PERTURBATION

In our examples so far, we have chosen the step direction and size intelligently. What happens, however, if these parameters are chosen at random? Would the method get anywhere? Remember, any point can be selected, which is then tested against the condition that it must result in a value of $f(x)$ that is larger than the previous value. Therefore it does not matter if the step size is random or not. Figure 4.7 shows what happens when the maximum of the parabolic function considered above is found using the same routine but with a random step size. The approach to the optimum is slower and more erratic, but the outcome is the same.

This result is significant enough to stop and recapitulate. The example found the maximum of the function $f(x) = 1 - x^2$ by starting at a random point that was a large distance from the optimum value $x = 0$, then moving by adding a random amount to the current value of x. If the resulting size of $f(x)$ was increased with the new value of x, the move was retained and, if not, the move was rejected. The process of adding a new random amount to x and checking $f(x)$ is then repeated until the algorithm has found the optimum. There is no intelligence involved here, just random changes and a selection process that only retains the beneficial changes.

Note that this method is a modified gradient method of sorts. Whereas the gradient ascent method we looked at above used the slope as the step (including the sign and the size of the slope), the step size and direction, in this case, is entirely random. If we assume the random step values are zero mean, which they are in this case, half of them will be in the correct direction and the other half in the wrong direction. This is okay as the selection process sorts and only retains those steps in the correct direction. The reader will have already noticed that this process is very reminiscent of random genetic mutation and natural selection. (A zero-mean random sequence spends equal amounts of time above and below zero.)

The next question would be, does this process still work as the dimension of the problem increases? Let's now look at finding the maximum of the function $f(x, y) = 10 - x^2 - y^2$. Again, it is obvious that the maximum is at $\{x = 0, y = 0\}$, but let's see if the same algorithm can find the maximum. Figure 4.8 shows what happens when this algorithm is applied to finding the maximum of this function. At each iteration, a random step value is found, alternating each time between the x and y axes, and if that step increases the value of the function, then the step is taken, and if the function is not increased, then the step is discarded. The result is an inexorable advance toward the function maximum, as shown in the figure. The open circles are candidate points, and the black circles are the points that were actually chosen by the selection method.

Figure 4.9 shows the trajectory of the optimization through the xy–plane. The light line shows every point considered in the optimization, and the heavy line with markers highlights the accepted points and the path the optimization took. The starting point is at $\{x = 3, y = 3\}$ with a cost function value of $10 - 3^2 - 3^2 = -8$, and the ending point is near $\{x = 0, y = 0\}$ with a maximum

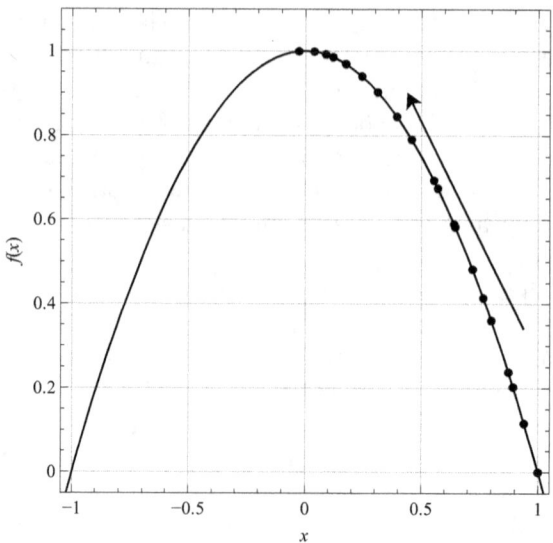

Figure 4.7: Finding the maximum of $f(x) = 1 - x^2$ using a random step size.

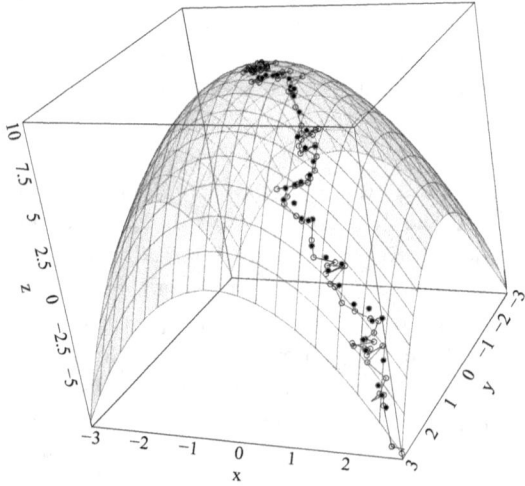

Figure 4.8: Finding the maximum of $f(x, y) = 10 - x^2 - y^2$ using a random step size. The open circles are all the points tested, and the filled circles are the points actually selected.

4.2. OPTIMIZATION BY RANDOM PERTURBATION

value close to $f(0,0) = 10$. Note, because random step values are being used to move the operating point, the algorithm never exactly finds the maximum and instead continues to get nearer to it over time. This is expected as a random step of exactly the correct size to close the gap to the maximum would be very unlikely.

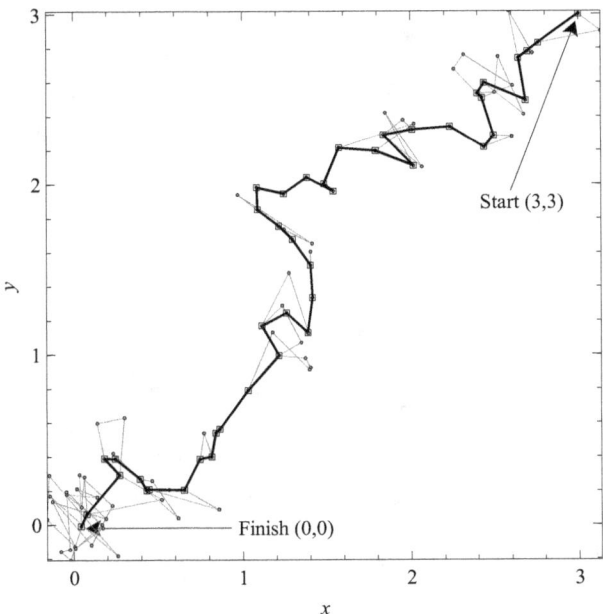

Figure 4.9: Finding the maximum of $f(x, y) = 10 - x^2 - y^2$ using a random step size. The lighter line with circle markers shows the random steps that were evaluated, and the heavier line with the square markers shows the steps that were actually selected by the algorithm.

The first step is from $\{x = 3, y = 3\}$ to $x = 3.1292$. The step to $y = 2.8991$ is not accepted because $f(3.1292, 2.8991) = -8.1967$ which is worse than $f(3,3) = -8$, meaning that the step is in the wrong direction. However, the second step moves to $\{x = 2.7583, y = 2.8296\}$ with a cost function value of $f(2.7583, 2.8296) = -5.6149 > -8$, and the move is accepted.

After 100 steps, the path ends at $f(0.0424, -0.00856) = 9.9981$, which is very close to the maximum of $f(0,0) = 10$. Again, finding this optimum only required random variation of the points in the xy-plane and the simple rule that a new point is accepted only if the resulting value of $f(x, y)$ is larger than the previous value.

There is not just one way to find the optimal in this example. Again, the step direction is random, and if the simulation is run again, a different route to the peak of the surface will likely be found. Figure 4.10 shows the results of four such simulations. In all four cases, the path to the maximum is definite even though the only two mechanisms being used are continual random stepping of the points in the xy–plane and the requirement that the value of cost function for the current step is larger than the previous. What should be clear is that there is no proper path to the maximum.

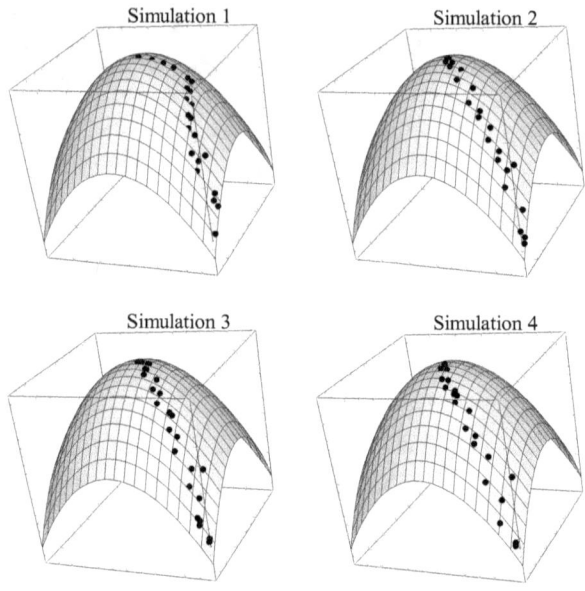

Figure 4.10: Four simulations of finding the maximum of $f(x, y) = 10 - x^2 - y^2$ using a random step size. Note the different paths taken.

As a more down-to-earth example, let's look at adjusting the firing angles of a cannon to land a projectile at a particular point on the battlefield. Figure 4.11 shows a cannon at $x = 0$ trying to hit a target on the field at $x = x_t$. Neither the angle of elevation (the angle of inclination of the cannon barrel) nor the azimuth (the angle of the barrel in the xy–plane) of the cannon required to hit the target is known initially. The starting angles are an elevation of $10°$ and an azimuth of $30°$. Similar to the examples above, the algorithm randomly steps these two angles and, if the next step results in the cannonball landing closer to the target, that point is accepted; otherwise, it is rejected.

The result of one simulation is displayed in Figure 4.12 in the the (x, y)–coordinates defined in Figure 4.11. The first shot lands at the point labeled in the figure and then inexorably tracks toward the target's center. I show this example to clarify that the simple optimization method we have been using can be used to solve practical problems despite the algorithm having no intelligence other than the ability to make numerical comparisons.

4.2. OPTIMIZATION BY RANDOM PERTURBATION

Figure 4.11: A cannon shoots toward a target on the ground to land a projectile on the bullseye.

The examples we have been looking at may appear trite at first glance, but they have quite a bit in common with biological evolution. (From this point on, I am going to use the terms biological evolution and evolution interchangeably. When other forms of evolution are considered, such as the time-evolution of dynamic systems, it will be made clear.) There are independent variables, genes written in deoxyribonucleic acid (DNA), which are contained within every biological entity on Earth, that determine where in the evolutionary landscape any such entity resides. Random genetic mutations continually modify the DNA. Genes are analogous to the independent variables in the xy–plane above. And like the variables in those examples, the genetic material passed on from one generation to the next is regularly modified or mutated during reproduction.

The genetic modifications express themselves as changes in the morphology of the entities that inherit these genes. And, in an ecosphere where survival depends directly upon the effectiveness of an entity's inherent skill set, a gene modification that introduces marginal improvement to an entity's performance provides a statistical advantage. The group that possesses this alteration tends to live longer and reproduce more often than others that do not have the same alteration. Hence this gene propagates throughout the species and a step is taken toward the optimum. On the other hand, if the mutation degrades a critical ability, the individual or group that possesses it is at a statistical disadvantage, resulting in shorter lifespans and reduced reproduction, hence the particular gene does not propagate. In this way, survivability acts as a selection function very similar to the mathematical functions in the examples above. A given gene modification is only retained if it maximizes the survivability of the group. Otherwise, it is rejected as a result of the individuals that possess it dying before they can reproduce and pass the modification on to their progeny.

Just like the previous examples, gene modification is random, but the optimization process that is called *natural selection* is absolutely not random. Natural selection is the process by which genetic variances are chosen for inclusion into the overall ecosphere because of their superiority in terms of reproduction or, conversely, their rejection because of a hindered ability to reproduce. There is no magic going on with evolution, and there is certainly no need to meet the impossible odds of the DNA of a new species being formed by a single event entirely

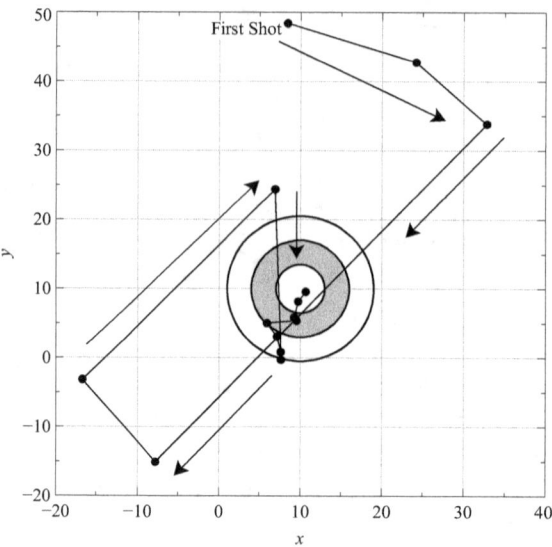

Figure 4.12: Random adjustments of the angles of azimuth and elevation to land a projectile on the bullseye. The cost function is the Euclidean distance between the landing point and the bullseye.

at random. This is just not what happens. The demonstrations above should at least provide plausibility to the claim that if a path to an optimum exists, there is continuous random excitation of the independent variables that define an appropriate cost function, and there is an appropriate selection function, then it is virtually certain that the optimum will be found in time.

There are much better algorithms that could be used to find the maximum faster or in fewer steps. These methods make use of gradient information to determine an optimal step size and direction. The gradient of a function computes the direction in which it is changing the fastest, and that is the direction one should step to make it to the peak the fastest. In addition, this information can determine how big of a step should be taken to reach the peak so as to not over- or undershoot it. Nevertheless, the point here is that even a relatively dumb algorithm can do just fine if the speed of convergence to the optimum is not the primary consideration.

Genetic mutation is, by definition, a random process. There is no intelligence dictating the changes made in the DNA of the subject. Therefore, more sophisticated gradient methods are not going to be discussed here. I bring the topic up only to emphasize that the algorithm is quite subpar to those constructed by most engineers. In addition, it needs to be clear that the selection of the end suitability of these changes to the benefit of the organism *is certainly not random*.

When the inputs to, or the parameters of, a system are continually modified, such as with the genetic mutations applied to DNA here on Earth, we call this persistent excitation. In other words, the input variables or parameters of the system are continually changing. This dynamism provides perpetual variation in the system that can be continuously evaluated and accepted or rejected in terms of the resulting optimality of the system. In other words, the operating point of all species is constantly being perturbed. The evolution of life on the earth depends entirely on this fact.

Interestingly, what we typically regard as a bad thing, such as genetic mutation, that in other instances leads to cancer and other problems, is the key to improving the human condition on Earth. Without it, we could not have adapted throughout history, or have been tailored, to the changing environment in which we live.

Every species on earth is continually moving toward an optimal position in an evolutionary sense. I say "an optimal position" because there are many points of optimality within the evolutionary landscape. Consider the function plotted in Figure 4.13 below. (For the curious, this function is $f(x,y) = \left| sin(x)sin(y)/\sqrt{|xy|+1} \right|$.) Four peaks of this function are shown in the figure, and several starting points have been chosen at the base of these peaks. The optimization algorithm described above is then applied to each of the starting points and continued for 100 iterations. Each optimization trajectory climbs randomly toward a peak.

It is important to note that starting in the same location does not guarantee ending on the same peak. For example, four of the simulations have been started in the center of the plot at $x=0, y=0$ and each, purely via random stepping, climbs a different peak than the others. It is easy to see why this happens. The algorithm that is being used is a sort of ratcheting process. Once the cost function value increases slightly, it will never come back down. The requirement that the value corresponding to the next step be greater than the previous value prevents retreating down a peak. If the maximum step size of the algorithm in the x and y directions is Δs, then as soon as the algorithm has moved to a distance $\sqrt{x^2+y^2} \geq \Delta s$, there is no coming back. There is no way to take a subsequent step large enough up an adjacent peak to overcome the requirement that cost function increase in magnitude.

The two-dimensional projection of the optimization process onto the xy–plane is shown in Figure 4.14. The start positions are labeled, the peaks are outlined with dashed circles, and the direction of movement is indicated with arrows.

In many instances of optimization, the cost function possesses multiple optima. This is generally the case as the function being optimized becomes more complex. Attempting to optimize systems with multiple optima can be problematic in the sciences, particularly engineering. Optimization processes like the one described above search for a point where the slope of the cost function becomes zero, indicating a peak or valley, but there is no guarantee that the maximum (or minimum) found is a *global* maximum (or minimum). It could certainly be

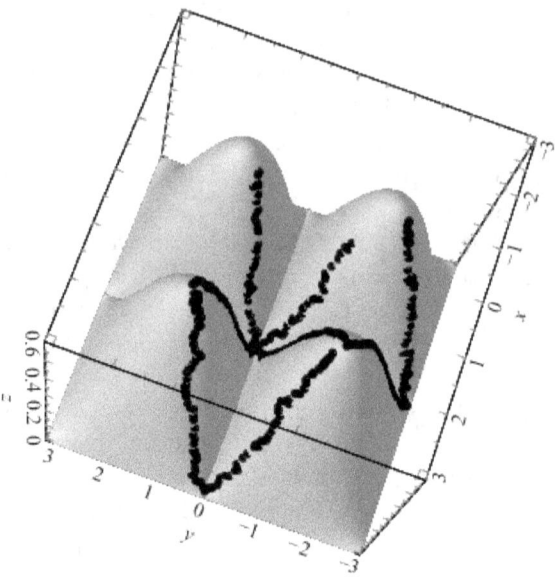

Figure 4.13: Several optimization runs, some starting from the same point, and others starting at different points. (All start from the bottom of plot.)

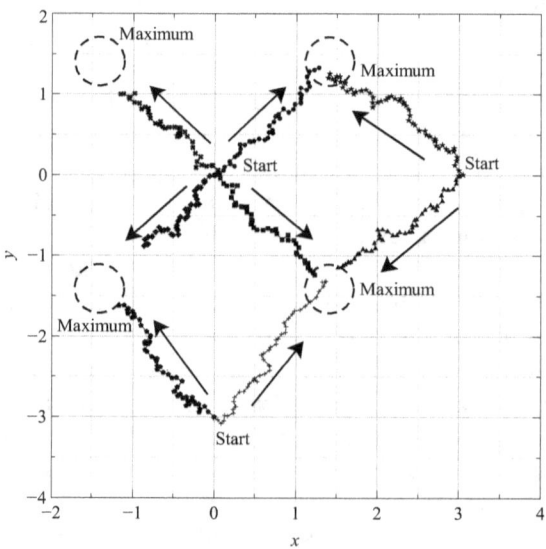

Figure 4.14: The 2D projection of the optimization paths of Figure 4.13. The peak locations are outlined with dashed circles.

4.2. OPTIMIZATION BY RANDOM PERTURBATION

the case that the solution found is a *local* maximum (or minimum). For clarity, if a function has multiple maxima, and one of them is larger than the rest, this largest peak is called the global maximum, and the remaining peaks are referred to as local maxima. It is impossible for an optimization method with a maximum step size smaller than the spacing between the multiple maxima to guarantee it finds the global maximum. The process will simply climb to the top of the first peak it finds and remain there.

It is important to point out that biological evolution does not find global optima. It is difficult to pin down even what would be meant by a global optimum in a biological context.

Optimization processes can only operate where there are differences, that is, where the derivative or slope is nonzero. The function $f(x,y) = 10 - x^2 - y^2$ in the examples above changes and has a maximum along both the x and y axes. Now, let y be a constant and equal to zero, $f(x,y) = 10 - x^2$. The plot of this function is shown in Figure 4.15. The function is parabolic in the x-axis and flat in the y-axis. As the optimization method climbs this particular function, as shown in the figure, there is definite movement in the x-direction and none in the y-direction.

Figure 4.16 shows the movement of the operating point in the xy-plane. It also shows the value of the cost function being optimized $f(x,y)$. The trajectory starts at $x = 3$ and moves left to the optimum of $x = 0$ and, in concert, the value of $f(x,y)$ moves from zero to a value of 10. The random points considered by the algorithm are plotted as open circles and never stray too far from the value of $y = 3$. In fact, the mean movement in the vertical direction is zero. This is because the derivative or slope of $f(x,y)$ in that direction is zero, thus providing no impetus to move along that direction. Stated another way, movement along the y-axis has no impact on the value of $f(x,y)$.

We have examined several optimization methods in this section, including the random search and the gradient ascent method. Evolution uses neither of these approaches. The first is inefficient, and the second requires some intelligence (or at least some luck in constructing the gradient calculator). However, the third method we looked at is similar to what is found in nature. This method has two components that implement the optimization process. The first is a means of randomly perturbing a set of independent variables (inputs or parameters). Changing these variables results in a change in the system output. The second component of the optimization process determines if the random step taken increases or decreases the optimality of the output value. If there is an increase, the change in the independent variable is retained; if there is a decrease, the change is discarded.

Biological evolution is similar to this process. Random mutations are made to the DNA of an individual during conception. If these changes improve the ability of the individual to reproduce, the genes will be passed on to the individual's progeny. If the changes are detrimental, presumably, the individual will have a hard time passing on these modified genes. Random mutation of DNA

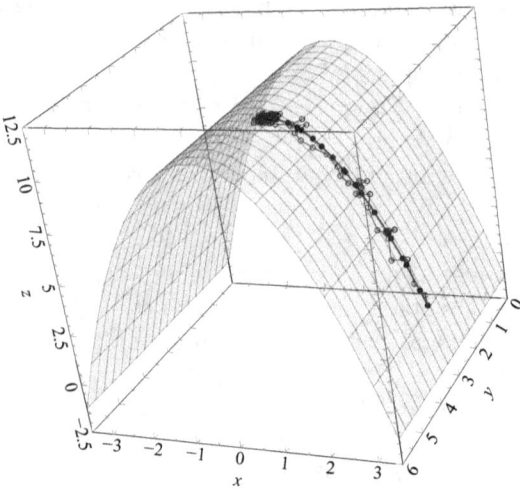

Figure 4.15: Optimization of the function $f(x,y) = 10 - x^2$.

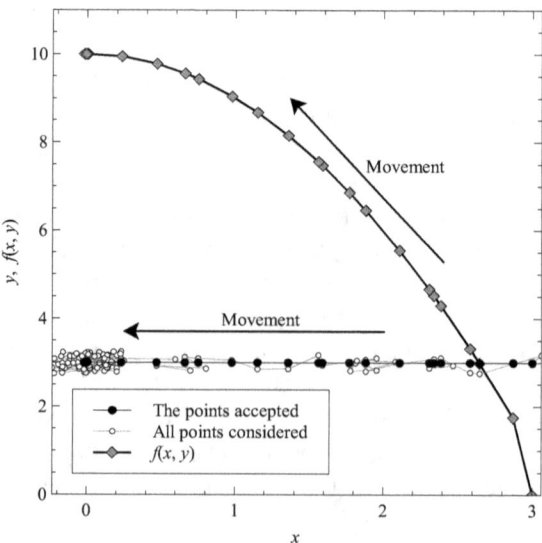

Figure 4.16: The projection of the optimization of the function $f(x,y) = 10 - x^2$ onto the xy–plane.

corresponds to random step changes in our examples above, and natural selection is akin to the greater-than selection used above.

Both the synthetic program that was used above and the natural evolutionary process rely on persistent excitation. This means that the independent variables of the system are randomly perturbed continuously. This requirement has to do with the fact that the cost function is generally dynamic as well.

Consider again the 3D function of Figure 4.13. If this were actually the landscape of optimality for a physical system, including the evolutionary landscape, the peaks would likely move dynamically. They would not remain constant. The only way to ensure the optimum is known at all times is for the optimization process to run perpetually, always driving toward the optimum.

In the next section, we will delve into how persistent excitation is used in engineering to find dynamic optima. The results will not be directly applicable to biological systems, but they will give the reader a clearer idea of how these processes operate.

4.3 Optimization using persistent excitation

Engineering can, in one way, be regarded as the practice of doing the most with the least. For example, consumers want higher performance at lower cost and more functionality in smaller packages. The engineer is constantly dealing with tradeoffs. The drive to continually improve system design provides an impetus to use what is already available instead of implementing unneeded, redundant systems. If a particular subsystem can serve dual functions, it will be required to do so.

Several ID authors claim that noise is not used in engineering because, well, it is noise and not valid for any practical purpose (of course, they do this to claim by analogy that randomness is also not useful in producing biological complexity) [42]. Of course, when we are talking about signals that must maintain their integrity, noise is a problem that must be dealt with. For example, noise in a communication channel is undesirable and may not be correctable. Noise superimposed on the output voltage of a DC power supply is not a good thing. However, as mentioned above, if a particular property can be co-opted for a practical purpose, it will be, and in the case of noise, this is done on a regular basis in electronic devices.

In many cases, there are parameters of a system that can only be measured when the system inputs, outputs, or internal states change. Making such a measurement can be accomplished, of course, by having the system software change the necessary input to the system while making the needed measurement. But this can lead to suboptimal system operation because the operating point moves away from the set point. However, if there is already noise present on the input signals—which there always is—why not just measure the change that it induces in the system to compute the necessary parameters? Contemporary designs make these types of measurements all the time.

In the remainder of this section, I will look at a few examples in engineering that may appear unorthodox to those outside the areas of physics and engineering.

The first topic we will look at is the characterization of the dynamic behavior of electronic circuits using white noise. A white-noise signal is completely random in time. Its value jumps randomly from value to value as shown in Figure 4.17. The amplitude of the noise is limited to a fixed value. The interesting thing about white noise is that it contains all frequencies. That is, the signal is constructed from sinusoids of all frequencies.

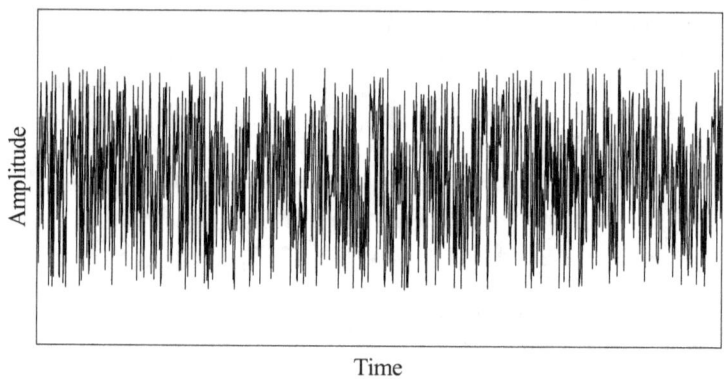

Figure 4.17: An example of white noise in the time domain.

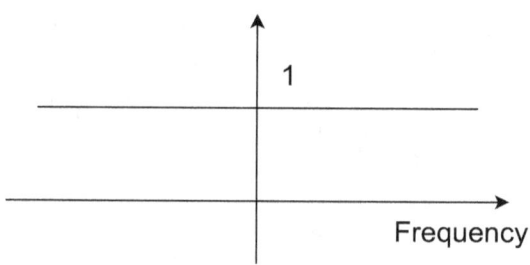

Figure 4.18: The frequency-domain representation of time-domain white noise. The spectrum for the time-domain delta function is the same.

As we learned in Chapter 3, the sum of sinusoids of all frequencies is a delta function at the origin. The frequency-domain representation of the delta function is shown in Figure 4.18; it is a horizontal line with amplitude of one. The delta function transforms to a horizontal line that extends across frequencies from negative to positive infinity. Hence, the delta function includes all frequencies.

4.3. OPTIMIZATION USING PERSISTENT EXCITATION

Interestingly, the frequency spectrum of white noise is the same as that for the delta function. That is, white noise contains sinusoids of all frequencies. Its frequency response is also that shown in Figure 4.18. I will not go through the details as to why this is the case as they are fairly mathematical, but it has to do with the autocorrelation of white noise being a delta function.

So, where would a signal that contains all frequencies be useful? One of the applications of white noise is in testing audio equipment. For example, an audio amplifier amplifies audio signals from a microphone or playback deck to a speaker. The characteristic of most interest concerning audio amplifiers is the flatness of the amplifier's frequency response. In other words, it is undesirable to have the amplifier increase the amplitude of the input signal at 2,000 Hz twice as much as that for signals at 200 Hz—the amplifier gain should be the same for all frequencies. This test is performed by connecting a white-noise signal to the amplifier's input and measuring the amplifier's output with what is called a spectrum analyzer. A spectrum analyzer measures a signal's amplitude across a specific frequency range.

An audio amplifier should work over a range of 20–20,000 Hz. If the gain of the amplifier is ten, then given a white-noise input of amplitude one, the spectrum analyzer should show a flat-line response between these frequency limits and zero elsewhere. However, this is not the case because there are always imperfections in real designs. Real amplifiers do not have flat responses, and there will be a wavy line between 20 and 20,000 Hz. The intent, of course, is to adjust the parameters of the amplifier to make this line as flat as possible.

I present this introductory example to show that noise is not always just something we want to be rid of in engineering. Noise can serve a useful purpose, as will become even more apparent in the following examples.

Solar power installations have become ubiquitous in today's world, and provide clean energy production. These silicon cells, which go by the name of *photovoltaic cells*, convert a percentage of the Sun's light energy into electrical energy. The conversion process is, however, not as straightforward as one might assume. Of course, it is possible to simply connect a solar cell to a load such as a small DC motor, and it will provide a DC current to drive the motor. Many toy kits are sold that do just this. But when used as part of the power grid, it is desirable to operate the cells at the point where they are transferring the maximum power possible to the grid.

The output power of a typical photovoltaic cell as a function of its output voltage is shown in Figure 4.19. The cell can be used to generate power at any point along this curve, but there is only one *maximum power point* as shown in the figure. The output voltage at which this maximum occurs is dependent on many parameters, including the intensity of light impinging on the cell and the cell's temperature. Therefore, it is advantageous to have some way of tracking this maximum power point and adjusting the load on the cell such that it operates at the optimal point.

This means that we can't just connect, say, a resistive load, to the cell and

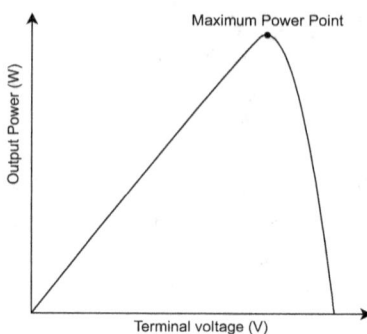

Figure 4.19: The output power of a photovoltaic array as a function of its terminal voltage.

expect the cell to operate at its maximum power point. The load will need to be able to adjust its resistance dynamically. Since real loads do not do this, we add a power-electronic converter between the cell and the actual load. This converter is capable of acting as a power transformer. It presents itself to its source, the photovoltaic cell, as a load that operates at a specific voltage and current and provides an output to the real load at a different voltage and current. Note that no rules are broken here as the input voltage and current multiplied together, or the input power, is equal to the output voltage and current multiplied together, or the output power. And, so, power is conserved as it should be. The converter simply acts as a means of matching the operating point of the cell to that of the load. In this way, the input voltage and current of the converter can be adjusted to the maximum power point of the solar cell, leading to maximum power being transferred to the load. However, to achieve this, the maximum power point must be found, as it is generally not known.

The simplest way to do this is using the power converter to make step changes in the cell output voltage (or current) and observe how the cell output power changes. For example, if the operating voltage increases by 0.1 V and the cell output power increases by 1 W, the controller knows it is moving in the correct direction. It keeps increasing the voltage until the power starts dropping off, at which point it knows that the maximum has been found. The downside to this type of control is that the converter is continually moved from its operating point to attempt to find the optimal operating point. However, there is another way to find the maximum power point that does not involve artificially changing the operating point.

The power-electronic converter matches its input voltage to its output voltage by chopping it into pieces and delivering only a portion to the output. There is a semiconductor switch that is used to connect the input to the output. This switch is turned on and off with a specific ratio of on-time to on-time plus off-time. For example, if the switch is on for 50% of the time and off 50% of the

4.3. OPTIMIZATION USING PERSISTENT EXCITATION

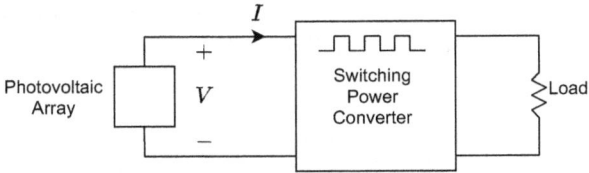

Figure 4.20: A switching power converter is used to maximize the power extracted from a photovoltaic array and delivered to the load.

time, the output voltage will be one-half of the input voltage. And, since power conservation must be maintained, the output current will be twice the input current (1/0.5).

These converters typically switch very fast, up to millions of times per second. The faster the switching, the easier it is to filter the signals back to DC. This switching noise is generally something we do not want. However, in this case, it can serve as the perturbation to the system that allows the optimality calculation to be made.

As shown in Figure 4.20, power out of the cell and into the converter is

$$P = I(V)V. \quad (4.1)$$

Taking the derivative of power with respect to voltage V, we have

$$\frac{dP}{dV} = V\frac{\partial I}{\partial V} + I. \quad (4.2)$$

As discussed previously, the power maximum will occur where the derivative of the power is zero. So, let's set $dP/dV = 0$ and rearrange as follows:

$$\frac{\partial V}{\partial I} = -\frac{V}{I}. \quad (4.3)$$

This equation says that at the maximum power point, the change in cell voltage with respect to the change in cell current is equal to the negative of cell DC voltage divided by the cell DC current. If the controller forces this equation to be true, it is ensured that the photovoltaic cell is operating at its optimal point [43]. The DC values can be found by simple filtering and measurement. The changing values stem from the converter switching and can also be found by filtering (high-pass) and measuring. The interesting thing is that we do not have to artificially perturb the system to find the maximum but can rather take advantage of the switching noise already available.

The maximum power point tracker above takes advantage of the fact that noise exists within the system and uses this noise to calculate the optimal operating point directly. Unlike manually stepped systems, this controller always drives the operating point toward the optimum.

Finally, let's look at placing the selection process of the previous section into the framework of a filter. This is a nonlinear filter that simply acts in accordance with the following rule

$$x_{k+1} = \begin{cases} x_k + \Delta x & \text{for} \quad f(x_k + \Delta x) > f(x_k) \\ x_k & \text{for} \quad f(x_k + \Delta x) \leq f(x_k), \end{cases} \quad (4.4)$$

where Δx is a random movement and $f(x)$ is an associated cost function. We will see how random perturbations converge on the function $\sin(x)$ using the selection function above. This sine wave is discretized into 32 points as shown in the curve labeled "ref" in Figure 4.21. We start with a horizontal line with 32 points that are all zeros. We then generate random-sized steps for each of the 32 points and add them to the initial values of the horizontal lines. If these new points are closer to $\sin(x)$, the selection function retains them. However, if the resulting sums are further from $\sin(x)$, they are discarded. We then repeat this process over and over. So, we move each of the 32 points by random amounts (if successfully selected) each time we iterate this process.

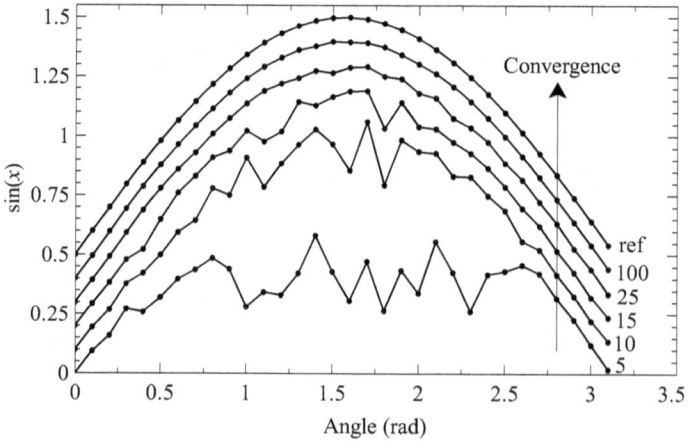

Figure 4.21: Simple example of the selection process acting as a filter on random excitation. The cost function is $\sin(x)$.

As shown in the figure, after only five iterations, the curve has risen but is jagged and is by no means a good approximation of a sine wave. With ten iterations, things are getting better, and after repeating the process 100 times, the curve is almost indistinguishable from the reference sine wave.

Now, this seems obvious. It is meant to be. Nevertheless, these processes are very analogous to what we will see in biology. The random excitation in the evolutionary process does not generate complexity as such; instead, it reveals the complexity already present within the system, complexity built by meeting the boundary conditions of a given species.

For instance, let's say that the environment of a given species experiences a step change, such as the introduction of a new predator. The number of ways the species could evolve to improve its chances against the new predator may be manifold, but the possible successful evolutionary paths already exist in potential before the prey follows any of them. Random mutation coupled with the selection function simply uncovers these paths similar to the process of finding the sine curve above.

This approach is simply another way of looking at things. But do not take this as though it is a simple static operation. Remember, since all species are evolving together, the boundary conditions for all of them are changing continuously, and therefore, what is optimal is also changing continuously. The selection process inexorably drives the system toward optimality.

These concepts will come up again in later chapters.

4.4 Genetic algorithms

Evolutionary programming is an optimization method that mimics biological evolution. Darwinian evolution, when synthesized programmatically, can solve optimization problems that are very difficult for conventional methods. The problem may possess multiple optima rendering gradient methods ineffective, or maybe the gradient is just too complicated to calculate, or possibly there is no mathematical representation of the cost function such as with aesthetic optimization. A very convenient characteristic of evolutionary algorithms is that they do not require gradient information, only a performance value, which is a consequence of their being derived from the theory of Darwinian evolution—survival of the fittest is the only performance measure.

In this section, we are going to look at an example of what is called a genetic algorithm. Genetic algorithms are a subset of the broader class of evolutionary programming. And let's be clear, genetic algorithms are not the perfect optimization methods in every situation. They certainly cannot compete with gradient methods applied to well-behaved, approximate second-order cost functions. For instance, Newton's method converges in one iteration for parabolic problems because it is optimized for that particular shape of cost function. However, genetic algorithms perform better than classical methods on specific problems, those, in particular, for which we cannot derive an analytic cost function.

Consider the traveling salesman problem. This is the problem that every delivery service faces daily: how, given many parcels that must be delivered to specific residences within a network of roads, does the delivery person choose the path that minimizes the distance traveled? In general, this problem does not have a closed-form solution. Calculating gradient information to solve the problem is not possible, thus making conventional optimization methods not useful. However, it can be placed in a genetic algorithm framework with the fitness metric being the inverse of the distance traveled for a particular route [44].

Another problem that conventional optimization methods cannot readily solve is in the area of antenna design. There are two characteristics of an antenna that are in direct conflict with one another, physical size and bandwidth. Physical size is, of course, the size of the antenna. Bandwidth refers to how wide a frequency range the antenna is capable of receiving or transmitting. In general, the wider the frequency range the antenna can accept, the larger its physical size. In the last two decades, genetic algorithms have been brought to bear to solve this problem [45,46]. The algorithms are allowed to adjust linear dimensions and angles of prototype wire antennas, using antenna performance as the optimization parameter. The result is a collection of utterly unorthodox-looking antenna designs that outperform conventional designs.

A flowchart of a simple form of genetic algorithm is shown in Figure 4.22. The first step is to create an initial population of individuals whose chromosomes are randomly generated (they are random because we have to start somewhere). Each individual possesses one chromosome that contains a number, bit string, etc. The chromosomes of the individuals define the parameters of the underlying process to be optimized. The initial random set of individuals will likely not do well at all, and if they do, it is simply by chance. The number of individuals within the population is somewhat arbitrary in simple examples but may significantly affect more complicated cases by ensuring more diversity in the group.

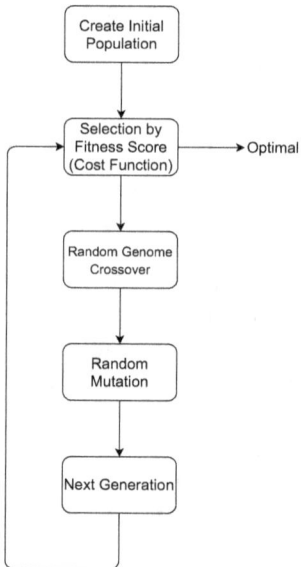

Figure 4.22: Flowchart for a genetic algorithm.

The second step is the start of a loop that repeats until a specific performance or fitness objective is met. Here the fitness of each individual is assessed via a cost function. The cost function simply takes the parameters provided by an

4.4. GENETIC ALGORITHMS

individual's genome and returns a number that corresponds to how well the particular genome solves the problem at hand. Note that the cost function is not required to be analytic; it simply must return a valid number related to performance. All individuals are assessed, and a small group of those who perform the best, usually the top two, are selected. The rest are discarded.

The next step is to replenish the population by creating a new generation from the chromosomes of these two top performers or parents. First, a genetic crossover is done by creating the chromosomes of the new individuals by splitting the parents' chromosomes at random demarcations, one portion going to each of the two progeny. At the same time, random mutation is applied to the chromosomes of the progeny with a specified probability. This involves randomly flipping bits in a bit string, randomly changing characters in a character string, etc. In other words, the chromosomes are modified randomly to some degree. This process repeats until the entire population has been regenerated to form a new generation.

Once the new generation has been formed, it is subjected to the selection routine to find parents for the next generation. This process repeats until a certain number of generations has been evaluated, or a specific fitness level (cost function value) has been reached.

It is important to note that genetic algorithms do not guarantee optimal solutions but rather attempt to find good-enough solutions, meaning that a certain fitness level is achieved. In fact, there is no way to determine in many cases if a solution is optimal or even close to optimal.

Let's look at a simple example that Richard Dawkins originally proposed in his book *The Blind Watchmaker* [47]. It starts with a random 28-character sequence and attempts to find the phrase "methinks it is like a weasel". The trick is that the only information available to the optimization routine is the output of a fitness function which returns an integer corresponding to the number of correct characters found.

Of course, there are many ways to approach this problem. If only simple random guessing is allowed, the number of possible combinations allowed in the 28-character phrase is $27^{28} = 1.2 \times 10^{40}$ (we are using the 26 characters of the alphabet and a space, yielding 27 total characters). Even using a fast computer that could make three billion guesses per second, successfully finding the correct phrase would likely take many lifetimes of the Universe.

A smarter approach would be to sequentially change the first character until the fitness function increases by one. In other words, if the initial value of the fitness function is zero, it will change to one as the first character becomes "m". If this process is repeated, the maximum number of letter flips that would be required is $(28)(27) = 756$, well doable on my computer. In fact, if the characters are modified randomly, with each character remaining unchanged after its correct value is found, convergence to the correct phrase will happen quickly. This is the approach that Dawkins took in his book, and it is repeated in Chapter 6 of this book. Here, we are going to apply a genetic algorithm to the problem.

The point of the optimization is, by starting with 25 initially random chromosomes, to get at least one of them to match the phrase "methinks it is like a weasel". Each of these chromosomes is length 28. The program is based on the flowchart in Figure 4.22 and was implemented in Octave. For this specific example, the initial fitness number—the number of letters that match correctly in any given chromosome—is three. That is, across all 25 chromosomes, three correct letters are in the correct positions in the phrase. When executed, the program went through 410 generations before one chromosome matched the target phrase identically. The ending population of chromosomes is shown in Figure 4.23 in descending fitness order. Note that the top entry matches the desired phrase.

```
methinks it is like a weasel
methinks it is like a welsel
methinks it is likt a weasel
methinkf it is likt a weasel
methinks it is like a welsrl
methinks it is like a wglsel
methjnks it is likt a wyasel
merhinks it is like afwelsel
methinks ut is like t wexsel
methinks it is liko a wexsep
mjehinks it is like a welsel
fethwnks it is like a welsel
methinks io is like avweloel
methinksfit rs likm a weaiel
jethinkstia is liks a welsel
mrlhinks ht is likt a geasel
metwinkv it is yike a zelsel
mevhinks it is likexa welqei
mytoinks itzis like aewellel
methinks itxpk lixe a welsgl
bethinks it ismlike    walh l
meqhinks it is ltktiabwblseg
uenawfirmjtlxsgitoaerumxspql
yabcbt wtjjeteicwbxhlsobaqbr
rmgksfezqsavlmdypephntz llhj
```

Figure 4.23: The 410th generation of the genetic algorithm solving Dawkins's "methinks it is like a weasel" optimization problem. There are 25 final-generation chromosomes that have been sorted from top to bottom in terms of their fitness in matching the desired phrase.

The genetic algorithm for this example is actually less efficient than that of Dawkins's original method of randomly modifying characters until each is matched. However, the genetic algorithm is closer to what is observed in nature.

Of course, this simple example in no way matches the complexity of that found in nature, but it demonstrates that it is possible, by using random mutation and chromosomal crossover, to extract information necessary to maximize a fitness function applied to a set of constraints. In this case, the phrase is analogous to a biological genome that determines the fitness of an organism. The fitness function, which measures the number of correct characters, is analogous to a biological fitness metric.

4.5 Conclusion

Optimization is a means by which a "best" solution to a particular problem can be found. In order to find the optimal point, the independent variables of the system being optimized must be varied in such a way that they pass over the optimum. Only in this way can the optimum be found. The best algorithms need only examine very few extraneous points of the function to arrive at the optimal point. However, in many cases, slower searches are the only ones available, the slowest of these being a purely random search.

The optimization method we are interested in here is one in which the independent variables are randomly perturbed, these new points being retained or rejected via a selection function. The selection function simply looks to see if the new point improves the cost function, and if there is an improvement, the point is retained; otherwise, it is discarded. In this way, the search leads inexorably to the optimum.

As we will see later, biological evolution uses a search method very similar to this, where random mutation perturbs the DNA of a species (independent variables), and the resulting outcomes are sorted based on survival of the fittest (selection function). Survival of the fittest represents a direct assessment of the corresponding cost function.

It makes no difference at all other than in convergence speed that the search uses random stepping. In fact, in some cases, such as with genetic programming, a search that is driven by random perturbation can prove to be more robust than conventional methods in the presence of cost functions with multiple optima.

It is important to understand that these methods have no goal that they are seeking. Humans tend to impose their own behaviors upon the subjects they study. These methods have a cost function $f(x)$ they are maximizing. The operating point is randomly perturbed on a continual basis. The selection function determines if the random steps are taken or not. This is a perfunctory machine; a board game where the moves are randomly chosen. There is no intelligence or meaning behind it.

We will see that the same is true with the evolutionary process. It is a mathematical optimization process written in the physicality of nature. The reason it behaves as it does is the same reason the optimization routines above do—although it may be driven by random variation, the optimization process is built of necessity from the physical laws of nature. Although more complicated, it is really no different than a rock rolling down a hill due to gravity.

As was discussed earlier, the fact that a rope suspended from both ends takes the shape of a catenary curve is itself an optimization process that minimizes the energy of the rope. These kinds of optimizations are ubiquitous within the natural world. It is not surprising that they also show up in biological processes as we will see in the next chapter.

Chapter 5

Evolution

> "One general law, leading to the advancement of all organic beings, namely, multiply, vary, let the strongest live and the weakest die.", Charles Darwin, *The Origin of Species*.

> "The two go hand in hand like a dance: chance flirts with necessity, randomness with determinism. To be sure, it is from this interchange that novelty and creativity arise in Nature, thereby yielding unique forms and novel structures.", Eric Chaisson, *Epic of Evolution: Seven Ages of the Cosmos*.

Evolution is undoubtedly one of the most contentious theories ever proposed in the scientific community because of its perceived conflict with the tenets of many religions, including Christianity. It proposes a direct and unintelligent means of explaining the diversity of life on planet Earth and, presumably, if there is life elsewhere in the universe, there as well. Religion has found this to be a very unappealing proposition because it conflicts with the idea of a god creating life in all its various forms. Evidence continues to mount for the theory, and several religious groups, including the Catholic Church, have yielded to its inexorable advance, but there remains a minority of scientists still attempting to argue the alternate viewpoint of intelligent design.

Intelligent design will be talked about in a later chapter. For now, it is enough to say that it rests on a notion coined by Richard Dawkins, which he called the argument from incredulity—that it is inconceivable that proposition A could be the cause of an effect; therefore, the cause must necessarily be proposition B.

In the case of ID, since it is unbelievable that the evolutionary process could lead to the vast number of complex morphologies seen on Earth, they necessarily must have been designed by some as of yet unknown intelligence, which many ID advocates outright admit is the Christian God.

I remember one particular instance of this type of reasoning when I was a graduate student at the University of Illinois at Urbana-Champaign. Each year the campus had a sort of open-house event and the engineering departments participated by setting up numerous exhibits demonstrating various aspects of physics and engineering. I was in the basement of the Everett Building, on the corner of Green and Wright Streets, staffing the event for a few hours. One of the exhibits was a motor created from an old-fashioned steel coffee can surrounded with bent iron fence wire with coils of copper wire wrapped around them. When this contraption was excited with three-phase currents, the can would rotate. It was, of course, a rustic demonstration of an induction machine. The induction machine produces a rotating magnetic field in its stator, which, in turn, induces a subsequent magnetic field in its rotor. The torque produced from the interaction of these two fields causes the rotor to spin. These machines are still used ubiquitously today in applications requiring rotational motion.

This setup was intended to engender curiosity in the visitors of the open house about the mysterious nuances of the electrical and magnetic fields, getting them to ask questions. However, at other times, as I found out, it brought out incredulous anger. I recall that one visitor became very agitated and explained to his group of students that there had to be magnets within the coffee can, despite my arguments to the contrary. The individual found it inconceivable that the simple motor could operate without using magnets. In the same way, the ID community contends that it is unbelievable that evolutionary processes can explain the diversity of life on Earth; therefore, a divine designer must be the real explanation.

I could have indeed been lying about magnets within the coffee can—I was not, but it was a viable possibility. Intelligent design goes one step further and introduces an alternate, non-viable solution. We will explain why intelligent design is not a satisfactory solution in a later chapter. For now, let's see if the coffee can of evolutionary theory really needs God's magnets to function.

5.1 What is the theory of evolution?

This chapter intends to examine evolutionary biology with a focus on the generation of complexity and diversity from a physical and mathematical standpoint. This is certainly not a biology textbook. One of the primary questions this book attempts to answer is whether or not there is a plausible explanation for how nature and the physical laws might produce the vast complexity observed in biological systems. The groundwork for talking about this subject has already been laid in the previous chapters and now will, in this and following chapter, be brought to bear upon the subjects of evolutionary biology and ID.

5.1. WHAT IS THE THEORY OF EVOLUTION?

The theory of evolution does not address the subject of *abiogenesis*, or the origination of life on this planet, and it is not going to be the main focus of this chapter.

In 1838 Charles Darwin tumbled to a somewhat novel explanation for the diversity and complexity of life [48]. Darwin's theory had two components: the change in heritable characteristics of organisms over time and the selection for retention of the most successful of these in aiding reproduction. Of course, Darwin did not know changes in the DNA of organisms caused that hereditary variation; he only knew that variation did occur, a good deal of this knowledge originating from his work with finches on the Galapagos Islands.

Darwin was well aware of the selective breeding of animals by human beings to enhance desirable traits and hypothesized that this same method could be employed by nature to improve the fitness of all species [49]. He reasoned that there must be large numbers of organisms that die in the struggle for life given that their respective populations do not overrun the area where they reside, and this competition for survival would sort out those in the population with advantageous genes from those with genes that did not perform as well. If the genes of an organism conferred a reproductive advantage, they would likely be passed on to the next generation, resulting in continual improvement. Darwin named this process of culling the lower performing members of the population *natural selection* as opposed to the artificial selection performed by humans in selective breeding.

That was Darwin's theory of evolution in its simplest form: there exists variation in hereditary forms, and those individuals with advantageous changes are the best at reproducing and passing on those changes to their offspring. Only later, with the discovery of DNA by Watson and Crick, was the rest of the story understood. It was found that the changes in the performance of individuals were due to random changes in an organism's DNA, or its *genotype*. The DNA of an organism directs its construction and hence determines its morphology and functioning, or its *phenotype*. Successful reproduction meant passing this modified genetic material on to the next generation. Genetic changes that increased the reproductive fitness of an individual were more likely to be passed on to its progeny than those that didn't aid reproduction.

The idea of organisms being modified over time was not unique to Darwin. In fact, the concept stretches back to the Greek philosophers and into Roman times with the poet and philosopher Lucretius. In the 18th and early 19th centuries, naturalists including Pierre Louis Maupertuis, Georges-Louis Leclerc, Compte de Buffon, and Jean-Baptiste Lamarck all suggested that there might be descent with modification from one species to another.

The British naturalist Alfred Russel Wallace, a contemporary of Darwin, discovered evolution via natural selection independently of Darwin. During his work in the Malay Archipelago collecting samples, Wallace concluded that life descended from previous species through a process of natural selection. Later, he would publish jointly with Darwin on the subject, spurring Darwin on to finish his book *On the Origin of Species*.

Although the theory of evolution provided a scientific framework explaining the diversification of life on Earth, it was met with criticism and resistance from many scientists of the time who leaned toward divine explanations of life. They believed that biological organisms needed a designer, leading Darwin to quip,"We can allow satellites, planets, suns, universe, nay whole systems of universe, to be governed by laws, but the smallest insect, we wish to be created at once by a special act."

Later discoveries, like that of the double helix of DNA by Watson, Crick, and Franklin, shed more light on the evolutionary process. Since then, what started as Darwin's theory of descent by natural selection has become the most successful explanation of the diversity of life on Earth and ties together all the biological sciences.

Over the 150 years or so since Darwin published his theory, the primary objection to it has been that a process as simple as natural selection could not lead to the complexity we see in the biological world. This view is still echoed by ID advocates today, just as it was in Darwin's time. However, as has been introduced and will be further expounded in this book, natural selection is no more than a guiding mechanism in the overall process. It is necessary but does not by itself generate complexity.

5.2 Random mutation and natural selection

Random mutation and natural selection are the two critical components of biological evolution that Darwin's theory so ingeniously used to explain the diversity of life on Earth. Although they have caused more confusion and controversy than probably any other two ideas, they are really quite simple, being no more complex than the optimization examples we previously looked at.

For simplicity, in this book, when we use the phrases random variation or random mutation, the intent will be to include all means by which the genetic material of offspring varies randomly from that of the parents. This includes formal random mutation where nucleotides of the DNA are changed or deleted during mitosis, chromosomal crossover, where an organism takes different combinations of its parents' genes during meiosis, lateral gene transfer, where one organism acquires genetic material from another, etc. It is not necessary for our purposes to look at these in detail. What we need, and what these processes amount to, is random variation of the DNA of individuals within a species. However, genetic crossover during mitosis and lateral gene transfer do serve to provide larger than usual, fully functional variations that probability rules out for straightforward mutations.

Chromosomal information contained within biological cells is akin to parameters or independent variables in the systems we considered in the previous chapter. This is, of course, an approximation because these genes do not necessarily act independently of one another and are subject to feedback in certain instances.

5.2. RANDOM MUTATION AND NATURAL SELECTION

For example, certain genes, or feedback from other parts of the cell, can be used to turn off the expression of other, specific genes. Nevertheless, there is no loss of generality in treating DNA as a set of independent variables similar to those in the previous optimization examples.

The DNA of an organism dictates the construction of that organism (although there are exogenous factors that come into play as well, such as biological interactions with the mother), and if the genes that are comprised of DNA are modified via mutation or another form of variation, the functions of the organism will change, potentially leading to a change in the organism's ability to reproduce.

The point is that the changes are very similar to those along the Cartesian axis of the simple optimization examples of the previous chapter. The main difference is that changes in the DNA are steps in the phenotypical space of the species to which the organism belongs. This is not a straightforward mapping such as that of a function like $f(x) = x^2$. In the case of this simple function, the mapping is one-to-one; that is, for each and every x, there is a distinct value of $f(x)$. This function is also continuous—intuitively, if x is changed smoothly, then $f(x)$ changes smoothly.

The mapping of DNA to a phenotypical function of an organism is different. Multiple configurations within the DNA space can map to the same changes in the morphological space. Moreover, the function that translates the DNA encoding to the morphological structure is inherently discontinuous because DNA is discrete—it is formed of gene sequences comprised of only four bases.

The genes within DNA encode for amino acids used in the assembly of proteins, which are then used directly in the construction of physical, biological structures. The generated proteins are responsible for generating structures that act as structural supports for cell shape and guide wires, acting as catalytic enzymes, the transport of ions and molecules across membranes, acting as signals, sensors and switches, and functioning as motors to move proteins and organelles. Proteins join into very large assemblies (the organism itself being the largest), the functionality of which can be changed to a great extent by a simple mutation within a single amino-acid defining gene. Hence, the effects on the functions of the organism due to genetic variation in any of its forms are completely nonlinear and discontinuous.

Despite the functional relationship between DNA and the phenotypical space being nonlinear and discontinuous, survival of the fittest still acts as a filter to sort out which variations are beneficial and which are not. Changes to an organism's DNA either result in improved reproduction, or they do not. If there is an improvement, more of the organism, and hence more of the modified genes, are produced, allowing optimization with reproductive performance acting as the cost function. This is evolution.

It is important not to anthropomorphize the process, however. This point was emphasized in the previous chapter with regard to numerical optimization and is again made clear here. The primary enabler of the evolutionary process is self-replication. The better organisms are at reproducing, the more they reproduce.

There is no intent here, however. Rabbits make more rabbits because they are instinctually driven to do so and capable of doing so.

Self-replication is an example of a positive feedback mechanism. To understand positive feedback, think of sitting on a motorcycle or four-wheeler and cranking the throttle. For those who do not ride these vehicles, the throttle or accelerator is activated by twisting the right handlebar grip backward toward the driver. This is similar to depressing the accelerator pedal in a car. Beginning riders often get a surprise the first time they twist back on the throttle of a bike. The bike lurches forward, and they are pulled back as it does so. They try to hold on to the handlebars, leading to them twisting the throttle even further as they fall backward, leading to the bike accelerating faster and their falling back further and twisting the throttle even further back, and so on. This is an example of positive feedback. The faster the bike accelerates, the harder the driver pulls backward on the throttle, making the bike accelerate faster.

Positive feedback is the mechanism that keeps evolution going. Variation and selection are actually secondary processes. Reproduction is a system with positive feedback because the more a species reproduces, the more of that species is available to reproduce, leading to more reproduction, and so on. Evolution adjusts the magnitude, or gain, of the positive feedback. If there is an improvement in a species' ability to reproduce due to an evolutionary change, then that species reproduces faster—an increase in the positive feedback. If the change is not beneficial, the positive feedback is reduced. In fact, the feedback can sometimes even become negative, the ultimate result of which is the extinction of the species.

Negative feedback is provided by the environment in which an organismal population finds itself. For example, predators act as negative feedback to a population of prey. As the prey population increases, the predators have better opportunities to succeed in capturing prey. The limited food supply of the prey can also provide negative feedback because more of the population will starve as its size increases. This negative feedback directly opposes positive gains that evolution provides, resulting, at some point, in equilibrium.

Of course, there is no guarantee of finding an equilibrium. Invasive species, for example, impinge on an area too fast for evolution to react. In the early 1950s brown tree snakes were accidentally brought to the island of Guam in the South Pacific. The snakes hunt birds and other small animals, and they have no natural predators in Guam. The snakes multiplied rapidly and are responsible for wiping out nine of the 11 of the island's forest-dwelling birds. The snakes had evolved for hunting birds, and no predator had ever evolved to prey upon the snakes on Guam. The result was that the snakes had an undue advantage, and the step change was too fast for the birds to adapt.

In summary, evolution is a process based on the random excitation of a system's independent variables followed by a selection function. It can be placed in an optimization framework, but this is not absolutely necessary. The genome is modified by random variation and, as improvements in phenotype performance occur due to this variation, they are retained via natural selection. This is akin to vibrating or tapping a pan of sand, causing it to settle. Similar to the way the sand settles in the pan to minimize the potential energy of the grains, random variation provides the motion necessary for the genome to settle into its optimal state, the optimal state being the one that allows maximum reproduction. There is no goal here; the improvement due to evolution is a purely mechanical process.

Evolution is not some magical, impossible process as many people unfamiliar with its operation claim it to be. It has no goal and does not create "winners and losers". It is simply a physical process that matches the boundary conditions between species and their environment.

5.3 From small changes to large

There is an old saying that originated from a Chinese proverb, "A journey of a thousand miles begins with a single step," implying that no matter the distance one must go, the process is carried out one step at a time. If we had to make the journey in one leap, it would be impossible. Fortunately, the sojourner need only ever be concerned with the one step he is presently taking.

Complex tasks are often broken into pieces this way. When a carpenter builds a house, he does not put the entire structure in place at one time. Rather, the foundation is placed first, then the frame, followed by the floor and walls, then the roof, and so on. In traveling to work, one does not just step out the front door of the house, take one big jump, and land at work. (I guess this might be the case for people who work at home.) One city block at a time is accumulated on the drive to work. The road segments add up to deliver you from your front door to your workplace.

Biologists use two terms to differentiate between short-term genotypical modification that stays within a species and that of long-term change that leads to speciation: *micro-* and *macroevolution* [50]. Intelligent design advocates are happy to admit that microevolution exists and leads to adaptive changes such as those in viruses to get around immunity in their hosts (this is why a flu shot is needed every year). They will not, however, accept that evolution can lead to changes in morphology or function in organisms.

The reason typically cited for disbelief in macroevolution is that the odds are so against it; that it would be extremely unlikely the large number of DNA changes necessary for speciation could occur naturally. Many ID advocates argue that the random formation of a specific sequence of 3×10^9 base pairs would have a probability of success of $1/(4^{3 \times 10^9})$. And, if this were true, it would represent an inconceivably small probability and would likely never happen. But there are two incorrect assumptions in arriving at this number. First, large evolutionary changes do not take place in single jumps. The string of human DNA did not form in a single act of random chance. It formed as a sequence of events. And, second, these events are dependent; that is, the probabilities of the events are not multiplicative.

Electrical engineers work with microcontrollers on a regular basis. These are integrated circuits composed of millions of microscopic transistors on a silicon die only a few square millimeters in size. The transistors and accompanying interconnects comprise a microprocessor (instruction processing unit, arithmetic logic unit, etc.), memory, and peripherals to the world outside the microcontroller, including communications, digital signals, analog signals, etc.

These devices provide the computing power for cell phones, video games, TVs, and all sorts of electronic devices. The machine code that runs on the average cellphone is generated from literally hundreds of pages of C, Swift, Java, or another high-level coding language. There is no way that a programmer could generate this large codebase in one straight shot. Instead, these languages are set up specifically to break the code into many functional subunits. Several subunits with specific functionality are written and tested in isolation and linked using a top code unit. Many of these top units, including their subunits, are brought together under another top unit to form a second-tier layer. This process is repeated over and over until the ultimate top unit is reached (there could be many of these top units, by the way, all of them running concurrently in an event-driven system, but that is not really important here).

The point is that the programmer cannot possibly create the entire codebase in one shot. That would be an impossibly difficult task. Fortunately, the programmer does not need to attempt this. Instead, if done correctly, the programmer never works with a block of code that is overly complex and yet, when combined with the other blocks, forms an overall program that is immensely complex.

This situation is analogous to the formation of a genome via evolution. The overall problem is reduced to smaller problems, which are then solved sequentially.

From a probabilistic view, assume for fun that a hiker must pass through ten gates that are interspersed between the beginning location of a trail and its ending location. Each gate has a preselected association with either tails or heads, of which the hiker is unaware. In order to pass a gate, the hiker must flip a coin until the coin lands on the side, tails or heads, that matches the gate's value. Now, the probability of matching all ten gates in one sequence of flips is $1/2^{10} = 1/1,024$, which are not very good odds. However, the probability of getting past the first gate is $1/2$ and, in a couple of tries, the gate's value has been matched, and the hiker is through. The same with the second gate: a couple of tries, and it is passed. And so on through all ten gates. If on average, it took a couple of tries to get past each gate, only 20 flips would be needed to pass through all ten. Compare this with trying to solve all ten gates at once. The latter method is the way evolution solves complex problems.

Consider the cost function shown in Figure 5.1. The intent is to make it from starting point A to optimal point B using the same optimization method. As can be clearly seen, the process is broken down into many smaller steps. If an attempt was made to randomly choose the proper length and direction of a

5.3. FROM SMALL CHANGES TO LARGE

single step that would lead from A to B, it would take a vast number of tries to be successful as the odds are very slim. However, if the problem is broken into small steps, like the example above of the ten gates, the process easily finds its way to the maximum.

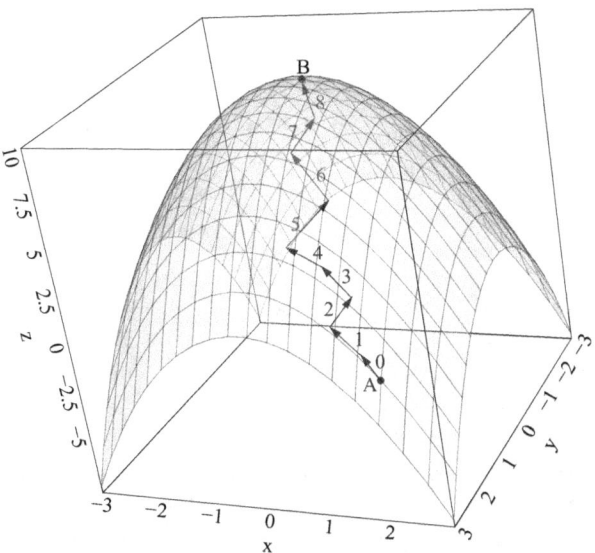

Figure 5.1: The climb to an optimum broken into many small steps.

There is one issue that needs some discussion: how can we guarantee that it is possible to break the path to the optimum into small, incremental changes? The answer to this, in general, is that we cannot guarantee it.

In the previous chapter, we discussed that two things would stop the optimization routine from reaching the maximum. The first is the presence of a local maximum, as shown in the left plot of Figure 5.2. If the optimization starts at point A and the step size is small, it will undoubtedly get stuck at B and never reach the true maximum at C. It will find the zero slope at the top of peak B and correctly decide that it has found a maximum.

On the other hand, if the optimization algorithm is started at point D in the plot on the right of the figure, it will also get stuck but for a different reason. In this case, the slope is zero on the flat portion between D and E. The algorithm stalls and never makes it to F.

It is quite possible that both of these situations can occur within biological evolution. The evolutionary cost function for any organism almost certainly contains multiple optima, and evolution will potentially drive toward a local optimum. This is just the way it is, and it is not a problem. In no way does

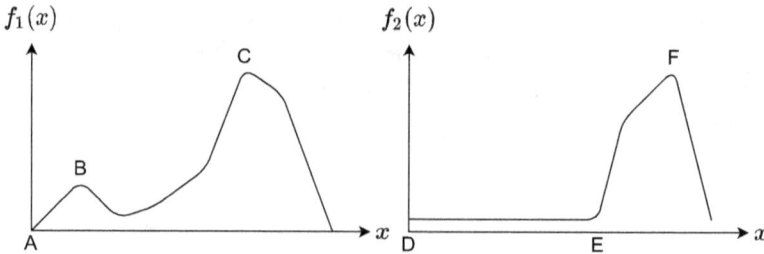

Figure 5.2: Two fitness functions for which the maximization routine will fail. If the routine starts to the left of B on the left curve, it will become stuck at the local maximum at B. If it starts at D on the right curve, it will not be able to move across the flat portion between D and E because the slope is zero.

evolution ever claim to be the "best". It only claims to optimally match an organism to its environment, and the match may well be suboptimal in strict mathematical terms. The second case of a flat cost function is also possible in biological systems.

Both points C and F could be referred to as irreducible in Behe's parlance [37]. As we will see in the next chapter, both C and F denote systems that are effectively unreachable by an optimization routine that relies on slope information and has an incremental step size too small to jump across the local optimal in the left plot or the flat portion of the curve in the right plot. Therefore, according to Behe, finding an organism at either point C or F should be inconceivable, as there would be no way for it to arrive at either point. And one would expect this is the case for a great many optima in the evolutionary space, that there are peaks that simply cannot be reached and, hence, are never realized.

However, this conclusion assumes that only small, linear steps can be made in the evolutionary process, and this is not the case. First, the process is not linear. Nature would never make exclusive use of linear processes because doing so would require the genome to be as large as the organism it generates. For example, if it were the case that an individual gene is present for each individual amino acid that went into the makeup of the fibers within a muscle, the DNA instructions would be as complex as the totality of muscle fibers. In other words, the total number of genes would be equal to the total number of amino acids present in the muscle fibers. This is not the case in real biological systems. Only one template gene is used for a particular amino acid within a muscle fiber. Additional genes instruct (kind of) how many fibers are present and how they are to be assembled. In this sense, the genetic instructions that define the morphology of an organism are nonlinear. A minute change can have a very large effect, and this is one possible means of getting past local optima.

It is not to say that small, incremental changes cannot be made in morphology; they can and do happen all the time. However, large changes are also

5.3. FROM SMALL CHANGES TO LARGE

possible. Nevertheless, even a large change will not result in a complete change in morphology—there was no single step between a bacterium and a mammal; rather, it was a long chain of steps both large and small that cumulatively added up to the differences between the two species.

Evolution is an ongoing process of continual change that we often think about in static terms as if every species is frozen at its current point in evolutionary history. Every species is changing in response to the changes in every other species and the environment. This continual evolution leads to the never-ending production of complexity.

We will talk about it more in the next chapter, but it is a strong claim to look at a complex biological machine and assert that there was no possible path that evolution could have taken to get there. The most famous of molecular machines chosen for this argument is the flagellar motor that some bacteria possess. It is an electrostatic motor powered by a proton pump that rotates a flagellum, providing propulsion for the bacterium. It is similar to human-designed motors in that it has a stator and a rotor. The irreducibility argument made by the ID community is that even if one piece of this motor, for example, the rotor, were missing, it would serve no function. Therefore, the flagellar motor could not have been constructed by the evolutionary process because, for example, the cost function would go to zero without the rotor, and without any slope information, the optimization algorithm would lose its way.

This conclusion is arrived at by assuming a linear, continuous path is required to get from no-rotor to rotor. Darwin made this assumption when he first proposed evolutionary theory, and the ID community still thinks in these terms. However, since then, evolutionary theory has expanded and shown that continuity in the functional biological landscape is not necessarily a fundamental requirement of the evolutionary process. Remember back to the behavior of a chaotic system where even minute changes in the initial conditions can lead to large changes in the final state. It would be an incorrect conclusion, for example, to assume that if changing one DNA base pair leads to the morphological change Δx, then changing two base pairs leads to a morphological change of $2\Delta x$. This is based on linear thinking, and chaotic systems are not required to behave in this fashion. It could very well be that changing one base pair leads to a change of Δx and changing two base pairs produces a change of $1,000\Delta x$. There exists no linear mapping from DNA to phenotype.

Furthermore, as shown in a later section, every point change in the genomic space (the DNA space) does not map to a different phenotype (the morphological space). There are large sets within the genomic space that are connected by single-point mutations that map to virtually the same phenotype. Moving along these networks via single-point mutations does not change the fitness of the organism—that is, a change in genotype results in no change in the phenotype. This property allows the genome of the organism to move over vast distances in the genome space without consequence, expanding the range of the accessible operating points within the genome space. (It is true that the gradient is zero along these paths, meaning this is simply random wandering through the genomic space.)

The main takeaway from this section is that evolution does not fight the losing game of probabilities. That is, it does not make large jumps across the genomic space as these have a high probability of failure, as many in the ID community would readily point out. A 747 jet is not created in midair flight due to a tornado passing through a scrapyard. Both evolution and the manufacturers of jets operate in a similar manner in regard to the fact that they build in steps. The execution of many assembly steps leads to the completed construction of a 747 airliner. Similarly, an almost uncountable number of evolutionary steps throughout the history of the Earth led from single-celled life to the diversity we see today.

5.4 The probabilities are too small

This is another one of the biggest arguments against evolution from the ID camp and is very similar to the topic of the previous section. They ask the question, how could there possibly be enough time to evolve DNA as a blueprint for constructing biological life, given that it occurs by random chance? When a question like this is asked about an apparent impossibility, it is not wise to just jump to the conclusion that the process must be impossible and a deity must have intervened. It is far better to step back and look at the situation again and make sure you have understood it correctly. The latter process has led to many breakthroughs in our understanding of reality, and the former has led to none.

In his YouTube video *Evolution and the Experts* [51], Douglas Axe explains the supposed flawed thinking of Richard Dawkins and other evolutionists when they claim that an astronomically improbable leap into the evolutionary landscape can be made more tractable by, instead of a single large step, taking many small steps. Axe confidently asserts that it is the same overall probability regardless of which path is taken. In other words, the individual probabilities of the small steps multiply to come to the same probability as the larger, single-step leap. Let us demonstrate the error in his thinking.

Consider a bicycle lock that has four dials, each capable of selecting the numerals zero through nine. In order to open the lock, the proper four-digit code must be entered. The enterprising thief who wishes to steal the restrained bike with this lock will have to decipher the combination by guessing. The most logical way of going about this endeavor is to first move all of the dials to zero and try the lock. Then move the first dial to one, leaving all others at zero, and try the lock again. Then move the first dial to two leaving the others at zero, and try opening the lock again. When every position on the first dial has been tried, she then moves the second dial to one and repeats the process of trying all the combinations on dial one again. This process is repeated through all of the dials giving a total number of combinations of $10^4 = 10,000$. Odds are, since every combination is equally likely to be the correct one, that she will hit the

5.4. THE PROBABILITIES ARE TOO SMALL

proper sequence within 10,000 tries. If it takes ten seconds to try a combination, it will take on average approximately 10,000 attempts multiplied by 10 seconds each, or 27.8 hours, to break the combination.

Now, let's modify this combination-guessing game a bit to give it a little more of a *Wheel of Fortune* flair. Let's say that when the thief finds the right number on a particular dial and tries the lock that its mechanical innards respond with an audible click. In this case, the process of guessing the combination goes differently. First, all dials are set to zero, and the lock is tried. If the click is not heard, the first dial is moved to one while leaving all the other dials at zero, and the lock is tried again. If no click is heard again, then the first dial is moved to two, and the lock is tried again. The moment the click is heard, the first dial is left where it is because that is the correct digit. The thief then goes through the process of incrementing dial two until the click is heard again. Each time, once a particular digit is found to be correct, it is left alone. The maximum number of combinations that must be tried to guess the combination of the lock, in this case, is 40; that is, ten on the first dial, ten on the second, and so on until the fourth dial. This is faster than the previous scenario by a factor of $10,000/40 = 250$ and takes a maximum of six minutes and forty seconds to open the lock.

These two situations of trying to guess the combination of the lock demonstrate the difference between randomly guessing at a particular sequence and using selective information along the way to decode the combination. Evolution is much more like the second case. But why are these probabilities (one over the number of tries) so much different if Axe is right that the number of small steps, added up, amount to the large leap? In the first case, each try at the lock combination is independent of all of the others. It does not matter when you try sequence 0109. The probability of it working is always $1/10^4$. However, in the second case, the guesses are not independent. If the first three dials have been positioned correctly, and five is guessed at for the last digit, the probability of success is not $1/10^4$ but, instead, is $1/10$. In fact, correctly guessing the appropriate digit of any of the dials in this way is $1/10$, but the probabilities do not multiply to arrive at $1/10^4$ because *they are not independent*.

It is only accurate to multiply probabilities of elements of a sequence together to obtain the probability of occurrence of the sequence when the statistical properties of the elements are independent of one another. Another simple example of this would be the probability of drawing an ace of hearts from a standard deck *without replacement*. The probability is 1/52, right? How about drawing an ace of hearts a second time? Is it 1/51? No, because the situation has changed. The probability in the second case is obviously zero.

Evolutionary theory is not so naive that it asserts that astronomically small probabilities are just ignored. A complete DNA molecule of three billion base pairs does not form out of thin air from a single guess at the appropriate sequence. This is just silly, and biologists are not stupid; they know this is not the case. That's why the argument is always brought up by those who do not understand the simple claims of evolutionary theory, mathematics, or both.

To drive this point home, let's look at two simple simulations. The purpose of each simulation is to match a bit string of lengths 2, 4, 6, 8, or 10 bits. The target series is a string of ones. The initial guess at this series is a string of zeros. The intent of both simulations is to randomly flip the bits in the initial guess until it matches the target series. However, there is a difference between the two simulations as well. The first simulation tries to randomly guess the entire sequence in one shot, whereas in the second simulation, whenever any bit in the guess sequence matches the corresponding bit in the target, that bit is locked from then on.

Each simulation is ran five times, once for each bit string length. During each of these simulations, the target bit string is found 2,000 times and the number of random guesses required for each run are averaged. A comparison is made between these two methods in finding an unknown bit string. In particular, the convergence time, or the number of guesses required by each method to find the target, is the quantity of interest.

Figure 5.3 shows the outcomes of the first simulation, where the routine attempts to solve for the entire bit string in one shot. For example, the lowest curve in the plot corresponds to a two-bit string length. The number of iterations of guessing at the target string (vertical axis of the plot) varies initially, but as it is averaged along the horizontal axis, it smooths out to four on the right side of the plot. On average, it takes four attempts to solve, in one shot, a bit-string target of length two. This result is not surprising because the probability of guessing the two-bit target string is 1/4 (one outcome in a sample space of four). This is the case for any of the other string lengths in the plot. The average number of guesses to hit the 10-bit target is 1,024 because the number of combinations 10 bits can take is $2^{10} = 1,024$.

Both axes of the plot are logarithmic. The horizontal axis is logarithmic to make the character of the curves more apparent, and the vertical axis is logarithmic to show that number of iterations required grows exponentially with the length of the bit string. We can see this by understanding that the probability of guessing a bit string of length of N is $1/2^N$. On the other hand, the number of combinations the bit string can express is 2^N. Since N is in the exponent, the number of guesses required grows exponentially with N.

The number of required iterations in these curves is what ID proponents claim of evolution; that the individual bits (base pairs) of DNA are treated as independent random variables and, therefore, the probabilities of success of the bits multiply, leading to exponential decrease in the probability as the number of bits (base pairs) grows.

In stark contrast, Figure 5.4 shows the same simulation except that, in this case, the bits are locked the moment they correctly match those in the target. For example, if the length of the bit string is 10, then if, in the first guess, two of the bits are guessed correctly, those bits are no longer allowed to change. From that point on, the problem reduces to solving for an eight-bit sequence. If, with the

5.4. THE PROBABILITIES ARE TOO SMALL

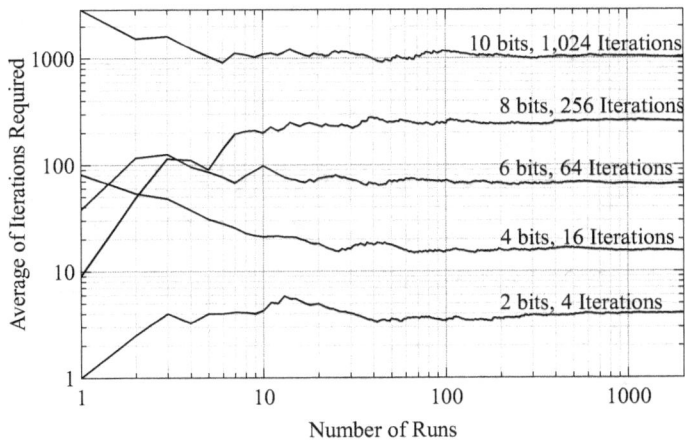

Figure 5.3: The average number of attempts to solve for a target bit string in one shot for bit string lengths of 2, 4, 6, 8 and 10.

next guess, two additional bits are guessed correctly, then the problem reduces to a six-bit string. And so on. This is similar to the way evolution works. Once a DNA base pair change has been made that increases the fitness function, that particular base pair will not be changed again. (If a change is made to a base pair that is already optimal, natural selection will filter it out.)

The curves get closer as the number of bits increases, implying that the convergence time is logarithmic with the number of bits, N. (In the previous case, it was exponential.) This can be shown with the following reasoning. As the transitions between zero and one are random, the expected value of each bit is $1/2$. Therefore, with N bits, the sequence of iterations goes like this: in the first iteration, on average, one-half of the bits in the guess sequence become equal to the corresponding bits in the target. In the second iteration, one-half of the remaining bits in the guess become equal to those in the target, and so on until all the bits are equal. The sequence of bits matching the target goes as $N/2, N/4, N/8, \ldots, N/N$ until all N bits have been found. The average number of iterations to accomplish this is $\log_2 N$ and matches the values found in the plot.

This latter simulation case is closer to the way the evolutionary process works than the former. Once an advantageous value of a particular bit of DNA is found by natural selection, it retains that value. There is no requirement that natural selection find a complete solution purely by chance, and certainly no reason to think that something as complicated as a string of DNA forms by complete chance, and in one shot. The small steps are far more likely to find a solution than one big leap.

Another misconception about the evolution of life on Earth is assuming it

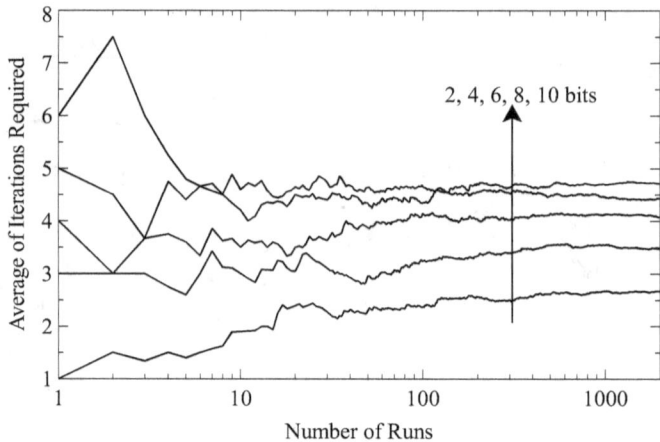

Figure 5.4: Simulation of guessing a target string of lengths 2, 4, 6, 8 and 10. In this case, once a bit is guessed correctly, it remains constant.

could have only gone one way. In other words, the lifeforms we see today are the only ones that would have been permitted to show up on Earth if any were going to do so. We tend to see ourselves as the pinnacle of evolution and want to believe that no other forms of intelligent life could have emerged. Hence, there is a problem with confirmation bias when calculating the odds of intelligent life appearing on the planet.

If one is not careful, it is easy to get confused about the actual problem that needs solving. The question is not about the probability of life showing up exactly as it is today, but, rather, the probability of life in any form emerging. These are two entirely different questions. The first leads to silly probabilistic non-sequiturs like conflating the probability of life with the probability of the three billion base pairs in the human genome emerging randomly. The second realizes that there may be many genome combinations that would produce life, and we as humans just happen to be one of them. The first case is impossibly improbable, and the second is not.

We have been over the fact that the Universe is inherently indeterministic. If reality's clock were wound backward by four billion years and released, life on this planet would not evolve exactly the same as it has. On the other hand, under the belief that there is only one possible solution, life would likely not emerge at all. But no evidence drives this conclusion. However, plenty of evidence suggests there are many different forms that life could've taken through evolution.

Evolutionary convergence is a topic we will take up in a later section. For now, understand that it is analogous to convergence in complex systems like cellular automata. As we saw previously, there are automatons that will always converge to one or a few definite patterns when started at very different initial

5.4. THE PROBABILITIES ARE TOO SMALL

conditions. It is the same with biological evolution. Many morphological and functional biological features have evolved multiple times on Earth [50, 52]. The best example is that of eyes. They have evolved several separate times on Earth, in particular in cephalopods, vertebrates, and cnidaria. Although camera-like eyes are incredibly complex devices, evolution has a good track record of bringing them into existence. And they are not simply repeated designs, as ID proponents might argue. Vertebrates, like us, have an optic nerve and blood vessels that enter through the back of the eye and branch out to connect on the inner wall of the eye, producing a blind spot in our vision. The optic nerve and blood vessels in cephalopods, on the other hand, do not pass through to the front side of the eye but, instead, connect behind it, resulting in no blind spot.

Other examples of convergence include flight, opposable thumbs, echolocation, and body plans. We will talk more about this subject in a later section. For now, convergence is strong evidence that there is no single path for evolutionary development. Convergence drives the development of common body plans and functions, but it by no means dictates a single viable evolutionary path. Therefore, simply calculating the odds of the assemblage of a string of DNA is not sufficient to determine the probability of life's development. Even conceding that DNA formation is an entirely random process, which it is not, we would need to know the size of the sample space of the trajectories that were possible for life on Earth to take before we could make any sensible probability statement.

Assuming that there is only one configuration for life on this planet is like saying there is only one configuration all the fundamental particles in the Universe could take and pondering the probability of the configuration we see today. If time were to play out again, the particles within the Universe would be in completely different positions than they were in the first run. I do not think anyone would argue this point, but they certainly do argue the point when it comes to evolution.

A prime example of the probability space admitting multiple solutions is given by the robustness of molecular networks as researched by Andreas Wagner, and others [53, 54]. Here, the term robustness refers to the ability of a genotype to experience point mutations and still express the same phenotype. That is, the genome's ability to change and yet still produce the same morphology and function. Specifically, in terms of RNA and proteins, many different versions of these molecules provide the same functionality.

Figure 5.5 shows a fictitious network of changes in three different proteins within the genomic space. Proteins are formed from amino acids that are stitched together end-to-end and used for many biological functions, including as catalyzing enzymes (enzymes speed chemical reaction times but are not altered in the process). Constituent amino acids are created from three-base-pair instructions from the DNA of an organism. Proteins can contain hundreds or more amino acids. The question is, how much of a change in the amino-acid sequence of a protein does it take to alter the protein's function? Is changing one amino acid enough to destroy the protein's functionality?

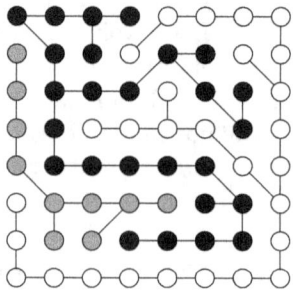

Figure 5.5: A depiction of constant phenotype networks in the genotype space. Each move from one circle to another represents a single-point mutation in the genotype. Moving on similarly shaded, connected paths means the genotype changes but the phenotype doesn't.

It used to be thought that any change in the makeup of a protein would substantially alter its function or at least make it unsuitable for the biological function for which it was needed. It has been discovered over the past two decades that this is not the case. Enter complexity into the picture once more.

It has been pointed out by ID advocates that proteins employed in the construction of life consist of strings of hundreds (or more) amino acids. They argue that the odds of the random construction of a protein composed of 200 amino acids, of which life makes use of about 20, would be on the order of $1/20^{200}$, which is a mighty small number. Once again, this argument falls into the trap of assuming there is only one way a protein could be constructed and still perform a specific function. As molecular biologists are discovering, this is not the case. There are many paths to the same functionality.

In a small digression, let's consider a somewhat analogous situation. Programming languages used today are extremely complex. More complex than most people realize. One of the most difficult aspects is ensuring that the language is *Turing complete*. This means that it is possible to implement any algorithm within the rules of the language. (It would be a very uncomfortable situation, as a programmer, to have to explain to your manager that the missile-guidance program you are working on cannot be completed because the programming language you are using does not allow a particular combination of arithmetic functions to be implemented.) At first glance, it might seem that ensuring Turing completeness would mean that only a few languages can be constructed. Now, keep in mind that new languages are not created just to see if they can be. There are various reasons as to why new ones show up, but it is not arbitrary. Surprisingly, there are an estimated 9,000 languages in use today. In other words, despite that Turing completeness represents a hard rule dictated by mathematics, apparently the requirement can be met in a vast number of ways.

The function of a protein is determined by how the large molecule folds on itself. This folding creates a distinct electrostatic bonding key. Since the protein is built by many amino acids, and the interaction of these elements with one another as the protein folds exhibits complexity similar to that of cellular automata, it is not surprising that there exists convergence in the resulting protein behaviors.

The shaded circles of the networks in Figure 5.5 represent genotype changes that result in the same or similar phenotype. The three different colors represent three different phenotypes. Each branch (solid line) in a network represents a genotype change, a change in the amino acid structure of the protein. Since they do not affect the phenotype or function of the protein, these changes are neutral in terms of natural selection—a protein, or organism for that matter, can move along these networks, and its fitness will not be altered.

We will talk about this subject in the later section on complexity. For now, we are only interested in one effect of the robustness of phenotype networks: the number of viable genotypes for the given configuration of an organism is not one; it is many. There is no longer a need to hit the exact recipe of bases within the DNA chain. There are a very large number of combinations that will work *without any phenotype change*. Therefore, even if we concede that only one form of life is possible, there are many ways to get there.

As we will see later, this interconnectedness in the genotype space also has strong implications for irreducible complexity. Irreducible complexity is a term used by ID proponents that says there must be operating points within the genotype space that are unreachable by natural selection. As we see from the networks in the figure, an organism (or a molecule) can cross vast expanses of the genotype space independent of natural selection.

In summary, evolution producing the diversity of life on Earth only seems to be a probabilistic impossibility when looked at from a naive perspective. Speciation does not occur in large strides (although this cannot be ruled out in certain instances). Evolution is the cumulative effect of many changes over long periods of time, breaking the large improbabilities associated with DNA formation into smaller pieces with higher probabilities that can be more easily overcome. In addition, the number of viable outcomes is not one—it is many. Nature did not have to hit a bullseye—maybe just getting in the inner circle was close enough.

5.5 All forms are transitional forms

One of the most cliché questions about evolution is where are all of the transitional forms? In other words, where are the forms that are, for example, halfway between one species and another? The answer to this question is, there aren't any. And, confusingly enough, the correct answer is also that all species are transitional species. This confusion comes from an attempt to shoehorn a word as a description of a process where it does not fit. It is a non sequitur.

Let's start with the simplest part of the dilemma: by definition, there can be no organism halfway between two different species. A species is defined as a taxonomic group whose members can interbreed. A member of a species cannot successfully reproduce with a member of another species. Therefore, there cannot, by definition, exist an inter-species organism.

If we look at this issue from another perspective, a transitional species could mean an evolutionary intermediate between an ancestor and descendant.

However, both of these views are misleading and are derivatives of the human mind's need to classify everything into distinct categories. Figure 5.6 shows a continuous curve $f(x)$ with three points marked A, B, and C. Is it helpful to say that point A is categorically different from point B? Is point B an intermediate form between A and C? In order to answer these questions, let's look at how we can move from point A to B.

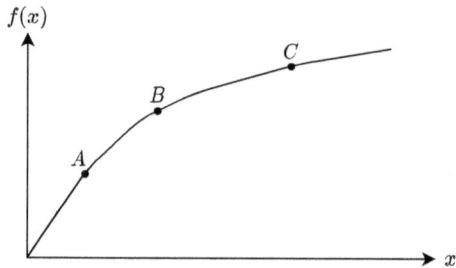

Figure 5.6: A continuous function with three marked points A, B, and C.

Starting from A, move an infinitesimal distance to the right, ΔX, and call that point A_1. As the distance between the points A and A_1 decreases to nearly zero, the two points become indistinguishable, yet they are different. Nevertheless, when used in calculations for engineering designs or balancing bank accounts, either one may be used with no discernible effect. This is one definition of continuity. If we pick another point A_2 an infinitesimal distance to the right of A_1, again, there is no real difference between A_1 and A_2. This process can be repeated until point B is reached. The crux of this entire exercise is the question of whether or not there were ever two adjacent points that were different from one another, and the answer is no. And, yet somehow, there was travel from point A to B.

The same is true for the evolution of biological entities. First, every form is in a state of continual change. Second, no species ever gave birth to a different species; their progeny were of the same species. And, yet, just like the movement of the point above, one species transforms into another over long periods of time. This is why a fossil that is half reptile and half mammal is never found. However, a continuum of morphologies between the two is found. All forms are transitioning to other forms. Contrary to ID and creationist claims that there is a lack of transitional fossils, it actually turns out that *every* fossil is a transitional form.

5.6 There are no superior forms

Evolutionary theory talks about natural selection through survival of the fittest. And this is an apt description of the process. However, it is misleading to many people who think that it implies the existence of animals that are superior to other animals, and that is entirely not the case. Once again, this misunderstanding is an example of the human mind anthropomorphizing unthinking physical processes.

Evolution has nothing to do with making an organism better than another organism. And it certainly has nothing to do with the argument that one group of people may be better than another. That is an ignorant point of view.

All animals, plants, and even viruses live together within a closed ecological system, and each one is dependent upon the others. It may seem, for example, like wolves or mountain lions are superior to deer as they hunt and kill the deer for food. These predators are faster, stronger, and also likely more intelligent than their prey. And from a human perspective, this must mean that they are better than the deer. This is a complete misunderstanding of how evolution actually functions and, frankly, an ignorant view of life itself. Humans see things like this as better or worse because we tend to anthropomorphize everything around us. It is not as if the wolf has an ego or thinks of himself as a better animal than the deer. There are no feelings involved. The wolf hunts, kills, and eats its prey. And eventually, *the wolf dies*, just like the deer. That's it.

Populations of these animals support one another without being aware of it. In fact, it runs deeper than just supporting one another. They both allow the other to exist. Remember, the existence of various lifeforms is contingent on mathematical and physical relationships between them. It is not like that the wolves win once they kill all of the deer—that would lead to the extinction of the deer and wolves. The reverse is true as well; if the wolves were not present, overpopulation and food shortage would eventually lead to suffering and extinction of the deer. There is a balance that exists between the two. We may not like it because it causes us pain to see a deer killed, but the fact is that we created those feelings. They do not actually exist in nature. This point of confusion has caused many problems.

Figure 5.7 shows the time evolution of the Lotka-Volterra predator-prey coupled-differential equations $\dot{x} = \alpha x - \beta xy$ and $\dot{y} = \delta xy - \gamma y$. The variables x and y represent the populations of prey and predators, respectively—the cross-terms on the right-hand sides of the equations model interaction between the two species. The assumptions have been made that the prey always have enough food and that the predators are always hungry.

Note that the predator population always follows the prey population in time. As the prey population grows, the predator population will also grow after a time lag. The same occurs when the prey population falls. In practical terms, as the number of prey increases, food for the predators is more plentiful, and their population also begins to grow until, finally, there is an overabundance of predators

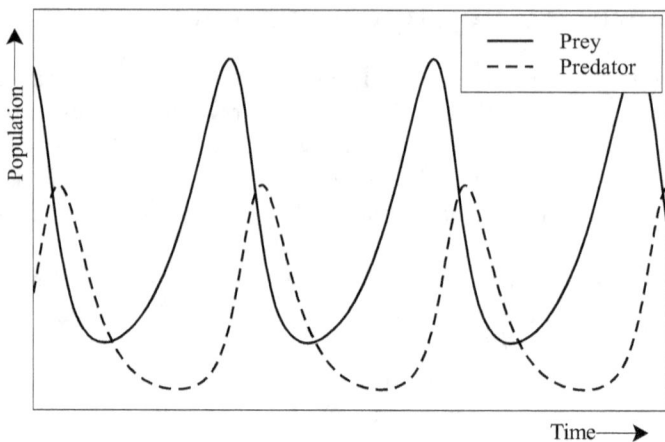

Figure 5.7: Solution to the Lotka-Volterra equations, $\dot{x} = \alpha x - \beta xy$ and $\dot{y} = \delta xy - \gamma y$. These equations describe the populations of predators and prey within an idealized ecology.

that culls the prey population. This leads to far too many predators for the prey population to support so it falls drastically, leading to the predators having very little food and their population falling off. This cycle repeats indefinitely.

If the prey population falls to extinction, the predators go with it. They are not superior. If there are no predators, the prey population grows without bound, leading to starvation and overall poor health in the group. Predators, despite appearances, are what keeps the prey population healthy.

There is no goal in the evolutionary process—that would be misleading as well. Here is an example to think about. A person steps up on a trampoline causing the fabric to stretch under her weight. As the springs and fabric stretch further, the force the fabric applies upward on her feet increases until it exactly matches the force of gravity pulling down on her mass. So, at some point, the fabric stops stretching, and she stops sinking, leading to a state of equilibrium between the force of her weight and the spring force of the trampoline. This is analogous to two species in equilibrium. One applies force (of a kind) to the other, and the other applies its own force against the first until they find an equilibrium.

There is nothing anthropomorphic here as can be seen by the force balance $-k_s x = mg$ that applies to this situation, where k_s is the spring constant of the trampoline, x is the displacement of the trampoline, m is the mass of the person, and g is the gravitation constant at Earth's surface. We can easily solve the displacement at which equilibrium occurs as $x = -mg/k_s$.

Of course, this example only considers two contributors to the equilibrium. Consider a silly experiment of filling a closet full of aired-up balloons. They are

5.6. THERE ARE NO SUPERIOR FORMS

all crammed in, and then the door is forced closed. Balloons in the middle of the closet are being pushed on all sides by other balloons. The closing of the door applies a great deal of pressure to the balloons, and their spherical shapes are distorted by the pressure of the balloons around them until the entire room is completely filled. The entire ensemble is constrained by the walls and door of the room. This is similar to the evolutionary landscape where species expand against other species and the boundary conditions to fill the entire landscape. However, there are no kings or queens here; all of the entities find equilibrium with one another to fill the entire landscape.

Humans have broken this equilibrium between species. It seems that no other species has ever been in the position we are in today, especially in our ability to control the conditions of the world around us. Moreover, although we no longer need the autonomous urges of animals like the wolves and deer above, they still exist in us from a time when they kept us alive. In particular, the urges for unrestrained reproduction and attempts to stave off death lead us to a potential over-population of the Earth that no previous species has ever experienced. We are out of balance with the rest of the world.

The primary advantage and disadvantage we possess is our enhanced ability to think and remember compared to other species on Earth. It allows us to construct a virtual mental world where we believe we are somehow better than or reside in a privileged position among all species, and we forget that this is simply our imagination. We create gods and proclaim them to be real. We become afraid of living in the world that created us and believe that the length of time we live is the most important thing. We worry about imaginary fears and attempt to control the world around us. In other words, our fear drives us to break the rules indiscriminately, and due to our intelligence, we have the ability to follow through with this desire.

The truth is that the natural world is always seeking balance. All its constituent parts find some near-equilibrium point while attempting to satisfy their boundary conditions (in the strict sense, equilibrium is never found, and change is perpetual). There is no morality involved in this process, and there are no organisms that are better than any of the others. We humans need to understand that when such ignorant concepts as superior forms enter the natural world, we brought them.

We believe ourselves to be the apex life form on the planet, and, in some ways, this is undoubtedly the case. We are capable of easily killing any other animal on the planet. Even killing every life form on Earth is not out of our reach. However, the most intelligent, most powerful species on Earth so easily overlooks the fact that it cannot survive without other plants and animals. The alternative case could be made that we are not the strongest species on the planet but the most vulnerable due to the inevitable instability we generate. If there is a belief in superiority, it was not created by evolution or nature but by the very human minds concerned about it.

5.7 Complexity and convergence

Arguably, the most important aspect of biological entities is their ability to replicate. As Richard Dawkins stated in his book *The Selfish Gene*, "At some point a particularly remarkable molecule was formed by accident. We will call it the *Replicator*. It may not have been the biggest or the most complex molecule around, but it had the extraordinary property of being able to create copies of itself." And once this ability was put into place, the door was opened up for the generation of unbounded complexity.

Although the origin of a self-replicator in biology is still an open topic for which no one yet has a definite answer, this will not affect the subject matter of this book. Abiogenesis is not a thesis point that will be taken up here. Instead, the concern will be how a fantastic level of complexity was generated after replicating processes arrived.

Once replication has started, assembly rules govern the types of complexity that emerge in biological organisms. Recall from the automaton simulations that there were no overarching rules that governed their behaviors. Instead, the complexity of the whole emerges from simple local rules that are only enforced at the level of a single cell. Now, think about this for a moment because it is a subtle, fleeting concept. *The complexity demonstrated in Figure 3.20 of chapter 3 is not produced by an overall design plan—it is created using four simple rules that are followed at the cell level.* The overall structure has no idea about these rules, and cells more than one cell away from one another do not know their respective states.

If one is not careful, it is easy to miss that automata produce global complexity using only local rules. That is very counterintuitive to the way the human mind thinks. We are top-down designers. We plan the overall picture first and then fill in the minor details afterward. Obviously, a house does not get built from the bottom up. The plumber and electrician do not come in and place the bathroom amenities and light fixtures *and then* the carpenters build the walls, floors, and ceilings around them. But in nature, things are done in the reverse direction.

Of course, one might suggest that complexity for the sake of itself, as is the case in the cellular automaton examples, is not what is seen in nature. Some argue that what is seen in nature is so-called *specified complexity* [55], that there was a plan all along. This is a very important distinction as unspecified complexity is not improbable, whereas specified complexity is highly improbable. But we will leave this topic for the next chapter. For now, let's get back to the question of whether or not complexity is even realizable in our understanding of biology.

Although cellular automata may be an intellectual curiosity, there is the question of what connection they really have to biological complexity. How do a set of objects, with simple local rules defining how they should assemble, appear in biology? This seems like a stretch.

As they are an ensemble of many atoms, most large molecules exhibit a sort of

5.7. COMPLEXITY AND CONVERGENCE

electrostatic fingerprint [56]. Consider first a small molecule like water. The two hydrogen atoms are positioned at one end with an angle of 104.5° between them and the oxygen atom at the other end. The result of this asymmetric positioning is what is called an electric *polarization*. The end with the two hydrogen atoms has a slightly positive charge as compared to the end with the oxygen atom. This polarization alters the way water molecules behave, how they behave in large groups, and how they ionically bind to other molecules like sodium chloride (table salt). When molecules like these bond, they do so such that a positive charge always links with a negative charge. Water molecules arrange themselves in a tetrahedral mesh solely due to polarization effects.

Larger molecules may have many sites of positive or negative charge as shown in Figure 5.8. There are two large molecules in the figure, one on the left and one on the right. In this instance, the two molecules have mirror-image charge distributions—where one is positive, the other is negative. This is called a *lock-and-key* configuration. When molecules have an opposite charge distribution like this, there is a strong *affinity* for them to attach to one another. Molecules bind to one another in only one or a few ways.

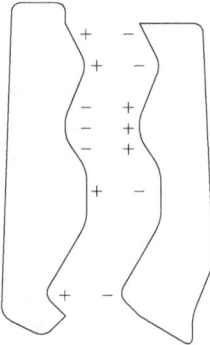

Figure 5.8: An example of an electrostatic molecular lock and key.

Molecular interactions and electrostatic lock-and-key configurations are not the only methods of interaction available. Nevertheless, all the methods available are very similar to the rules used in our simple automaton simulations. These biological interactions are also a simple set of localized rules, and they also serve to enable self-assembly.

I do not want to give the impression that complexity spontaneously arises from self-assembling molecules or autocatalysis. If these processes function appropriately, they reliably produce repeatable structures every time. However, the assembly rules and their initial conditions are dictated by other processes.

A catalyst is a molecule that enables or enhances a chemical reaction, yet it is not modified permanently by that reaction. If you've ever had a minor scrape or cut, you may have applied hydrogen peroxide at some point. When applied, hydrogen peroxide releases oxygen that kills bacterial infection in your wound. Normally, this release of oxygen is slow. However, in a high-school chemistry class, you may have performed the experiment of adding the catalyst of manganese dioxide to hydrogen peroxide. Manganese dioxide is a black powder that, when added to hydrogen peroxide, speeds up oxygen release significantly. Very rapidly, the hydrogen peroxide will release oxygen and only water *and* the manganese dioxide will remain. The interesting thing is that the manganese dioxide can be used again as it is not altered. This is a catalyst.

Autocatalysis is the process by which a molecule catalyzes its own formation. If sufficient raw materials are available, once a small amount of the molecule appears by a slow reactive process, from that point on the production of the molecule increases in speed due to it aiding in the formation of copies of itself. This is a positive-feedback process. The more of the molecule that is present, the faster more of it is made (so long as the raw materials are available).

The primary source of instructions for creating assemblies of amino acids is DNA. It is the blueprint from which the amino acids used to form proteins are built, and it is subject to change. In our cellular automaton analogy, not only does DNA prescribe the form of the molecules to be used in assemblies and, hence, the rules that dictate how the molecules interact, but it also determines the initial condition from which the automaton starts. In this way, the resulting complexity (if any) becomes dependent upon the structure of DNA.

It is not my intent to argue that complexity necessarily shows up in this way. I do not think anyone could successfully make that argument today. The intent is to show that it is plausible and possible that biological complexity derives from such a process.

Just as one cannot determine the outcome of a simulated cellular automaton without actually simulating it, in the same way, biological complexity cannot be predicted. Some biologists use the term historicity to denote that such processes are unpredictable and that the entire space of possible outcomes can never be discovered. For example, the total number of possible proteins that could be generated in the Universe will never come into existence because of sheer improbability—there are just too many of them. Stuart Kauffman refers to this as the non-ergodic Universe [38].

The term *ergodic*, when used to describe a dynamic system or random process, indicates that every possible state of that system or process will be visited. Consider a gas in a container. Any and every gas particle in the container will eventually pass through every point of the space enclosed by the container. This property is important in physics and the analysis of random signals as it allows the development of statistical properties that describe these systems.

When Kauffman describes the Universe as non-ergodic, he is calling out the fact that the Universe will never go through all of its possible states and that there is no way to predict the exact form of the biological complexity that will emerge. He terms this phenomenon as being beyond physics due to its unpredictability. I cannot entirely agree with that claim. For example, even if we could not define a statistical variable called temperature that defines the mean kinetic energy of a

5.7. COMPLEXITY AND CONVERGENCE

group of particles, it would not mean that the physics that describes the behavior of these particles was broken. The processes are still characterized adequately by physics, to the degree even that we know that we cannot predict them. Physics does not break down here any more than it does in using probability as an integral part of quantum mechanics. We cannot pin down exactly the properties of quantum mechanical particles, but the physics still works.

That said, the large number of possibilities these systems provide speaks to the robustness of biological systems finding a way to produce an ecosphere. Earlier, we said that we could not compute the specificity with which life showed up on Earth. We have no idea how many different paths life could have taken that would have led to entirely different biological functions and morphologies than we see today. We can assume that the number of possibilities is quite large due simply to the number of species we see extant today and within the fossil record. It is not difficult to speculate that the possibility space is far larger but, of course, only one solution emerged.

There would be extant today only a subset of the complete ensemble of possible outcomes that would be capable of surviving on a terrene landscape. These possibilities were filtered through the properties of the environment in which they found themselves. Only those that could survive did. This leads to the emergence of the specified complexity we will talk about in the next chapter. In a callback to the anthropic principle, we find ourselves perfectly tailored to the world we live in because we represent the set of most optimal possibilities as determined by natural selection.

This understanding also implies there are two aspects to biological convergence. Remember, convergence is the tendency of certain biological functions like eyesight evolving multiple times throughout the history of life on Earth. The first aspect is that specific functions are highly desirable within the environment on this planet. For example, the value that eyes provide to an organism is well worth the energy costs by that organism to possess them (there are instances where this is not the case, where animals who had eyes have transitioned to living in the dark underground and have lost them). Many lifeforms have developed eyes, including mammals, reptiles, fish, etc. The abilities of tactile sensing, hearing, temperature sensing, smell, and taste fall in the same category of valuable functions that many species developed. The convergence to these specific capabilities is a direct consequence of the characteristics of the world in which we find ourselves. If early life had evolved in an atmosphere replete with radio waves, perhaps evolution would have developed a biological receiver for sensing such signals. (It is not far-fetched; sharks use lateral lines to detect electric fields, and some birds make use of magnetic materials to act as compasses.)

We have already seen the second form of convergence from the cellular automaton simulations. Some complex systems naturally converge to a particular outcome or set of outcomes despite their starting condition. As we discussed earlier, the eye has evolved several times on Earth, and each time resulted in, a not identical, but similar form to the others. This is true of many aspects and functions of animals, including body plans, flight, echolocation, opposable thumbs, etc., all of which have shown up multiple times in evolutionary history.

This should sound familiar after our work with cellular automata. The plots in Figure 5.9 show the evolution of cellular automatons with the following rulesets:

$$
\begin{aligned}
18 &: \{(100),(001)\} \\
22 &: \{(100),(010),(001)\} \\
82 &: \{(110),(100),(001)\} \\
90 &: \{(110),(100),(011),(001)\}.
\end{aligned}
\tag{5.1}
$$

Remember, these three-bit rules are used to match patterns in a row to determine the state of the center cell in the next row. If any one of the rules makes the cell active, then it remains active.

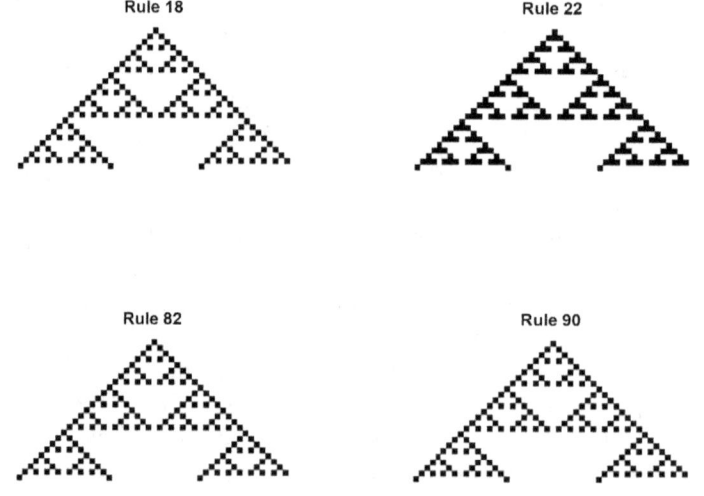

Figure 5.9: Convergence of several cellular automata using the rule sets {(1 0 0),(0 0 1)} (upper-left), {(1 0 0),(0 1 0),(0 0 1)} (upper-right), {(1 1 0),(1 0 0),(0 0 1)} (lower-left), and {(1 1 0),(1 0 0),(0 1 1),(0 0 1)} (lower-right).

The patterns resulting from these rule sets, starting with the initial condition of the center cell in the top row as active, are shown in Figure 5.9. This is an example of the convergence of complex systems. The patterns are not necessarily exactly alike, but they are very similar. In the same way, genetic configurations that lie on any one of the networks shown in Figure 5.5 are convergent in that they produce the same phenotype. That is, each of the points in a network composed of circles of the same color represents rule bases that will converge to the same outcome in the phenotype space.

5.7. COMPLEXITY AND CONVERGENCE

Again, proteins in the human body are constructed from a base set of 20 amino acids—other animals use a similar set. The acids are stacked sequentially to form a protein molecule, and the protein folds on its self, producing a very complicated electrostatic key that can bind to other molecules. Since proteins can be comprised of thousands or more amino acids, these lock-and-key patterns can be quite complicated.

The way proteins fold is determined chemically by the ordering of the amino acid sequence that makes up the protein. The amino acids act similar to the blocks in cellular automata, the chemical properties of each acting as the ruleset. The assembly of these amino acids under this ruleset can produce complex and convergent behavior in the resulting protein structure. Although the structure may be slightly different between similar varieties, the electrostatic binding pattern need be only similar enough that the phenotypical behavior is the same for many different proteins in the set. The result is a huge number of distinct protein functions or phenotype networks, each comprised of a large number of genotype neighbors linked by single-point mutations. Each member of a particular phenotype network, when the corresponding protein is assembled, converges to the same chemical functionality.

As was briefly mentioned in a previous section, recent work in the area of the mutation of RNA and proteins by the likes of Andreas Wagner [53] has explicitly demonstrated the power of convergence in complex biological systems. For example, as shown in Figure 5.5, configurations of proteins and RNA molecules with the same or similar phenotypical behaviors are topologically connected, all such states being accessible via single-point mutations. In other words, many different genomic configurations exhibit the same phenotypical behavior, and each of these configurations are connected to one another through a network of single-step changes, meaning that an organism can move throughout these paths without its fitness changing. This means that selective pressures will not prevent the organism's genome from traveling freely through the genome space on these paths, resulting in a cost-free method of moving through the phenotypical space.

Again, this should also sound familiar. The phenotype networks distributed throughout the genomic space represent basins of attraction for the cellular automata that assemble the proteins, and they are most likely fractal in form. This conjecture, I cannot prove here, but for our purposes, it is enough to know that there are large, interconnected phenotypical sets within the genotype space of biological assemblies, including RNA, proteins, and organisms. The structure of these phenotype networks is dictated by the chemical properties of the molecular blocks that make them up, just like the overall structure of a cellular automaton state depends on the ruleset that dictates its evolution.

One of the primary takeaways from all of this is that an organism's genome does not need to be exact to generate a specific phenotype—these assemblies are robust in the presence of genotypical variation. This, in turn, implies that the probability of evolution getting it "right" is increased substantially. Evolution never had to find a single, exact solution; any solution that was close would do.

It turns out the seemingly impossible precision needed in the construction of organisms was not so impossible after all.

The genomes of organisms are free to move along these phenotype networks because there is no selective pressure as they are traversed. Remember, each node on the network results in the same function in the phenotype space; therefore, as far as natural selection is concerned, there is no change when moving between them. (Actually, there is likely some change, but it is slight enough to be undetectable in selection.) The mutations that comprise the nodes of these networks are typically called *neutral* as they cause no selective effect. This interconnectedness allows undisturbed movement throughout the genome space as long as the organism moves on one of these networks, which opens a door leading to vast possibilities.

Figure 5.10 shows two genome spaces. Circles of the same color indicate different genomic points in the space with the same phenotype. Each step between adjacent circles represents a single-point mutation. The problem here is to get from point 1 to point 2. In the genomic space shown in the left of the figure, the only way to do this is to make a move with simultaneous multiple mutations. Assuming the white circles are of lower fitness than the black circles, any single step away from point 1 in the left grid of the figure results in a detrimental phenotypical change, and is denied by selection. In other words, it is highly improbable to move from point 1 to 2.

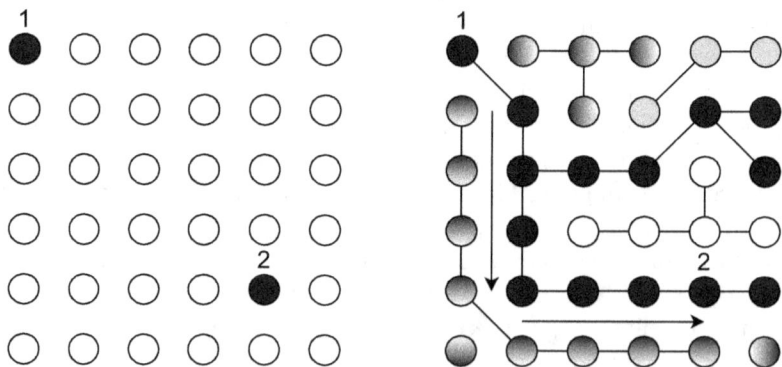

Figure 5.10: Depiction of two genomic spaces. The space on the left is unconnected, and that on the right is characterized by constant-phenotype networks.

The genomic space on the right side of the figure is a different story. The operating point can easily move from point 1 to point 2 using only single-point mutations and following the path delineated by the arrows in the figure. This path can be followed because there is no evolutionary pressure to prevent it as the path is on a constant-phenotype curve (the network of black circles). Hence, point 2 is easily reached from point 1. The question is, what benefit is the connected topology to the organism in question?

5.7. COMPLEXITY AND CONVERGENCE

Consider the two genes shown in Figure 5.11. The top gene moves two steps along a constant-phenotype network via neutral single-step mutations to arrive at the gene on the upper-right of the figure. The lower gene takes a single step to arrive at the new gene in the bottom-right of the figure. The functions of these genes remained the same through both mutative processes, so selection never came into play. However, one consequence of these mutations is that the two genes have moved to a new location in the genome space where perhaps something more interesting is possible. Although these two genes originally performed functions 1 and 2, with maybe a final mutation elsewhere in the genome, the two genes come to work together to perform function 3. Function 3 is a phenotypical response that would have been unreachable from the original positions of genes 1 and 2. Only after a series of neutral gene mutations did the potential arise to select an entirely new function.

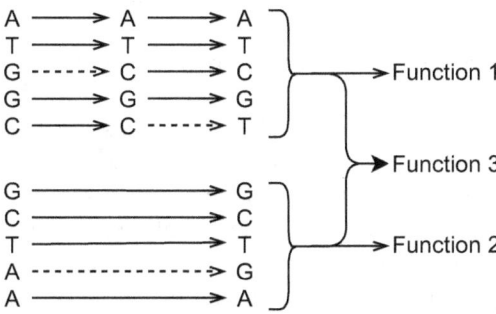

Figure 5.11: Movement of functions 1 and 2 through a constant-phenotype space. The combination of the genomic configurations of these two functions, with perhaps a last mutation, result in a function 3 that could not have been reached from the starting configurations of 1 and 2.

As Wagner pointed out, a single node in the genomic space may have access to a few phenotypically different nodes, its neighbors, via one-step mutations. However, a large constant-phenotype network has access to a vast number of neighbors. No longer is it necessary to perform a probabilistically-prohibitive jump in genome space—the same jump can be accomplished by taking small steps along a network of neutral nodes to ultimately arrive in the neighborhood of the genomic state of interest.

Constant-phenotype networks permeate the genotypical space. However, their branches lie very close to the branches of other networks as they thread their way through the space. In other words, they harken back to the extreme sensitivity of chaotic systems. It is quite possible to move along a path that has zero selective pressure to arrive at a far-away position in the genotype space. From this new position, it is possible to take a single step that substantially changes the organism's fitness, even though all along and surrounding this path, there may have existed steps resulting in a detrimental effect.

Complex behavior, including convergence, is not only displayed in the robustness of molecular chains but also in the robust wiring of gene circuits [54]. Gene circuits are just as the name implies, circuits that regulate or control the expression of genes. These circuits can be programmed to turn a particular gene on or off. They take inputs such as molecular concentrations in a given area and generate protein signals that are used to determine the active state of a particular gene. The intent of the protein signal is based on how the protein folds. Gene circuits are very similar to electronic circuits as they can perform boolean logic to produce an output based on the state of a set of inputs. These circuits can be immensely complex and, hence, run afoul of the same question of how nature came to evolve such a complex system correctly.

The answer is the same as for the molecular problem above: gene circuits turn out to be very robust in the face of alteration—they are convergent. Many different gene circuits provide the same logical behavior [54]. Again, there is no need for extreme precision. It is estimated in some studies that gene circuits could differ by as much as 90% in their connections and still perform the same function.

In both cases, the implication is that there exists a highly interconnected, likely fractal, underlying structure (current research is on track to answer this question). The consequence of the genomic space possessing this structure is a reduction in the specificity required by biological systems and, hence, an increase in the probability of evolution producing such systems.

We will talk more about this behavior and its consequences in the next chapter in reference to irreducible and specified complexity.

Before finishing up this section, it is important to briefly talk about autocatalysis one more time. The generation of complexity and subsequent application of evolution likely began at the molecular level before life appeared on the scene. As mentioned above, autocatalytic molecules are positive feedback systems. With sufficient raw materials, exponential growth is possible. As this occurs, there are occasionally slightly altered molecules that are generated due to external influences. If these new molecules are better at catalyzing their own construction, they will eventually dwarf the populations of the other molecules in solution that are slower in catalyzing themselves.

The idea that evolution started at the molecular level is gaining in popularity in scientific circles, and we'll talk about it again. However, there is one point that I wish to make about this topic: we are talking about evolution in what I think most everyone would agree are nonliving systems. This should make it abundantly clear that evolution is not an intelligent process; it is an emergent functioning within the natural world. It is driven by natural forces to form complexity from the ground up.

5.8 The fractal nature of biological systems

Fractals are employed by biological systems on a regular basis, including the structure of blood vessels, lung tissue, neurons in the brain, tree limbs, capillaries in tree leaves, flower seed arrangements, and many others. There are two reasons for this. The first is that fractal geometries allow for constructions that seem to be barred using conventional methods. For instance, the fractal design in lung tissue provides a very large surface area compared to the volume required. As a consequence of a fractional dimension, the surface area of an adult human lung is on the order of that of a tennis court, all squeezed into a relatively small volume. The human circulatory system provides oxygen to every cell in the human body but takes up only 4% of the body. Remember from chapter 3, a fractal line can have an infinite length yet reside in a finite area. Biology uses the peculiar quality of fractional-dimensional fractal shapes to accomplish tasks that would otherwise be extremely difficult.

Fractal geometries are very efficient for functions like maximizing fluid-flow efficiency for a given amount of piping. An example of a simple tree fractal is shown in Figure 5.12. In this case, the width of the branches scale as $(N-n)/N$ where n is the current branch evolving from the bottom and N is the total number of branching steps. (The bifurcation angle is 30°.) The particular width scaling in this example was used for ease of viewing, but in a real-life fluid-flow application, the bifurcation ratio would likely be closer to $1/2$—that is, the subsequent lines from any branch would be half the size of the previous line. A ratio of $1/2$ would equalize the flow cross-section at the bottom of the tree with that of the sum of the branches at the top.

In addition, all of the paths in the figure leading from the tree trunk to any of the very top branches are equal in length. The primary advantage of this architecture in terms of fluid dispersal is that it approaches an optimum in terms of the most area reached by the upper branches compared to the total length of branches needed for the task. Biological systems like blood vessels and lungs exploit this characteristic to reduce the energy expended to acquire oxygen from the atmosphere and disperse it throughout the body.

Another reason biological systems use fractal geometries is these structures require a minimal amount of specification for construction. The creation of a fractal tree like the one shown in Figure 5.12 requires only two numbers, one that specifies the ratio of diameters in branching and a second that determines the angle of branching. Building the structure of the lung does not require a DNA instruction for every branch in the airway. Rather, a relatively simple set of recursive instructions specifying *how and when* each bifurcation is to occur is enough. This represents another reduction in the amount of information required for constructing an organism.

However, it is still a matter of debate as to how much the splitting of, for example, blood vessels is specified by the genome and how much is simply a matter of the dynamics of the biology involved. Both neurons and blood vessels

Figure 5.12: A simple tree fractal whose branch widths scale as $(N-n)/N$, where n is the current branch evolving from the bottom, and N is the total number of branching steps. The bifurcation angle is 30°.

in the human body are indeed guided to some extent by growth cones for neurons and tip cells for blood vessels that follow chemical markers. However, part of the process that determines the final physical structure of blood vessels may include the dynamics of assembling the vessels. For instance, the stability of the blood vessels may be length-dependent, and when the limit of that stability is reached, the vessel branches in two, not because it is guided to do so, but, rather, because of the physics involved in the construction. It would not be too much of a stretch to claim that the resulting fractal shape is the basin of stability for the blood vessel dynamics.

Authors Professor Adrian Bejan and J. Peder Zane have proposed just such a theory that Bejan calls the *constructal law* [57]. As he frames it, the law is the tendency of flow systems to evolve in such a way as to enhance flow. For example, the structure of the network of channels carved during the evolution of a river basin produces the most efficient removal of water from the land. Rivulet flows are reinforced by erosion of the soil, allowing more water to flow. The more water that can flow, the faster erosion takes place, leading to even more flow. This is an obvious example of positive feedback. The shape that evolves from these flows is fractal.

This same behavior potentially governs a wide range of processes, including blood vessel formation, lightning patterns, snowflake construction, and many others, and may be a general property of dynamic systems. Bejan gives a particularly lucid description of the formation of tree-root patterns and of bones. Systems, via this constructal law, tend to add material along high-stress paths.

For example, the human femur (thigh bone) experiences severe stress when walking and running and, therefore, is heavily reinforced by adding copious amounts of bone. It is the stress in the bones that leads to the ossification.

As formulated by Bejan, the constructal law can always be construed as the tendency of nature to maximize flow. River basins form the structures they do to maximize the flow of water from one region to another. Trees grow where there is more water in the ground than in the air and facilitate the transport from the former to the latter. Lightning takes the shape it does to maximize charge transfer from regions of high charge density. The shapes of all of these transport mechanisms are fractal.

Chaotic systems possess fractal stability basins that define the behavior of their associated dynamics. It seems more likely that, for example, a river basin forms in the shape of a tree-like fractal as a derivative of the stability region of the dynamics involved. In other words, the shape of the basin is not determined in some special, meaningful way. Instead, it is the only shape the basin could have taken in the same way a ball rolls downhill and not up. (We'll discuss this further in a later chapter.)

Biology's use of fractal designs for construction is another example of how nature unknowingly uses very simple and efficient processes that reduce information and energy-expenditure requirements. Genomic instructions that define the location of every cell of the circulatory system are not needed. This amount of information could not be stored in the genome regardless. The circulatory system, and many other systems in the body, are built from a small set of recursive instructions. These instructions act in a relative, not absolute, context. For example, generating a fractal tree requires only the ratio of sizes of branches and the angle of bifurcations.

5.9 Extending dimension of the function space

We have been looking at genes as if they generate specific functionality, but how did they come to be connected to that functionality? For example, there are genes that dictate the formation of eyes, but how was the function of seeing brought about when neither the organism nor its constituent parts had any idea what vision was? Despite the many species extant today that possess eyes, there is only one that has some understanding of what "seeing" is.

Biological machines are far different than the machines humans create, and the levels of complexity between the two are not really the main differentiating factor. Many electrical devices such as microprocessors are very complicated in their own right, but they do not even approach biological hardware in terms of interconnectedness and the way it interfaces with the outside world.

Human-built hardware has access to the world around it only if an electrical engineer designs it that way, and it is limited entirely to the scope of access that its design allows. For instance, a flight computer in a jet has access only to the sensors the design team deemed necessary for successful flight capability.

These may include gyroscopes, temperature sensors, wind-speed sensors, altitude measurement devices, etc., but the hardware of the jet is limited to these once it is built. (Although reconfigurable hardware exists today, it is very limited in its capability.)

Biotic hardware is not limited in this way. This stems from the way biological organisms are constructed versus electrical circuitry. Printed circuit boards and their associated silicon integrated circuits are built in a factory. Biological machines are self-assembled devices. They do not require sensors to be attached to them as *they are the sensors*. They are also the computational hardware and the actuators.

An electrical machine cannot evolve on its own or self-manipulate its form and function. As they build themselves, biological machines can. If the engineer wants to add a vision sensor to a particular system, she must explicitly add the camera, wire in the connections, and provide appropriate firmware. Biological entities, under certain circumstances, can simply rewrite their internal instruction set to generate the proteins necessary to construct an eye. Of course, this is not a simple process and requires the proper environmental stimulation and sufficient time and a vast number of generations to complete. But it can be and has been done.

With an electrical system, adding a camera would include adding additional memory locations that correspond to control variables, image pixels and status signals. The firmware would need instructions that determine how these memory locations should be manipulated to control the camera and receive images.

There is no distinction in how the memory and control circuits are built in a biological system—they are all constructed from proteins synthesized from DNA instructions. In other words, whereas computer instructions have no meaningful relationship to the outside world without the associated hardware and firmware to provide meaning and context, and firmware only instantiates abstractions within the computer's memory as the code runs, biological instructions actually *instantiate objects in the physical world*. This is a phenomenological difference of the highest magnitude.

So, how does a species go about expanding the dimensionality of its set of functionality by adding a new piece of hardware? In the simplest terms, the DNA instruction set of the organism simply writes the necessary organic hardware into existence. Of course, this is an immense understatement of the actual processes involved, but it does put on display the wide gulf that exists between human-constructed electrical hardware and biological hardware.

But how does a species know it needs an eye in order to survive better? And how does it know what an eye is? As we mentioned above, it does not, in both cases. Let's look at a hypothetical example.

Suppose there was an ancient animal with its brain close to the surface of its skin, close enough that photons of light from the Sun could penetrate the skin and impact cells on the surface of the brain. Since neurological systems use electrochemical communication pathways, it seems not much of a stretch of the imagination that these neural pathways could be modulated or stimulated by light energy. How photoelectrical transduction occurs in organisms through the use of light-sensitive proteins called retinal is understood well today [58–60]. The possibility of these proteins being introduced into an organism's makeup falls directly under the umbrella of change through random mutation of DNA.

Once this mechanism was in place, optical signals could be routed into the brain. Of course, the organism possessed no viable context from which to interpret these disturbances, and, from the organism's perspective, these signals were originally just noise, in the same sense that electromagnetic noise is coupled onto sensitive signal lines within the devices we engineer today. Conceivably, when light impinged on this sensitive area, it might have been registered as pain or some other sensation by the brain. One thing is sure: if such a signal were referenced to a context it would have been to a preexisting context.

The provision of an accurate context and, hence, an appropriate reaction to the newly acquired sensor would have been provided via *neuroplasticity*—correlating the new sensation with other events. Neuroplasticity is the method by which the brain rewires its neuronal structure based primarily on correlates. For example, in the case here, if the organism we have been discussing correlated this new sensation with the presence of a potential predator (maybe predators would block some of the light when they came near), a novel behavior might emerge that dictates the organism remain motionless when it felt the odd sensation associated with the fluctuation of a light source. This behavior might eventually be written into the neural pathways of the brain as a defense mechanism.

And just like that, a new sensing ability has been developed and assimilated. Note that the sensory hardware will be passed on genetically, and the contextual use of the information relayed by the sensor would need to be a learned behavior. This is not a novel concept. Mammals, for example, rely on excitation of their optical and audio pathways when they are young for the brain to appropriately wire up the connections to the eyes and ears and to plastically develop the appropriate neural processing to decode the signals [61–63]. Without sensory input, the brain will not come to use these functions properly. It is an interesting method the brain uses to unravel both the sensory and motor pathways. One is correlated and validated by the other. Independent senses and motor functions are tested and measured against one another, and the control interfaces for each of them are solved. In addition, the process relies on feedback from other human beings.

The hypothetical story of the generation of a primitive eye is not far-fetched. It is actually pretty close to how biologists believe eyesight has started several times throughout evolutionary history. Similar stories describe how the other senses and motor functions came about. *The functional space of biological entities is extended by accident, and context is only established after the fact.*

This implies a second characteristic of the expansion of the biological function space. The first we discussed above: the instructions contained within DNA are different from electrical firmware in that they actually write physical structures into existence. The second is that the set of senses we possess are constrained by the physical reality in which we reside but, in some ways, are arbitrary. The fact that eyes evolved to detect electromagnetic radiation is not shocking as the world we know is permeated with it. Conversely, life did not evolve a means of detecting gravity waves as the sensitivity required to do so is likely beyond what can be constructed biologically. That said, it was somewhat arbitrary that our eyes sense only a very narrow region in the visible spectrum (note it is only referred to as the visible spectrum because we called it that based on the limits of our perception). This region of the spectrum was not chosen through design but, rather, was a curious accident of the proteins that were co-opted for vision. The same is true for our abilities to hear, taste and smell. Many other animals have much wider sensory spectrums in these areas.

When we are very young and are looking at the world around us, no one explains what we are doing is "seeing". Oh, they may say something to us about it but we do not understand because we do not yet know how to talk. And, yet still, everything works. We calibrate our eyes against the movement of our hands and the movement of our hands against what our eyes register, but we do not really know what either is doing in the same context as an adult. Nevertheless, we do learn how to use these features and how and when they can benefit us. The evolutionary acquisition of a new function can occur in the same way.

We are not just limited to the set of features with which life started on this planet. Evolution has the ability to expand the dimensionality of that function set.

5.10 Evolutionary pressure

There is a common criticism of evolutionary theory that today we do not see macroevolution in action, nor a continuous spectrum of extant morphologies. For instance, why have we never observed a reptile evolve into a mammal? Also, if humans descended from a primate ancestor, why do we not see half-human, half-primate-ancestor intermediates today?

The evolutionary fitness landscape or cost function presumably has many local optima. Let's look at the cost function and put this in terms of a minimization problem—it will become apparent why in a moment. The function of the evolutionary optimization method is to find minima, be they local or global. The many stable life forms we see today represent minima within the evolutionary landscape.

Evolutionary change occurs only under what biologists refer to as *evolutionary pressure*. For instance, in a predator-prey situation, if the predators experience a genetic change in their population resulting in the average predator speed increasing by one percent, this will drive the prey to evolve until equilibrium is again reached. Obviously, if the predators become faster on average, this applies pressure to the prey to move within the biological topography via natural selection. There are multiple ways for this to happen. Perhaps the prey becomes

5.10. EVOLUTIONARY PRESSURE

faster as well, or maybe it moves slowly toward another ecological niche out of the predator's environment, etc. There is no way to ascertain what route will be taken as we cannot compute these routes. But one thing is certain: any modification within the prey group that prevents an individual from being eaten by the speedier predators is sure to be amplified within the reduced prey population. Whereas previously, a large percentage of the prey could outrun the predators, now only a small percentage who have also garnered a speed advantage (or some other advantage) through genetic modification have a good chance of outrunning the predators. The rest of the population is at a significant disadvantage. This is selective pressure, and in its absence, a population will remain at an equilibrium point within the biological landscape.

Of course, evolutionary pressure can come in other varieties as well. For example, a climatic change that results in the average temperature dropping will potentially drive the animals in that area to develop heavier fur coats. As a second example, people moving to the northern or southern latitude extremes where there is far less sunlight can experience pressure for the eyes to get larger to be more sensitive to ambient light conditions [64]. Along with larger eyes, there is an accompanying increase in the size of the brain, providing the processing power necessary for the increased visual data.

There is always a variance in the genome of a species, and this is a key component in the evolutionary process. The variation in the genome is caused by various mutations, copying mistakes, and many other factors. Now, if a change is made in a species' environment, selection will become active, and those individuals with a slight advantage in the face of this environmental change due to genetic variance will propagate faster than others. In a moment, we'll see what this looks like. For now, understand that environmental change can make the genetic mean of a species move in time. The more significant the change is, the more intense is the pressure for genetic change. The characteristics of the environment is what defines the evolutionary landscape for a species. It is not as if the organism is seeking out a path through this landscape toward an optimum, rather, it is the slopes of this landscape that drive evolutionary change. If there were no evolutionary pressure from the environment, a species would remain right where it is because the landscape would be flat—it would have zero slope.

When an organism lies in a valley of the evolutionary landscape, similar to point C in the graph of Figure 5.13, the tendency is for it to stay there just like a ball will not roll uphill. It is at a stable equilibrium point that biologists call *stasis*. It is where there is no evolutionary pressure for a species to change. Most of the species we are familiar with today, including humans, are in a state of stasis.

On the other hand, if the organism's environment changes, resulting in its evolutionary cost function changing such that the organism lies at either point B or D in the graph, evolution will become active to push the operating point to C. (Note this will not be the same point C as above as the curve changed shape due to the landscape changing.) The evolutionary pressure is proportional to the gradient or slope of the cost function at a particular point.

Point A is an unstable equilibrium point. On one side is a gradient that will push it toward the left, and on the other, a gradient will move it to the right. Of course, it is difficult to imagine an organism residing at an unstable equilibrium for long as there are too many environmental influences that will inevitably move it off to one side or the other.

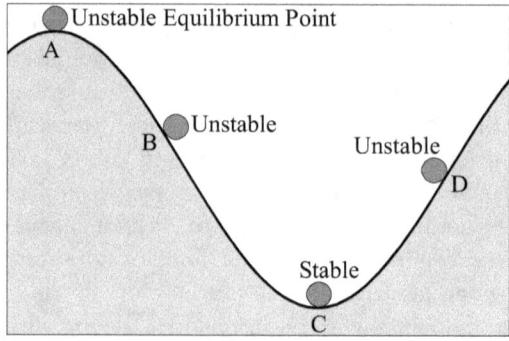

Figure 5.13: The stability of a ball placed at different points on the sides and top of hills and in a valley. Positions B and C are obviously unstable as the ball will tend to roll down the hill. The ball is stable at C. And, although A is an equilibrium point, it is unstable.

The gradual evolutionary change that many people think about when they hear the word evolution is called *phyletic gradualism* [65]. Darwin most likely was thinking of the slow progression of evolution when he developed the theory, although some scientists, such as Richard Dawkins, have suggested that Darwin did not require gradual change. Phyletic gradualism is akin to optimization with a smooth cost function on which evolution slowly moves in small steps from its current point to the optimum.

The fossil record shows that any given species is generally in stasis or equilibrium with its environment. On the other hand, there are cases where an evolutionary ancestor is in a layer of rock with its descendant(s) in a higher layer with no layer between that demonstrates speciation taking place. It appears at first blush that the predecessor morphed into its descendant(s) in the blink of an eye. That is, in *the blink of a geological eye*. The time spans represented by the rock layers are vast. The story that the layers are telling is that a predecessor lived happily in equilibrium or stasis for an extended period of time, and then its environmental constraints changed in some way, resulting in a corresponding change in the topology of its evolutionary cost function, leading to its change. The animal found itself on a gradient of the cost function, and, hence, evolutionary pressure caused it to change. This change happened rapidly compared with the amount of time it had been in stasis. Therefore, it appears that the change happened instantaneously between two rock layers.

5.10. EVOLUTIONARY PRESSURE

What has just been described is a case of so-called *punctuated equilibrium*, a term made popular in a paper by Niles Eldredge and Stephen Jay Gould [66]. Punctuated equilibrium declares that most of the time, species are in a state of stasis and, when pressure is applied, that is, they are pushed out of equilibrium due to the fitness landscape being altered, they begin evolving rapidly as compared with stasis until equilibrium is once again found. At that point, they will remain in stasis again until a change pushes them to evolve again.

The fossil record shows evidence for both phyletic gradualism and punctuated equilibrium, but there is disagreement still over the speed at which evolution acts and the prominence of stasis in this picture. There are many examples of fauna and flora remaining in the same phenotypical state for vast amounts of time, demonstrating that stasis can be long-lasting. Thus, the debate raged over the best way to describe the evolutionary process as slow and gradual, static or punctuated by bursts of activity, or a combination of all three modes.

There seems to be a desire to qualitatively characterize evolution in terms of its behaviors instead of the underlying process. As stated in his book *The Blind Watchmaker*, Richard Dawkins contends that punctuated equilibrium is an "interesting but minor wrinkle on the surface of neo-Darwinism" and "lies firmly within the neo-Darwinian synthesis" [47]. This seems to be the most fitting answer to the dispute. Evolution is, at its heart, an optimization method, and the gradient of the cost function it acts upon has a direct impact on its convergence rate and, once an equilibrium has been found, there is no slope impetus to push it out of stasis. At least conceptually, it is not difficult to imagine an organism remaining at a stable equilibrium point for an extended period of time. When a change increases the evolutionary pressure it experiences, it is driven rapidly to a subsequent equilibrium.

Even in stasis, there is a measurable variance in the DNA of a population. For example, human DNA today varies by about 0.1% across the population. This variance can be modeled as a statistical distribution about the mean. The mean is what might be called the reference or average human being. Due to mutation and other factors, humans around the mean experience varying degrees of genetic difference. This is simply a fact and has been confirmed over and over through genetic testing of the population. There is no denying that genetic change still occurs today.

The genetic variation around the mean can be plotted, similar to any other probability distribution, with the frequency of particular genes on the vertical axis and genome space on the horizontal axis. Now, if we want to show how the genome evolves through time, we must also add a time axis as shown in Figure 5.14. The genetic distribution has a typical Gaussian shape that moves from the lower-left to the upper-right of the plot through both the genetic space and time.

As this distribution moves, genes leave the population on the left, and new genes, created through mutation or other processes, enter on the right. Hence, there is continuous movement from left to right through the genomic space and, presumably, through the phenotypic space. The force that causes the distribution

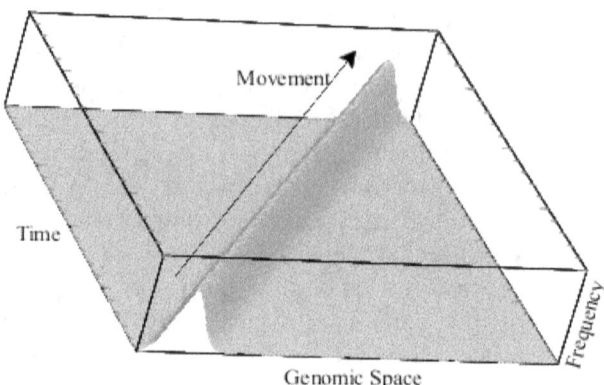

Figure 5.14: Movement of the DNA group of a particular species as it moves through both the genomic space and time.

to move along the genomic axis is evolutionary pressure—there exists a gradient in the fitness function associated with the organism. It is advantageous for the distribution to move to the right.

If the sine function used in the example of Figure 4.21 were moving in time, say a function like $\sin(t - x)$, the result would be a convergence of the random excitation to the shape of a moving sine wave. Of course, there would be a time lag between the actual cost function $\sin(\cdot)$ and the random envelope tracking it because it takes some time for the random process to converge to the sine shape. The same is true in the biological case: the fitness function is changing, and the actual genomic distribution is lagging—it has to lag in order to be driven by a gradient. There would be no evolutionary pressure to drive its change if it were perfectly in sync with the fitness function.

So, there is no difference between phyletic gradualism and punctuated equilibrium. They are qualities of the same underlying process. When a species resides at an equilibrium point, and there is no evolutionary pressure from its environment, its motion through the genomic space ceases. There is no gradient in the fitness function to force it to move. On the other hand, when there is a change in the species' environment, particularly a step change, the slope of the fitness function can increase significantly, driving the species' genomic distribution rapidly through the phase space. And so, a period of very fast evolution ensues until another equilibrium or stasis point is found.

Phyletic gradualism and punctuated equilibrium are not distinct operating modes, and we do not need to decide which process nature follows in the course of evolution. It follows both of them when the appropriate environmental fitness landscape presents itself. In a similar fashion, a ball rolls down a hill when the slope is steeper, but we do not need to qualify fast and slow rolling as distinct processes. In other words, evolution is simply a dynamic process, and its behavior is describable like any other dynamic process.

Before we leave this section, the process of evolution as presented here describes why we do not have multiple versions of human beings and do not actually witness a reptile turning into a mammal today. These are questions often asked by those critical of evolutionary theory, and there is a two-pronged answer. The first part of the answer is that the species we see in the world today are in stasis; they are in equilibrium and have been for a long while. With no impetus to change, life tends to remain the same.

The second explanation rests on the fact that as an organism tracks through the genomic-time plane as shown in Figure 5.14, it does so with minimal variance about its mean. In other words, the DNA of the group moves as a whole. The DNA space does not smear behind it, leaving a trail of intermediates. This would make little sense from an evolutionary standpoint: assuming that the individuals are the latest and greatest developments in terms of survival, any stragglers would be eliminated by natural selection. Hence, there is no string of partially evolved organisms between a predecessor and its descendants. (That being said, it is certainly possible to regard the 0.1% variance around the mean as containing intermediates.)

If it seems strange that there would not be a string of intermediate forms between our primate ancestors and modern humans in existence today, let's consider an analogy where it will be more clear why this is not the case. Consider the cars on lots and the road today. Today, in 2021, it is not uncommon to see 2018-2021 model year cars on the road—they are very common. And, slightly less common are cars that are a decade or two old, but they are on the road. On the other hand, cars that are 40-50 years old are very uncommon to see, but they do still show up occasionally. For example, my son and I rebuilt a 1978 Trans Am that we take for a drive once in a while. If one were to plot the number of cars on the road versus model year, a guassian distribution would emerge—a bell curve. This curve maintains its shape and moves through time. It also moves through the vehicle design space: newer cars possess newer and different technology than older cars. What emerges is something very similar to the movement of species through the genotypical/phenotypical space.

5.11 Paleontological evidence for evolution

As I stated previously, this is not a biology book and is more concerned with the physical and mathematical aspects of life and its evolution, but I would be remiss if I didn't include at least a few examples of evidence from the fossil record of the evolutionary process in action. Although there are a vast number of examples, here only a few will be included to confirm that the effects of evolution do indeed make their way into the world at large.

A fossil unearthed in the Canadian Arctic on Ellesmere Island by researchers Neil Shubin, Edward Daeschler, and Farish Jenkins became somewhat of a media superstar when its discovery was made public in 2006. It came to be known as *Tiktaalik* and was a fish from 375 million years ago [67, 68]. But it was a special fish. Just like any normal fish it had scales and gills, but it also had many anomalous features. Its head was flat, very much like that of a crocodile. And, most striking of all, its front fins had transformed into what appear to be semi-functional legs, allowing it to prop itself up in shallow water.

Tiktaalik is an intermediate form between ancient fishes and vertebrates. It lived in shallow water and grasped prey with its jaws similar to crocodiles today. Although it had flat fins that were spanned by thin bones at the front, the stalks of these fins possessed bones that were substantially thicker for supporting the weight of the fish. Oddly enough, Tiktaalik also sported a neck that allowed it to turn its head to some degree, a feature innate in later vertebrates but nonexistent in fish. Its ribs were also reinforced as compared to fish in order support its weight as it hunted in the shallows. And, as Neil Shubin, a member of the team that discovered Tiktaalik, notes, "The shoulder, elbow and even parts of the wrist are already there and working in ways similar to the earliest land-living animals." [69].

Tiktaalik represents a very convincing example of evolution in action in the fossil record. It clearly sits halfway between ancient fish and later tetrapods.

Another example of evolution in action in the fossil record is *Archaeopteryx*, a dinosaur that lived 150 million years ago that had many features in common with birds [70], including feathers and wings. It lived in what is now southern Germany at a time when most of Europe was an archipelago of islands. The first fossil was discovered in 1861, a few years after the publication of Darwin's *On the Origin of Species*. At the time, it was received as confirmation of Darwin's evolutionary theory.

Archaeopteryx was about the size of a raven, maybe 20 inches in length, and weighed around two pounds. Its feathers were very similar to those of today's birds. It had many features in common with dinosaurs, particularly raptors, including sharp teeth, a long bony tail, and a hyperextensible second toe (the claw that raptors have for slicing their prey).

However, it is known today that Archaeopteryx is likely not the missing link between dinosaurs and birds. Other candidates have been found such as Xiaotingia. Despite the fact that Archaeopteryx is not a direct ancestor of birds, it is still interesting in its own right in demonstrating the changes that evolution can make. Archaeopteryx is a dinosaur that evolved wings and was likely capable of flight.

The last example to be considered here will be human evolution because it is the one for which ID and Christianity claim there is no evidence. However, this not being a book on biology, we will not go through in detail the literally dozens of links in the fossil record that show a line of evolution from a hominin ancestor common with the great apes. This evidence can easily be found on line or in many books on human evolution. But I do want to point out the well-established evolutionary line from this common ancestor to modern-day humans and attempt to answer the question as to why ID and Christianity are so averse to recognizing it.

The overall order primates contains lemurs, lorises, tarsiers, tree shrews, monkeys, apes and humans. *Sahelanthropus tchadensis* is the oldest known hominid, which lived about 7 million years ago (mya). Chimpanzees and human ancestors diverged from this branch about 5 mya. The line that paleontologists believe led to humans starts with *australopithecus anamensis* and leads through *australopithecus afarensis*, *australopithecus africanus*, *homo habilis*, and *homo erectus*. Modern-day humans, *homo sapiens*, as well as the neanderthals, *homo neanderthalensis*, descended from *homo erectus*.

The actual evolutionary path is far more complicated than the brief outline that has been given here, and there is still debate going on about it as the fossil record is not complete at this point. However, despite an incomplete record, one thing is absolutely clear: humans were crafted in their present form by an evolutionary process. It's either that, or a god created an entire line of animals that appear to have evolved over time, including the appropriate dates in the timeline. Occam's razor rules out the latter case.

It has been pointed out by others many times that either evolution is real or the supernatural designer *made it look real*. There is a voluminous amount of evidence that points in the direction of evolution, and no evidence that humans were created by a god. Therefore, the argument from the Christian side is that we abandon logic and follow a path that has no physical evidence.

As we'll talk about later, one does not choose what they believe. They are driven to their beliefs by causes that are essentially out of their control. Once one has seen and understood the evidence for evolution, how could they just turn that off and say they believe that a god created man 6,000 years ago? Of course, they can say it, but that does not make it so. Whether or not we like it, all the evidence available points in the direction of a world where all life, even humans, was not created, but rather evolved over a long period of time. And it is misleading for ID to advocate a different point of view simply because it is what they wish to be true.

5.12 Genetic evidence for evolution

The fossil record is not the only physical evidence for evolution. We each carry around the accumulated changes of evolution over the past eons within ourselves in our genetic sequences. The difference between an evolutionary predecessor and its descendant is the changes that were made in the predecessor's DNA to arrive at those of the descendant. In other words, in order to get from predecessor to descendant, the DNA of the predecessor is changed slightly. The unchanged DNA, which is usually the larger portion, declares that the two organisms are related. The DNA change, which is typically a smaller percentage of the total DNA, says something about the difference between the two organisms. In this way, the common genetic material in the evolutionary tree leaves footprints so to speak.

Consider the hypothetical evolutionary tree shown in Figure 5.15. The ancestor common to the entire tree is labeled A, and it gives rise to descendants B, E, and F. These organisms then give rise to their own evolutionary descendants. Organism A is assumed to have a total number of DNA base pairs of N. For simplicity, and without loss of generality, the assumption is made that the DNA of each descendant is changed by the factor k. (If you want, think of k not as a constant, but as an operator that modifies N in any particular way.) In the real world, the genome would be altered by variable amounts between antecedent organisms and their evolutionary descendants. Nevertheless, our assumption of an equal change across all evolutionary links does not change the crux of the matter.

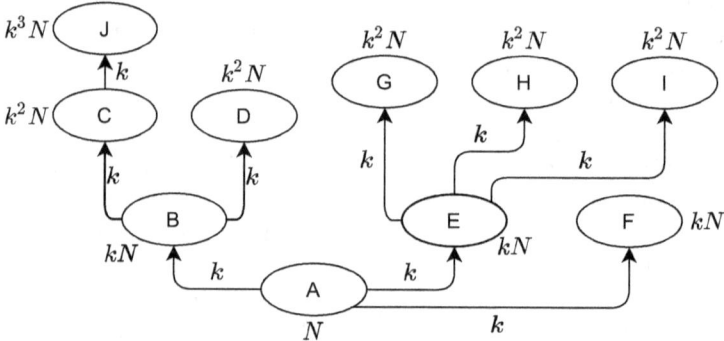

Figure 5.15: A hypothetical evolutionary tree. For simplicity, the modification of the genome from ancestor to direct descendant is assumed to be k. The DNA of the organism at the base of the tree is assumed to have N DNA base pairs.

In each of the evolutionary paths, the ellipses represent species that reside at an equilibrium point. Their descendants have split off and derived their own species. This means that all of them are extant together—for example, B did not become extinct just because C and D are on the scene.

In each step of the process, the genome of the ancestor is altered by the multiplier k. So, as we go from A to B, the genome transforms from N to kN. And in the transition from B to C, once again the genome of B is modified by k, resulting in the genome of C becoming $k^2 N$. And finally, in the transition from C to J, the genome transforms to $k^3 N$. Using this information it is possible to track J back to C by solving for $1/k$. In the same way, the entire chain can be traced back to A. Biologists trace the relatedness of organisms within the evolutionary tree in a similar way. (Of course, this description is oversimplified. For example, subsequent changes can also affect portions of DNA that were already changed in a prior change. The point is made, regardless.)

In the same way that the system $\dot{x} = f(x)$ can be integrated forward or backward in time, the evolutionary process can be worked backward in time

5.12. GENETIC EVIDENCE FOR EVOLUTION

using the data we have available today. These two cases are not exactly the same since evolution is a stochastic process, but it can still be characterized as with any other random process even if it is only in the simplest of terms as in Figure 5.16.

The plot in Figure 5.16 displays the percentage of similarity between humans and other animals. Primates are closest to human beings in terms of DNA similarity. Of course, this does not discount the likeness of our DNA to that of pigs—that match is also very close at 98%. This close match is what makes possible xenotransplants, such as the heart valves already taken from pigs and used in human patients. In the coming years, it is projected that full heart transplants will be performed.

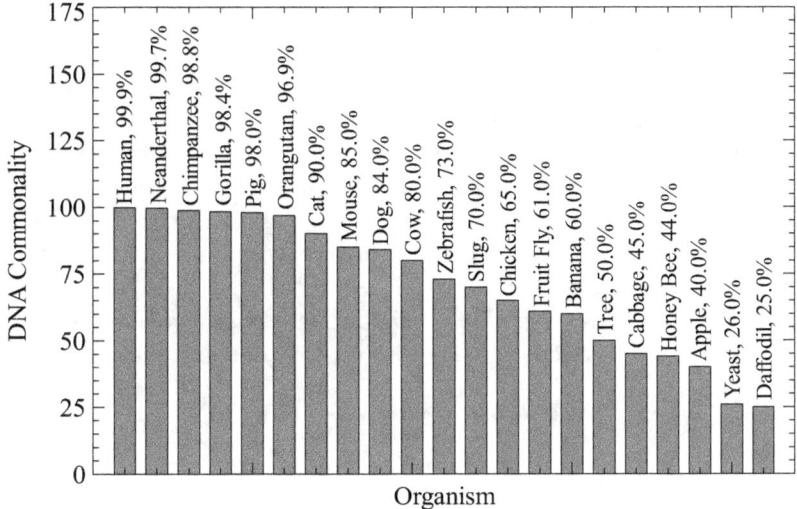

Figure 5.16: The similarity of human DNA when compared to other animals. (Note the first column states that humans share 99.9% of their DNA. This results from an observable 0.1% DNA variance across the human population.)

So, how do scientists know that the correlations they find are trustworthy? Couldn't it be that the DNA just happens to make it appear that the animals evolved over time, originating from a common ancestor? It could be possible, but it is so unlikely as to be completely ruled out. There are billions of base pairs within the DNA. I'm assuming that a god has no need to reuse designs as evolution is required to do, which means the DNA of all lifeforms need not be correlated at all in that case. Therefore, if a god did create all lifeforms, one could assume the chances that any DNA would be correlated at all is astronomically small. But that is not what is found in the DNA. Instead, extremely strong correlations are found. Once these DNA correlations are compared to the fossil record, the evidence for evolution becomes even stronger.

Let's look at one last example within the DNA of humans which is very much an oddity if a designer created it.

Chromosomes are strings of DNA base pairs that encode information needed for protein synthesis. The long stretches of encoded information are delineated by markers to help in processing and replicating that information. At each end, there is a *telomere*, a short string of base pairs with a specific sequence that indicates the end of the chromosome. Also, each chromosome has what is called a *centromere* in its center. The centromere is basically an anchor point. When a cell divides during mitosis, each of the new cells attach fibers to these anchor points to pull one-half of the chromosomes to each new cell.

We have already talked about humans and the great apes being descendants of a common ancestor. A piece of direct evidence for this is found in the fact that great apes have 24 chromosome pairs while humans have only 23 pairs. However, we know that two of the chromosomes that the great apes possess have actually become conjoined to form chromosome 2 in humans [71]. Whereas chromosomes normally have only two telomeres total—one at each end—the human chromosome 2 also has telomere remnants near its center. Also, normally chromosomes have only one centromere, but human chromosome 2 has an extra vestigial centromere. And, finally, the closest living relative to humans, the chimpanzee, has nearly identical DNA sequences to human chromosome 2, but they are found in two separate chromosomes in the chimpanzee. This is extremely strong evidence that somewhere in our evolutionary past, two chromosomes were merged, leading to an extreme distinction between humans and the great apes.

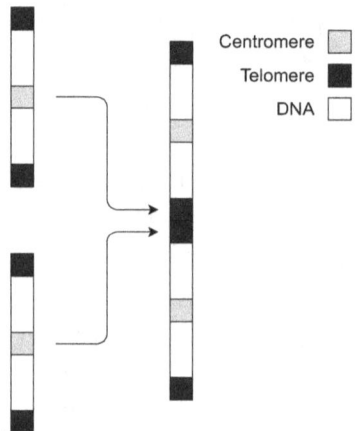

Figure 5.17: The formation of human chromosome 2 from two other chromosomes. This led to humans only having 23 chromosome pairs whereas other extant primates have 24 pairs.

Once again, if life on Earth was created in such a way as to make it appear to be evolved, why would a creator do such a thing? This seems to be an incredibly silly notion. If we give up rationality here, we will be forced to give it up everywhere else as well.

5.13 Conclusion

The theory of evolution is the single most successful explanation of the diversity of life on Earth. It is based on an immense amount of observed data and does not rely on preconceived notions of how the world should operate. Based on the deceivingly simple notions of random variation in the genomes of organisms and subsequent natural selection, it inherently leads some people to incredulity, particularly those trying to promote alternatives with no evidence, such as ID. This was the case in Darwin's time and persists to today.

Rejection by the theologically inclined is further exacerbated by the belief that evolution necessarily goes against the Biblical view of God as the creator of humankind. Furthermore, this is undoubtedly the case when referencing a literal reading of the Genesis account. However, many other things are ignored when taking the Bible literally, including the age of the Earth being greater than 6,000 years, as inferred from Genesis.

There are a number of religious organizations that have accepted evolutionary theory, including the Catholic Church. The usual means of reconciling evolution and faith is to assert that God guided the process behind the scenes. Although certainly unprovable, the notion can live comfortably with science precisely because it makes no measurable impact on reality. It is a "believe it if you want to" kind of thing. However, there are some religious groups, like those who endorse ID, that cannot live with this situation, and want people to believe (because it can't be proven) that a god runs the entire show and that evolutionary theory is a physical impossibility.

So far, ID has made no headway whatsoever in terms of displacing evolutionary theory. The biggest reason for this is that evolutionary theory fits the data so well that it is extremely difficult to argue against. Another reason, as we will see in the next chapter, is that ID is not really a theory of its own that makes positive claims, but, instead, it is only an objection to the validity of the evolutionary theory.

Chapter 6

Intelligent Design

"We can allow satellites, planets, suns, universe, nay whole systems of universe, to be governed by laws, but the smallest insect, we wish to be created at once by special act.", Charles Darwin, *Notebooks*.

"I do not feel obliged to believe that the same God who has endowed us with sense, reason, and intellect has intended us to forgo their use.", Galileo Galilei.

The fundamental thesis of ID is the negation of the scientific claim made by biologists that the evolutionary process is sufficient to explain the ecological diversity we see today and in the fossil record. Although it originally stems from *Creationism*, which makes unabashed claims that the world and life within it were created by a god (specifically the Judeo-Christian God), ID applies a technocratic, pseudo-scientific emphasis to its predecessor in a misleading effort to achieve scientific legitimacy. The primary example of this chicanery is detailed in the case of *Kitzmiller v. Dover* [1] that centered upon preventing the Dover school board from introducing ID concepts into the teaching curriculum due to their non-scientific nature.

Intelligent design is not a new concept, however. Even in the time of Charles Darwin, there were advocates for ID. In particular, William Paley, the late 18th century philosopher and Christian apologist, wrote the book *Natural Theology* near the end of his life that presented the main idea of contemporary ID [72,73]. In the opening pages of this volume, Paley presents the watchmaker analogy. To understand the gist of the argument, assume that one is walking along a beach and sees a rock embedded in the sand. This observation results in no surprise and warrants no further investigation as to its origins. However, if a watch is spied lying in the sand while walking on the same beach, Paley's argument is that the observer automatically knows that the watch had a designer. This conclusion is due to the watch's intricate construction, the complete interdependence between its multiplicity of internal components, and the obvious intelligence required in its design.

Intelligent design makes the first claim that, like with the watch, there is a certain level of complexity beyond which we should infer that a device or system was designed. In other words, a watch is too complex to be formed by random, natural processes, and this is so much more the case with the immense complexity involved in the construction of, say, a human eye.

Another observation of Paley was that the watch would not function without all its internal components. That is, every single component is necessary for the device to perform its intended function. This concept would later be co-opted and renamed irreducible complexity by Michael Behe in his book *Darwin's Black Box* [37]. Using an analogy, ID claims that just as a watch can't function without all of its internal components, neither can a complex biological system function without being formed in its entirety, meaning that for some biological systems, no gradual evolutionary path exists for their natural development, hence barring phyletic gradualism.

In this chapter, we will look at these arguments and their premises in detail and show that, although they seem legitimate at first blush, they are ill-posed notions.

Further complicating the issue, ID itself provides no theory by which biological complexity could arise. Instead, it simply attempts to show that such complexity cannot arise solely through evolution. As will be shown, in this regard, it indeed cannot be considered a theory. At best, it is an objection to a theory. At worst, it is an entirely unsubstantiated hypothesis. For example, suppose one proposed that the sky appears blue not because of the scattering of light in the atmosphere but is instead due to the action of magic fairies that cannot be detected. In that case, no one could refute this hypothesis, but since it has no backing in measurable facts, it is not a serious contender to explain why the sky is blue. It will become clear before we're done here that ID is attempting to hypothesize in precisely this way.

Despite all these other pitfalls with ID, its most subtle and insidious quality is that it supports forces or entities that are not observable or measurable in reality. This idea embarks on dangerous ground. Believing in only unobservable concepts would be less hazardous—these concepts would make no difference at all. For example, a divine entity, which is not detectable, that dictates the behavior of the tides, is not an issue. Since it is not observable and does not interact with reality in any measurable way, it is the same as the laws of physics, and the conclusion will be that the Universe behaves this way on its own. But there is another problem. The problem is that these ideas do not show up in isolation.

They tend to be like the in-laws showing up for a weekend stay with enough baggage for a month-long stay.

The smuggling in of religious ideas tied to the ID hypotheses is evident in the ID community's actions. While they are in the public spotlight, ID advocates tend to refer to the intelligent designer they have in mind in bland, innocuous terms as an unnamed force in the Universe. However, one does not have to look too hard to discover the real identity of this designer: the Christian God.

The Discovery Institute was founded in 1990 in Seattle by Bruce Chapman and George Gilder as a nonprofit think tank, and it has become a leading champion of the ID hypothesis [74]. The Discovery Institute initiated the Center for the Renewal of Science and Culture in 1996 with the intent of providing a place for scientists working on ID and challenging evolutionary theory.

In 1999, a five-year plan authored by the Discovery Institute was released onto the internet [75]. It came to be called the "Wedge Document" and outlined a detailed plan to replace scientific materialism with a theistic-based world view. Its stated goals were:

1. To defeat scientific materialism and its destructive moral, cultural and political legacies.

2. To replace materialistic explanations with the theistic understanding that nature and human beings are created by God.

And the strategy to be used was also provided in the document as follows:

> However, we are convinced that in order to defeat materialism, we must cut it off at its source. That source is scientific materialism. This is precisely our strategy. If we view the predominant materialistic science as a giant tree, our strategy is intended to function as a "wedge" that, while relatively small, can split the trunk when applied at its weakest points.

Over the past two decades, the Discovery Institute and the ID movement in general, have pushed several campaigns to undermine the credibility of evolutionary theory. The "Teach the controversy" campaign argued that evolutionary theory has many weaknesses and that ID is a valid contender that should be placed on equal footing. The "Free speech on evolution" campaign claimed that scientists, teachers, and students across America were prevented from having dissenting views on the validity of evolutionary theory by so-called "self-appointed defenders of the theory of evolution." There were also campaigns portraying books, such as the creationist book *Of Pandas and People*, as being banned, and that college professors and students were being oppressed for criticizing evolutionary theory.

I have noticed in reading ID materials that they tend to argue as if the choice between an intelligent designer and evolutionary theory is a matter of opinion. It is not. There is only one truth, and despite our feelings, we do not get to choose our own version of reality. Scientists are human, but generally, they are not swayed by feelings or preferences. And, if they are, the matter will be sorted out in peer-reviewed discourse just as ID has been weighed and measured and rejected.

But this subterfuge on the part of ID is not legitimate reason enough to reject the ID hypothesis. The only thing necessary to justify rejection is consideration of the claims of ID. These claims are what will be our concern in this chapter.

6.1 Is intelligent design a legitimate theory?

There has been longstanding debate over whether the idea of ID stands on the same footing as evolutionary theory, or any other scientific theory for that matter.

The thesis of ID, in its essence, states that since human scientists in the 21st century can't fully demonstrate that life could have come about by natural means, then a supernatural designer must have brought it about. There are several problems with this approach. The first is that it is an attempt to prove a negative, specifically that nature cannot produce the kind of complexity that is observed in life.

There are two ways to demonstrate that a cause leads to an effect. The first is to show that the effect is directly contingent upon the cause. The reason that the tides come in and go out on Earth is directly due to the gravitational pull of the Moon on the oceans. That it would lead to this effect is easily demonstrable through direct calculation: one can use Newton's law of gravitation along with the mass of the Moon to calculate the force exerted upon the oceans. The effect can be simulated using computers to further confirm that the Moon's gravity is indeed the cause.

There is another way to demonstrate a theory, however, and it is much more involved. Say, for some unknown reason, that you want to prove the tides are caused by the Moon's gravity *without using Newton's law of gravitation*. That is, you want to demonstrate that the tides are caused by gravity without using gravitational force in your theory. This is a silly thing to do but stick with me for a moment.

The way you might engage in this endeavor is to create a set of all the causes in the Universe. Of course, this is not possible, but for the sake of argument, assume that you can list all the infinite causes. One of the causes in this set would be the Moon's gravity. It has to be there because *it is the cause*. However, given that the set of causes contains every possible cause for anything, we don't have to directly demonstrate that the Moon's gravity is responsible for the tides. Alternatively, we can show that every other cause in the set is *not* the cause of the tides. Since this set is complete, if none of those causes are the reason for the tides, the only cause leftover, the Moon's gravity, must be the actual cause.

Effectively, this is what ID is attempting, although they are doing it half-heartedly. It seems to be an expectation of the ID community that if they can demonstrate that biological evolution cannot be the cause for the diversity of life on Earth, the default, the only remaining choice, must be an intelligent designer. Not only is this an ill-posed dichotomy, but even if it were possible to show that

6.1. IS INTELLIGENT DESIGN A LEGITIMATE THEORY?

all alternatives to ID were infeasible, it would be exceedingly difficult to carry out in practice. There is no way of knowing all other possibilities and most likely no way of verifying their inadequacy even if one could know them.

It's not that such an approach cannot work. Consider the following thought problem: a roll of two dice has a sum of five, and the combinations that form a sum of five are $\{1,4\}$, $\{4,1\}$, $\{2,3\}$ and $\{3,2\}$. Thirty-six combinations may occur from a roll of two dice. These are listed in Table 6.1.

The first way to demonstrate the solution is to consider the bold lines 4, 9, 14, and 19 in Table 6.1 and directly confirm that they add up to five. The other way is by using the contrapositive and ensuring that all the lines that are not bold *do not add up to five*.

The point is that the direct path in proving a cause-effect relationship involves considering one or a few possibilities. The second path, of demonstrating that no other cause is valid, can require showing that a possibly infinite number of possibilities do not solve the problem.

Intelligent design takes the latter path and claims that there is no way nature alone could construct the complexity observed in the biological world. Proving a simple proposition like the dice example above is one thing, but doing the same for a complex problem like the diversity of life is an entirely different matter. Whereas proving the dice example required analyzing thirty-six outcomes, doing the same for a problem like the evolution of human DNA requires examining a near-infinite number of possibilities. Science typically does not approach problems this way. Claiming a solution is valid because no other alternative is available without being able to demonstrate that *all* alternatives fail is an argument from ignorance fallacy.

In defense, the ID community would likely claim that they only have a theory and are not claiming a solution to the problem. This is a valid objection, but the point still stands that attempting to prove a negative is generally not a good approach. And this is the primary motivation for rejecting ID as a valid theory: there is simply no way to demonstrate its validity directly. Although a theory can possibly stand on the notion of proving every alternative is false, ID doesn't even go this far. It simply asserts that evolutionary theory is wrong and the only other viable candidate is ID. It leaves out things like UFOs, magic, and a lack of understanding on both the part of evolution and ID.

The second problem with ID is there is no direct evidence for it. In fact, as we will see later, there is really no way to obtain any direct evidence. Short of proving the contrapositive, which we discussed above, it is not possible to justify a theory without evidence. As explained in Chapter 3, a theory is an explanation of observed evidence. For example, observations relating to Newtonian gravity include that massive objects experience a force between them proportional to their masses and inversely proportional to the square of the distance between them. Data from measurements made on massive objects is wrapped under the umbrella of a theory of gravitation, and the behavior of these objects is expressed mathematically. Intelligent design provides nothing like this.

Table 6.1: Outcomes of a roll of two dice.

Line	Die 1	Die 2	Sum
1	1	1	2
2	1	2	3
3	1	3	4
4	**1**	**4**	**5**
5	1	5	6
6	1	6	7
7	2	1	3
8	2	2	4
9	**2**	**3**	**5**
10	2	4	6
11	2	5	7
12	2	6	8
13	3	1	4
14	**3**	**2**	**5**
15	3	3	6
16	3	4	7
17	3	5	8
18	3	6	9
19	**4**	**1**	**5**
20	4	2	6
21	4	3	7
22	4	4	8
23	4	5	9
24	4	6	10
25	5	1	6
26	5	2	7
27	5	3	8
28	5	4	9
29	5	5	10
30	5	6	11
31	6	1	7
32	6	2	8
33	6	3	9
34	6	4	10
35	6	5	11
36	6	6	12

There has been a longstanding argument that ID, or something akin to it, should be taught alongside evolutionary theory in public schools. The reasoning is that all such theories stand on equal footing as they all provide possible explanations of reality. The logic goes that doing so expands the ideological ground

6.1. IS INTELLIGENT DESIGN A LEGITIMATE THEORY?

to which the students are introduced, and that is a good thing in that they can use their critical thinking skills to arrive at their own opinion. But is this a good thing?

Students at all levels in the educational system have very limited time to examine various subjects and contemplate them critically. School curriculums are chosen intelligently and judiciously with this constraint in mind. Not all topics can be taught, nor should they be. There is no valid reason to teach children astrology as a possible alternative to astronomy as it simply muddies the water and takes up time. I can't prove with certainty that being born as a Virgo does not provide one with an enhanced ability to communicate with animals telepathically. Still, I'm pretty sure that the possibility is remote enough that it doesn't merit consideration.

A good number of children in the US are told at a young age that a man named Santa Claus exists and delivers Christmas presents to all the good children around the world on Christmas Eve. And I'm not dissing Santa Claus. It is cute when the kids are young, and they know they're playing a game. Still, suppose we continued teaching Santa Claus's existence when they started their sophomore year of high school as a potential explanation of how Christmas presents arrive under the tree Christmas morning. In that case, I imagine that many parents would have a problem with this equality of all theories proposition.

And as ridiculous as all educated people know the Santa Claus story is, we must admit that we can't disprove it. If there is a failure to detect Claus's arrival at any house worldwide on Christmas Eve, maybe it is not because he doesn't exist, but rather because every child in the world was naughty that year. Or perhaps he possesses some form of supernatural magic that prevents his detection.

I'm being a bit facetious in talking about Saint Nick, but I don't think it's too far off-topic. All theories are not created equal. There must be some epistemological rules that dictate how we determine what is real and what isn't. This goes straight to the heart of the problem with the ID theory: it is a hypothesis with no evidence at all.

A similar argument of a theological philosophy bent illustrates the point quite well. One of the most often used claims from this arena is that the logical absolutes can't be accounted for without the supposition of a god who defined them into existence. It is true that the laws of logic are axiomatic and cannot be derived or proven. Then why do we use them? I think Matt Dillahunty expresses the answer best when he states that we continue to use the laws of logic due to their continued effectiveness in producing reliable results. In other words, we cannot verify that they are indeed correct, but we can demonstrate and recognize their functional consistency within reality. It is not perfect, but it works.

However, the entire argument is rendered moot when one realizes that the Christian worldview cannot verify or justify the laws of logic either. If one is not careful, the notion that a god can justify these laws slips through the door with no one noticing. Just because an apologist claims that a god provides for the logical absolutes does not make it so. It is just a claim and is a claim with no evidence whatsoever. This situation is very similar to the ID argument in that it simply asserts (truthfully) that we human beings can't justify the laws of logic and (falsely) that the notion of a god can.

6.2 Probability

Before jumping into ID and all that subject brings with it, let's first look a little more at probability theory and some of its implications that may not be all that obvious to those who don't use it every day. Once that is done, we can look at concepts like specified complexity and how ID attempts to use a lower probability bound to infer design.

The foundations of probability theory are based on what is called the *law of large numbers*. An example is probably the best way to introduce this idea. Consider flipping a fair coin. The adjective "fair" here means that after being tossed into the air, the coin is equally likely to land on heads or tails, and it cannot land on edge (perhaps the edge has been rounded to avoid this possibility). Therefore, whenever the coin is tossed into the air, it must land completely randomly on one of its two sides.

Suppose this coin is flipped many times, and the percentage of tails is plotted on a graph. What do you think that will look like? Fortunately, we can see for ourselves with a little Octave code (assuming my computer has a true random number generator...). Figure 6.1 shows the fraction of tails obtained as a function of the number of up to 100,000 tosses of a fair coin. (Note that the plot has a logarithmic horizontal axis so that the salient features of the graph can be easily seen.)

The first toss is heads, resulting in the proportion of tails being zero. The second toss is tails, and the proportion becomes one-half. The third flip is heads, and the proportion becomes one-third. The fourth flip is again tails. And so on. The earlier flips have more of an impact on the overall proportion of tails, as expected. (Adding one to the denominator of, say, one-half has a far larger impact on the result than adding one to the denominator of, say, 1/1000.)

So, the start is a little rocky, but as the number of tosses increases, the proportion of tails approaches exactly one-half. And, if there's any concern that this may be a quirk, Figure 6.2 shows five more runs of 100,000 tosses, and each one of them converges to one-half.

This is not unexpected based on our everyday experience, but where does the one-half come from?

The number of outcomes we are testing for is one: the coin coming up tails. The number of possible outcomes is two: heads or tails. The probability of the coin coming up tails on any toss is the ratio of one desired outcome to two possible outcomes, or one-half. In probability theory, any possible outcome or set of outcomes is called an event, E, and the set of all possible outcomes is called the

6.2. PROBABILITY   193

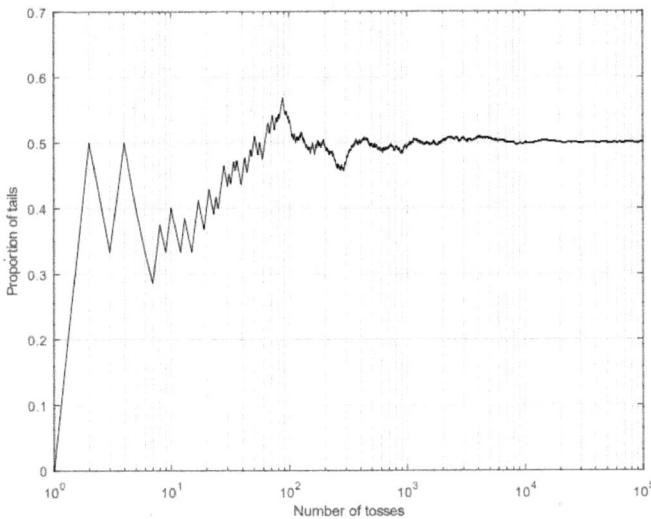

Figure 6.1: The cumulative proportion of tails as a fair coin is tossed between one and 100,000 times.

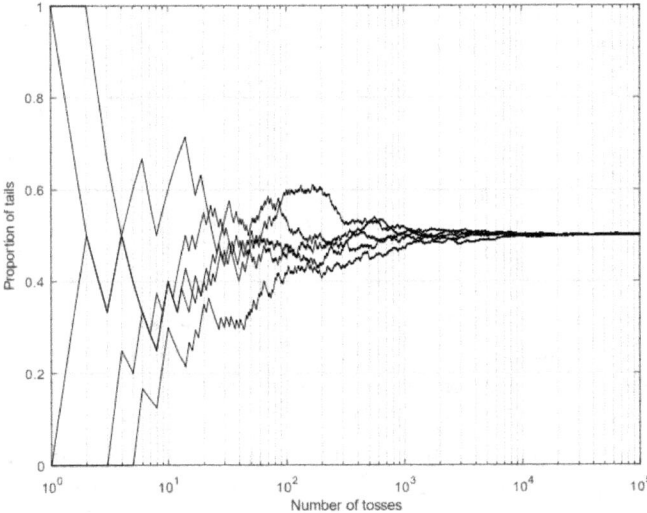

Figure 6.2: The cumulative proportion of tails during five runs of a fair coin being tossed between one and 100,000 times.

sample space, Ω. The law of large numbers states that with a sufficient number of trial runs, that the expected proportion of the occurrence, or probability, of a particular event E is the ratio of the number of outcomes contained in E to the total number of outcomes contained in the sample space Ω.

As a second example, consider rolling a six-sided die and getting a five or larger number. The sample space is the set $\Omega = \{1, 2, 3, 4, 5, 6\}$ and contains six outcomes, and the event set is $E = \{5, 6\}$ and contains two outcomes. Therefore, the probability of rolling a five or six is the ratio of two to six or one-third.

Similarly, the probability of rolling a number that is two or less with a six-sided die is also equal to one-third because the event set in this case also contains two outcomes. In fact, choose any event space containing two numbers between one and six, and the probability for each of them will be one-third. That all of these probabilities are equal doesn't seem strange, but let's look at another example where it may well seem odd.

Out of a deck of 52 cards, what are the odds of drawing the four aces, one after another? In drawing the first ace, there are four possibilities, four aces of four different suites, out of a sample space of 52, yielding a probability for the first card being 4/52. After this first card is drawn, there are three aces left and a total of 51 cards. Therefore, the probability of drawing one of the three remaining aces is 3/51. That for the third card is 2/50 and for the fourth 1/49. And the probability of drawing all four of these cards in any order is $(4/52)(3/51)(2/50)(1/49) = 3.694 \times 10^{-6}$. The odds of drawing four aces from a full deck is about one in 271 thousand.

This result is not surprising at all. Every card player knows that drawing four-of-a-kind in aces is extremely unlikely. Drawing a worthless hand like $\{2\clubsuit, 5\diamondsuit, J\heartsuit, 9\spadesuit\}$ is far more likely than four aces, right? Let's calculate the probability for this hand. Just like above, the probability of choosing one of the four from a deck of 52 is 4/52, drawing the second card is 3/51, the third is 2/50, and the fourth is 1/49. The probability of drawing these four cards is $(4/52)(3/51)(2/50)(1/49)$, identical to what we obtained for the four aces. In fact, the likelihood of drawing any combination of four cards from the deck is equal to this calculated probability. How can that be?

The result is surprising because, as card players, we imbue the four aces with a special status and regard other combinations of four cards as less significant or junk hands. Understanding this idea will be very important later.

Let's go back to the tossing of a six-sided die example for a moment. The probability of rolling a number equal to one or less is 1/6 because there is only one outcome in the event set and six in the sample space. The probability of rolling a number two or less is 2/6 because there are now two outcomes in the event set. The probability for rolling three or less is 3/6, that for four or less is 4/6, that for five or less is 5/6, and for six or less is 6/6 = 1. The point is that as the number of outcomes in the event space grows, the probability grows with it until it is equal to one. The probability of any event in the sample space occurring is exactly equal to one.

It is essential that we correctly state which outcomes are contained within the desired or successful event space; otherwise, the probability estimates we calculate will be erroneous. For example, drawing an ace from a standard deck of cards is not 1/52; rather, it is 4/52 because four outcomes fall into the desired event. Defining the event space of a particular probabilistic calculation is the equivalent of describing the context of the problem it models. This will be very important in later discussions about the probabilities involved in the generation of biological complexity.

6.3 Irreducible complexity

I was once sent to Louisiana to debug the electronic controller of a prototype grain dryer because it was malfunctioning. I liked Louisiana a lot, particularly the wildlife, the bayous, and the lack of concrete. I also enjoyed talking to the people as they are so down-to-earth and friendly. I was only about thirty miles from the coast and figured I would go and see the ocean, so I asked the owner of the farm on which I was working how to get there. He had a very thick Cajun accent, but I could understand him reasonably well as I grew up in the backwoods of Illinois. The man's response to my question was very simple and definitive, "Oh, you can't get there from here," which took me aback for a moment.

As it turned out, he actually meant that I couldn't drive south to the coast as it was all swampland. This is a kind of silly story, but it definitely captures the point of this section. The term *irreducible complexity*, coined by Michael Behe, refers to the impossibility of reaching evolutionary state B starting from A and, thus concludes, that there are biological structures, such as the eye, that could not possibly have evolved.

Phyletic gradualism requires a fitness or cost function that has a slope that is monotonic and smooth. The term monotonic increasing refers to a curve that only increases without any downturn. In this case, the slope of the fitness function on the left would be positive all the way to the maximum, and on the right, it would be negative all the way to the maximum. As we have seen, local maxima caused by dips in the cost function on either side of the maximum will capture the optimization process and prevent it from finding the global optimum (refer back to Figure 5.2). And if the cost function has a flat portion, the routine will get stuck because it has no slope information to determine a direction to move. Intelligent design proponents would classify these functions as irreducibly complex, meaning there is no feasible means of evolution crossing a valley or flat region in the fitness function.

What would cause the fitness function to have local maxima or have a flat region? Assume moving toward a more optimal point on the fitness function requires the operating point to first pass through a region of lower fitness. In that case, there will be no impetus to move toward the maximum. Optimization routines that only consider local gradient information cannot move across regions of zero slope. Furthermore, if the organism's operating point is already at a peak in the fitness landscape, and its overall fitness has to decrease to move toward a higher peak, it will stay on the current peak. In this case, the organism resides on a local maximum. These concepts are best illustrated through an example.

The bacterial flagellar motor is a flagship example of supposed irreducible complexity. Therefore, we will start with it.

The flagellar motor is a microscopic motor that some bacteria possess to propel them through their watery environment. Like motors that humans make use of every day, it has a stationary annulus-like stator in which a cylindrical rotor is positioned. This motor does not apply torque to the rotor using a magnetic field, as most of the motors we are familiar with, but it is, instead, an electrostatic motor—the interaction of electric fields produces the rotor torque. Remember from high-school physics that opposite electric charges attract one another and like charges repel one another. Everyone who has emptied a clothes dryer and had their socks stick to a sweater has witnessed the force of an electric field.

Before going further, it will be beneficial to discuss how a motor works by considering a conventional motor used in human applications today. The drawing of a typical two-pole DC motor is shown in Figure 6.3. It comprises four stator poles on the outer edge (numbered 1-4) and a rotor in the center with two poles that can rotate on a bearing. The four stator poles are distributed around the circumference of the motor at 90° intervals. They are actually steel (which is magnetic) extrusions in the radial direction with coils of wire wrapped around them. That is, they are electromagnets. The rotor, shown in the drawing as a bar magnet pinioned at the center of the motor but allowed to rotate freely, can be either a permanent magnet or an electromagnet.

If one again remembers back to high-school science, it is known that magnets have two poles: a north pole and a south pole. Also, opposite magnetic poles (north-to-south) exhibit an attractive force, and like poles (north-to-north or south-to-south), exhibit a repulsive force or repel one another.

The second concept may be a little more removed from most people, but it is just as simple. When a coil is wrapped around a piece of magnetic steel, as with the stator poles of the DC motor, and a current is made to flow through the windings of the coil, as was observed and described by Michael Faraday, a magnetic field is generated in the steel.

Note that there are two types of magnetic steel. One type can enhance a magnetic field created by a current-carrying coil of wire and will lose its magnetism after the magnetic field produced by the coil is removed. The second type retains its magnetism permanently and is often referred to as a permanent magnet—these are the refrigerator magnets with which most people are familiar. The poles of the motor's stator use the former type of steel; that is, they exhibit a magnetic field only when a current is applied to the coils. In this case, the rotor is a permanent magnet. It retains its magnetic field permanently.

Each stator pole shown in the motor drawing is made of wire wrapped around a chunk of steel that is radially directed toward the rotor, as shown. It is impor-

6.3. IRREDUCIBLE COMPLEXITY

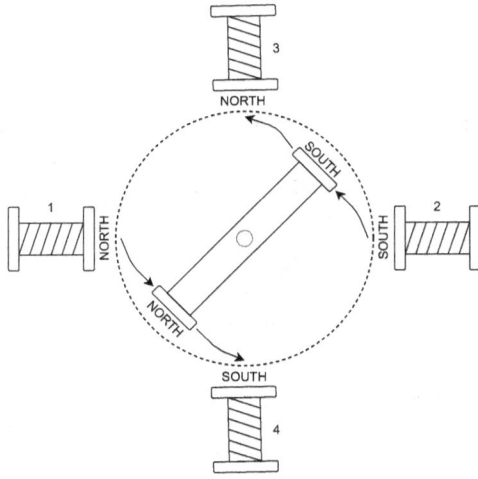

Figure 6.3: Diagram of a four-pole motor. The rotor is magnetized through its diameter, and coil sets 1 and 2, and 3 and 4, are series electromagnets driven by controlled currents. Coil 1 is wound opposite of 2, and coil 3 is wound opposite of 4.

tant to note that windings 1 and 2 are connected backward in this setup, and so are winding 3 and 4. Since pole 1 is wound opposite of pole 2, when a current flows through the wire comprising the windings, these two poles express opposite polarities of magnetic field. The same is true of windings 3 and 4. Windings 1 and 2 are always active together, as is the case with windings 3 and 4.

Now, as shown in the figure, the rotor position and the currents in the windings are such that the rotor is at 45 degrees, pole 1 is north, pole 2 is south, pole three is north, and pole 4 is south. The result is that stator pole 1, being opposite polarity to that of the nearest rotor pole, produces a torque on the rotor, making the rotor move counterclockwise. On the other hand, stator pole 4, being a south pole while the nearest pole of the rotor is north, pulls on the rotor and produces a torque in the counterclockwise direction as well. Examining both of the remaining pole combinations, it will be apparent that they also produce a counterclockwise torque. Therefore, the rotor will turn in the counterclockwise direction, but what happens after it turns such that the rotor south pole is aligned with stator pole 3, and the north pole of the rotor is aligned with stator pole 4? The rotor will stop at that point unless something is changed.

The trick is that as the rotor becomes vertical, the currents in all the windings are switched such that they flow in the opposite directions. It is important to understand this as it is the basis for motor operation. Since the currents change directions, the magnetic polarities of the stator poles also change direction, resulting again in a counterclockwise torque on the rotor, forcing it to continue to rotate counterclockwise. Then, as the rotor rotates another 90 degrees, the currents change direction again, leading to the magnetic poles changing polarity again, and so on.

The first DC motors used mechanical contacts to change the direction of the currents in the windings. Modern motor controllers use transistors to do this. In addition, modern motors use far more stator poles than four, resulting in a motor that operates in the same way but has increased and smoother rotor torque.

I want to mention one more thing about the operation of modern motors before we go on. In the DC motor described above, the torque comes in pulsations as the currents in the windings change direction. This is still the case for many lower-cost motors in today's world, but the majority of motors used today are AC motors that respond to sinusoidal currents instead of switched currents. If you remember your high school mathematics, sinusoids are the basis for creating a circle. In other words, if we take a two-dimensional coordinate system and plot $\sin(\omega t)$ on the horizontal axis and $\cos(\omega t)$ on the vertical axis, the result will be a vector rotating around the unit circle centered on the origin (this is usually called a *Lissajous figure*). AC motors, such as the induction motor, produce an actual rotating magnetic field that drags the rotor around with it. No longer needed is the trick of switching the stator pole polarities as the rotor comes close. Instead, in this case, no switching is required. Instead of approximating a rotating field by switching the polarities of stator poles, the polarities of the poles are varied as sinusoids producing a smooth, rotating magnetic field that leads the rotor to rotate smoothly.

I bring up this last point because biological motors likely operate more like our AC machines than conventional DC machines. However, there is one significant difference. Electric motors that we use on a daily basis to drive fans or power electric cars produce torque using magnetic fields. Biological machines, on the other hand, make use of electrostatic fields. There are many reasons for this, the primary reason being that electrostatic machines are more efficient than magnetic versions. We use magnetic machines on the human scale because they are practical at that size and electrostatic machines are generally not. However, at the microscopic level, machines based on electrostatic forces become practical.

Biological motors operate similarly to the motor discussed above by using two sets of coupled electrical charges. The rotor is built from proteins that contain negative charge distributions due to their assembly structure (remember the electrostatic lock-and-key discussion).

The biological cell uses a proton (positive hydrogen ion) gradient across a membrane to store potential energy in both a concentration and as a voltage. That is, a high concentration of protons is created on one side of a cellular membrane. If these protons are then allowed to flow through the membrane barrier, they will gain kinetic energy from an electrostatic potential and also from diffusion from a higher to a lower concentration. These two forces impart kinetic energy to the protons from the potential energy stored in the concentration gradient across the membrane. As the protons cross the membrane, they are guided such that their movement, through electrostatic coupling, imparts a

6.3. IRREDUCIBLE COMPLEXITY

portion of their kinetic energy to the negative charges attached to the flagellar rotor, causing it to rotate.

These biological structures are incredibly complex, to such an extent that we don't fully understand them today. Various proteins are constructed by the cell and used to assemble the rotor of the motor, its bearings, a type of clutch by which the rotation of the shaft can be started and stopped, the flagellum itself, and many other components. Many complexities have been left out of this description. For example, the rotor of the flagellar motor does not only have four poles as with the DC motor described above; it has many more.

With the complexity of the biological flagellar motor in mind, we can now give a first, simple definition of irreducible complexity: a system is called *irreducibly complex* if the removal of any one of the parts of the system causes it to stop functioning. This definition seems to make sense. For example, a car has four tires, and removing any one of them will prevent the car from being useable as a mode of transportation. As a second example, removing one of the keys on a computer keyboard prevents the ability to type a book, right? The computer can no longer perform its function as a word processor. Finally, a gun is only useful so long as one has bullets to use with it. I think most people would agree that removing bullets from a gun defeats its functioning as a weapon.

Intelligent design applies this same logic to many biological systems. It asserts that they could not have formed by gradual evolution as the lack of any piece of the system, for example, one of the poles of the flagellar rotor, would prevent the entire system from working. What good is a flagellar motor if it won't run? None, and therefore natural selection would never have gone down the path of evolving a motor of this kind. As the man said, you can't get there from here.

The process of evolution can be modeled as an optimization process in which there is a general movement in the direction of the gradient of a specific cost function by a selection function acting upon random perturbations of the independent variables. (Refer back to the chapters on evolution and optimization.) The gradient always points toward the direction of the greatest change of a function. Therefore, if the method follows the direction of the greatest positive change, movement will be toward a peak of the function. This allows a maximum to be found using an unintelligent algorithm. However, this method is not foolproof.

Consider the plot shown in Figure 6.4 which is formed by multiplication of a sine function and a function that decreases as it moves away from the origin. As shown in the plot, the optimization method we've discussed throughout the book begins in the lower left of the plot and starts moving toward the maximum. Now, remember, it is only moving in the diagonal direction toward the peak because the local slope of the surface is positive. The optimization method only takes into account local slope information. The consequence of being limited to local information is that the method stops at the top of the ring around the peak because the local slope goes to zero. And, if the local slope is zero, the method cannot proceed any further.

The ring situated around the global maximum acts as a barrier to the opti-

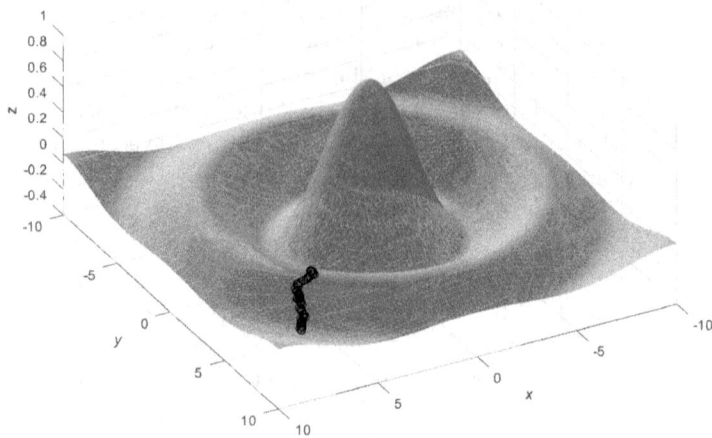

Figure 6.4: The surface of the sinc function $z = \sin(r)/|r|$. Starting at the initial condition $\{x = 7.5,\ y = 7.5\}$, the optimization method goes about finding the local maximum.

mization method. Once started outside this barrier, the method will never be able to cross it to gain access to the global maximum. What ID claims is that the biological entities we examine today are located at the global maximum. More to the point, the claim is that many biological systems presently reside at a maximum they could not have reached, and ID provides the alternate solution that a designer must have created them at this optimum since they couldn't have gotten there any other way.

The basis behind this claim is that natural selection will never select a less advantageous state than the current state. In other words, it cannot cross a boundary where a mutation results in a low fitness score. Therefore, natural selection, by definition, in attempting to find the global maximum, cannot cross a region where the slope of the cost function becomes negative.

The simple version of irreducible complexity goes like this: since evolution must move in small steps, and given that nature does no planning in what it brings about in the end, a flagellar motor would need, perhaps, the stator to be the first component constructed. However, a stator without the remaining components does not result in a functional motor. Therefore it is of little use to the organism and, hence, will be immediately selected for removal if for no other reason than it requires cellular resources to build and provides no benefit in return. Other members of the species that are not required to pay the energy expenditure for a useless piece of hardware will possess a survival advantage. The only other option is that the organism must evolve the entire structure in one generation. We know this is highly unlikely due to the probabilities

6.3. IRREDUCIBLE COMPLEXITY

involved. Evolutionary theory is supposed to remove the need to overcome such improbabilities. So, how does evolution bridge this gap?

The first item that needs examination in further detail is the claim that there is no monotonic path from no motor to motor. Remember, this is simply a claim, and it certainly hasn't been proven due to the complexity involved.

Assume that the motor of Figure 6.3 is missing pole 4; is the motor then completely nonfunctional? No, it can certainly still function. In fact, the motor is run in the same way as before. The only difference is that the maximum rotor torque has been reduced because of the missing pole. However, the motor would still be beneficial, only less so, and this is precisely the requirement of an increasing fitness function for evolution to work.

How about if it were missing poles 3 and 4? In this case, it is easy to see that if the rotor stopped aligned with poles 1 and 2, it might never get started again. If it were exactly aligned with these two poles, there would be no torque created at all between the stator and rotor poles because the angle between them is zero. However, how often would this happen? Often enough that we don't build motors like this today, but the question is, would it confer some advantage to an organism. Yes. Nine times out of ten, the motor would start if the rotor were off-axis of the stator poles, and this would undoubtedly be advantageous compared to no motor at all. Most of the time an organism employing such a motor would be able to move—only in rare instances would it lose this ability.

Would pole 1 be enough on its own to provide benefit? Maybe. Rotation would no longer be possible, but the rotor could be made to oscillate back and forth by continually changing the polarity of pole 1. Would this oscillatory motion be enough to move the organism slightly? Probably. And if so, again, this satisfies the cost function requirement.

In our motor example, we have shown a partial progression from a nonfunctioning motor to a fully functional motor that corresponds to a monotonically increasing fitness function, but does this carry over to the bacterial flagellar motor?

In the 2005 Kitzmiller v. Dover trial, the expert witness group for the defendants argued that the bacterial flagellar motor could not function with even one constituent piece missing from the assembly. If only one protein were missing from either the stator or rotor, or if the flagellum filament were not there, the result would mean that Darwinian evolution could not find a path to the motor because the fitness function would be flat or negative up until the point there was a complete assembly. Simply put, they argued there would be no use for a motor that was not fully assembled and presumably could not function.

Ken Miller played the role of an expert witness for the plaintiffs. He explained how Behe's irreducibility argument failed when considering three biological systems: the bacterial flagellar motor, the blood-clotting cascade, and the immune system. In each case, he described in detail a path evolution could have taken to arrive at a functional system.

A simplified diagram of the bacterial flagellar motor is shown in Figure 6.5.

It comprises a rotor concentrically placed in a stator structurally supported by the cell's double membrane. The stator and rotor include proteins at the bottom of the figure that use pumped hydrogen ions to generate torque on the rotor assembly. Multiple rings act as thrust bearings to allow the rotor to rotate and to prevent axial movement. The rotor is ultimately connected through a right-angle cylinder, called a *hook*, to the flagellum at the top (only a portion of the filament is shown). Harvard biologist Howard Berg has referred to the flagellar motor as "the most efficient machine in the universe." Although this may amount to speculative hyperbole, electrostatic machines built at this scale are indeed extremely efficient. Hydrogen ions are pumped through the rotor and stator assemblies at the bottom of the figure to bring about motion. As previously explained, this stream of ions is generated by a concentration gradient across the cell membrane, which provides kinetic energy to the ions, both from the associated electric field due to the gradient and from diffusion across the membrane.

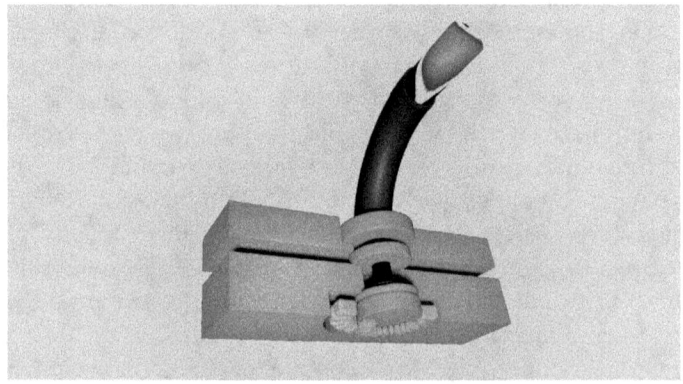

Figure 6.5: Simplified structure of a flagellar motor. The two cellular membranes act as stator and rotor support. At the bottom, rotor and stator proteins use hydrogen ions to generate electrostatic torque that is applied to the rotor. The rotor is connected by a shaft to the hook and flagellum at the top.

The question that concerns us here is whether this machine has to be formed in its entirety to benefit an organism or if there are graduations from a simpler function leading up to a fully-functional motor. Although the scientific community has not yet completed the picture of the evolution of this machine, there are strong indicators that a path of gradualism does indeed exist.

An example of how the flagellar motor might not be irreducibly complex was given by Ken Miller in the Kitzmiller trial, where he showed a deconstruction of the motor by taking away 30 of the constituent protein parts, leaving behind what biologists recognize as the type III secretory system that some bacteria use as a sort of syringe to inject materials into other cells. It is important to note that even in the flagellar motor, the shaft is used to funnel proteins from the interior of the cell to its exterior for the construction of the filament. In other words, the support structure of the motor, the shaft, and the flagellum are retained in this secretory system to perform an entirely different function from that of a rotary motor. Although not conclusive, this is certainly indicative of evolutionary co-opting of biological components as species move through the genomic/phenotypic space.

Since the trial, much more evidence has come to light showing that gradualism is possible. First, many types of *flagellin* proteins are used to construct flagellar structures within the bacterial world. Some flagellum forms are constructed in a right-handed way, some in a left-handed way, and some are rigid, and some are not. Mark Pallen and Nicholas Matzke state in a 2006 paper in Nature Perspectives, "In *Escherichia coli* alone there are over forty antigenically distinct flagellins, with good evidence that variation is driven by diversifying natural selection." [76].

The authors go on to claim that a conservative estimate of the number of different bacterial flagellar systems must be in the thousands or even millions. They note that there must have been thousands of creation events, or, in line with Occam's razor, this is evidence that these systems evolved from a common ancestor.

They also present evidence of flagellar systems rendered non-functional when they were no longer needed. Remnants of flagellar genes have been discovered in the genomes of bacteria that do not possess such systems. In addition, the various examples of flagellum seem to pick and choose from a set of 40 or so proteins, further indicating a common ancestor.

Some bacteria produce conventional flagellum but lack the biological components for *chemotaxis*, or the ability of a bacterium to move with or against a chemical gradient.

And finally, there are *archaeal* organisms or cells that possess flagellar motors like bacteria. Although ion pumps also drive the archaeal flagellum, they are different from bacterial flagellum in their construction and protein composition. (Archaea are prokaryote single-celled organisms distinct from bacteria.)

This is a summary of the evidence for the evolution of the bacterial flagellum, but it is not proof. Intelligent design proponents refer to these as "just so" stories, which is a demeaning way of saying that they are contrivances of biologists with active imaginations. However, when looking for the simplest explanation, do we regard them as evidence for thousands of interventions by a supernatural designer or the functioning of the natural world through evolution? There is absolutely no evidence, even anecdotal, of divine intervention, which makes ID the granddaddy of all "just so" stories. On the contrary, there is a myriad of evidence for biological adaptation and genetic modification within nature and the lab, regardless that biologists have not exactly mapped the evolution of the flagellum.

There are many other systems that ID claims are irreducibly complex, including the mammalian eye, the blood-clotting cascade, cilium, and the Bombardier beetle's defense mechanism. All of these have their "just so" stories that biologists have developed and, just like that for the flagellum, are not complete. However, the points of evidence do fit.

It has been proposed that the eye has evolved independently as many as 40 times [77], and about ten different types of eye still exist today [78]. As the different types of extant eyes are surveyed, a hierarchy emerges, ranging from simple to complex. We have already covered this in the previous chapter. Here, we are concerned with the validity of the notion that an eye is irreducibly complex.

Is it enough to demonstrate that eyes of varying degrees of complexity exist to refute them being irreducibly complex? In other words, we know that the eye of a nautilus is simpler than the mammalian eye, but is this enough to show that it is likely that evolution moves from simple to more complex eyes? Of course not. But the large number of different types of eyes confirms that the corresponding fitness landscape has at least as many viable operating points. Claiming that all these operating points are irreducibly complex implies they are all distinct optima and unreachable by a gradient ascent method.

Of course, there is no way at this point to answer this question with certainty. It isn't easy to assess the fitness landscape during the time of evolution of the eye. It is continually changing, and its shape is dependent upon a vast number of parameters. However, as evidence continues to roll in, none of it favors irreducible complexity. It all continues to fill in the gaps in the trajectories of gradualism. Consider the plot of a simplistic, hypothetical evolution curve shown in Figure 6.6. On this curve are various points in the evolutionary spectrum of the eye that we actually observe in reality. One leads to another in terms of complexity and performance.

Although biologists have not filled in the areas between the points on the curve (and there are far more known points than are shown in the plot), the points do fit on the curve. In other words, even though all the evolutionary steps between a simple and complex eye have not been explicitly observed in existing animals or the fossil record, a specific set of points on the transitional curve have been observed, and they line up.

The same is not true for ID. There is no predictive function with which to fit the observed points. The different varieties of the eye are simply the forms that the intelligent designer chose to implement. And that is as far as the investigation can go. Fortunately, science continues to investigate the evolution problem.

Multiple local maxima in the fitness landscape are a problem to gradient methods using a small step size. As we learned in Chapter 4 this is because the technique can get stuck in a flat region or at a particular local maximum. Another way of saying it is that the evolutionary method, starting at a specific fitness level, cannot cross over a region of lower fitness, even if it means getting to a higher overall fitness. The reason is the method relies on local information and has no idea of the global situation. However, this constraint only holds absolutely in the idealized case when the optimization step size goes toward zero.

6.3. IRREDUCIBLE COMPLEXITY

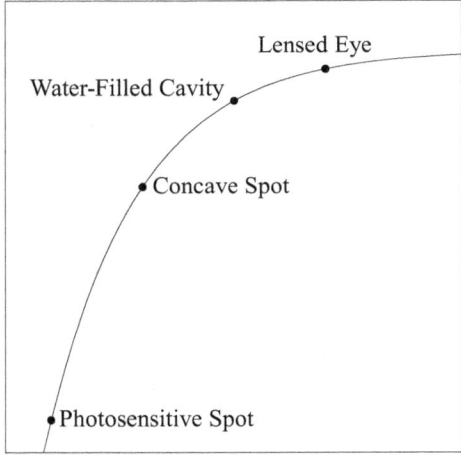

Figure 6.6: Varying stages of the evolution of the eye. The simplest version of the eye is just a photosensitive spot, followed by the same spot situated in a concave recess. Next is a water-filled cavity for focusing and, finally, a complex eye with lens.

In reality, the step size of any optimization method, including the evolutionary process, is finite. If it is large enough, it can cross over a valley that isn't wider than the step size. An example of this is shown in Figure 6.7 where the operating point of a particular species starts at the local maximum A and jumps over a suboptimal fitness valley to B. From there, it can make its way to the global optimum at C.

How does the process go about making the jump? It seems that would imply it was not done by a single-point mutation but rather by a low-probability, multi-point mutation. In other words, as was discussed previously, a small step in genomic space is far more probable than a large step. As the genomic step becomes too large, the probability will tend to forbid that step.

The answer to the problem is that we are not looking for large step changes in the genome. Rather, what is required is a large step in the phenotype. It is the phenotype that determines fitness, not the genome. And, since there is a nonlinear relationship between genome and phenotype, a small genomic step can result in a large phenotype change. As was discussed in the previous chapter, the designation of the phenotype by the genome is not like specifying the building of a house from a blueprint. As the house gets more complicated, with more rooms and amenities, the blueprint becomes commensurately more complex. This correspondence does not necessarily hold between the genome and phenotype.

Houses are built from blueprints that have a one-to-one mapping to the house.

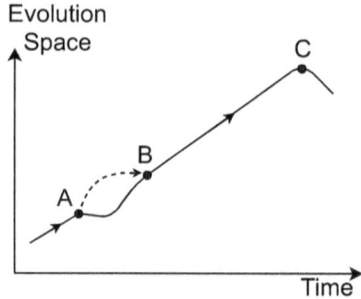

Figure 6.7: Evolutionary optimization with a step size large enough to jump over a fitness valley and progress from a local to a global maximum.

As you will recall, structures like cellular automata that rely solely on local rules do not share this one-to-one correspondence. This is why it is impossible to predict the shape of the outcome of a cellular automata simulation based solely on its ruleset. The same is true for biological complexity. There is no overall, organizing plan for an organism's construction. The whole process is a bottom-up endeavor. Since biological construction is performed from the bottom up based on a (relatively) simple ruleset, a single change to one of the rules can result in a system-wide change in the corresponding phenotype. It is this behavior—essentially a high gain between genome and phenotype—that makes the evolutionary process so efficient at generating biological complexity.

Irreducible complexity certainly becomes a problem when considering optimization on a smooth fitness function as the method can become trapped by local optima. However, with a biological system, the function of mapping the genome to the phenotype is not one-to-one and is not continuous. In addition, as we've already discussed, there exist many selection-neutral network paths through the genomic space that allow regions of that space that would be otherwise inaccessible to be reached. It is precisely these characteristics that give evolution the ability to overcome local optima.

6.4 Specified complexity

In colloquial terms, the ID argument states that natural processes alone cannot explain the biological complexity we see on Earth today. Furthermore, ID proponents claim that the only rational explanation is that the origin of such complexity must be the handiwork of a supernatural, intelligent designer.

Approaching this argument means quantifying complexity in such a way as to be able to decide on nature's ability to craft it. In other words, in mathematical terms, what does it mean to say that the biological complexity we see in the world is too complex for nature to have created it? William Dembski, in particular,

6.4. SPECIFIED COMPLEXITY

has developed a formalism, which he calls *the design inference*, specifically to answer this question [79]. At the heart of the formalism is what he calls *specified complexity*. We will attempt to explain these two concepts in this section and the next to see if they hold up under scrutiny. More specifically, we want to answer whether this formalism applies to the problem at hand.

Intelligent design proponents use the term specified complexity to measure the complexity of information set within a particular context [55,79]. The name has two parts, each of which applies to a particular component of specified complexity.

Let's look at the second part of the term first, that being complexity. It is defined here similarly as in conventional information theory, that is, as $I = -log_2(P)$, where the function $log_2()$ is the base-two logarithm, and P is the probability of a particular outcome or set of outcomes, comprising a probabilistic event. It's probably easiest to show what this quantity means through an example. Consider a standard binary, 8-bit sequence such as $x = 00110011$. It is easy to show that there are 256 different combinations of ones and zeros that eight bits can represent. These combinations could, as they do within computer operating systems, correspond to alphanumeric characters (the ASCII code). Or they could represent some other encoding.

If we ask the probability of a particular 8-bit sequence occurring purely by chance, the answer is $P = 1/2^8 = 1/256$. If we take the base-2 logarithm of $1/256$, we get an answer of -8. When we include the minus sign in the definition of complexity above, we get $I = -log_2(1/256) = 8$. Thus, the complexity or information carrying capacity of an 8-bit sequence is 8.

Intuitively this makes sense. As the number of degrees of freedom of a system increases, its information-carrying capacity increases proportionately. For example, the logarithm of an exponential, such as x^n, is $log(x^n) = nlog(x)$. The logarithm of an exponential is proportional to the exponent or, equivalently, the number of degrees of freedom.

Alternatively, consider a one-dimensional coordinate system, or number line, where a variable x is free to move in only one dimension. The logarithm of x is $log(x^1) = (1)log(x)$. The multiplier in front of the logarithm is one, and, hence, the information capacity of this system is proportional to one. A two-dimensional system can represent an area and is characterized by x^2 with a logarithm of $log(x^2) = 2log(x)$. Hence, the information capacity of a two-dimensional system is proportional to two. Similarly, for a three-dimensional system we have $log(x^3) = 3log(x)$ and the information capacity is proportional to three. Again, the information capacity of a system is proportional to the number of degrees of freedom that it possesses. (I took some leeway with the coordinate axes analogy. If the axes represent real numbers, the number of degrees of freedom is infinite.)

As another example, the human DNA base-pair sequence is approximately three billion pairs long. Each position in the chain can assume one of four chemical bases: adenine, guanine, cytosine, and thymine. The complexity content of this DNA chain is $log_2(4^{3 \times 10^9}) = (3 \times 10^9)log_2(4) = 6 \times 10^9$. This is an incredibly large number compared, for example, to the 8-bit sequence above.

The amount of information a particular sequence or system is capable of representing is a measure of its complexity (in the sense we are using the term complexity here). This makes intuitive sense. The more possibilities there are, the more complex a given situation becomes. There are far more possible moves in a game of chess than there are in a game of checkers, and hence, most people would regard chess as the more complex game. The same holds true here, but the expression for complexity, $-log_2(P)$, formalizes this intuition. Let's now move to what the word specific means in specific complexity.

Consider rolling a six-sided die. There are six possible outcomes when the die is rolled. The probability of the die coming up one is exactly 1/6. In fact, the probability of any digit between one and six coming up is 1/6. In other words, all outcomes are equally likely. But when someone makes a bet on a particular number, say three, coming up, it feels different. This is because the gambler has partitioned the event space. The odds of the desired number three being the result of the roll is 1/6, whereas the probability of a number other than three coming up is $1 - 1/6 = 5/6$. The probability of the gambler losing is five times as likely as winning. That is precisely why it is so surprising when the winning number comes up as a result of the roll. But it is not because the probability of rolling the winning number is any more unlikely than rolling any other number. It is the significance assigned to the winning number by the players that makes it unique.

The uniqueness of the winning number is what is referred to with the term "specified." There are many different examples of specifications. The book *The Design Revolution* provides the example of an archer shooting at a target. (I am going to modify the example slightly.) Assume the target is 1 ft^2 in area and is attached to a wall that is 10 ft by 10 ft square with an area of 100 ft^2. Further, assume that when the archer launches an arrow that it must hit the wall somewhere. Also, assume that the arrow has an equal probability of hitting the wall anywhere after being fired (the archer is a bad shot). Therefore, there is a $(1 \text{ ft}^2)/(100 \text{ ft}^2) = 1/100$ chance of hitting the target, and a 99/100 chance of hitting anywhere else. But what is to be concluded if the arrow does hit the target?

The arrow hitting the target is an improbable event, and reality doesn't favor unlikely events. If all events from a certain process are equally probable, reality doesn't favor any of them. It is exactly the same case as the gambler tossing the die in the example above: all outcomes are equally likely. Consequently, when an event is specified to have some relevance ahead of time, when that event occurs, it seems significant. Just like with the die, if the probability of an event within an event space containing N equiprobable events occurs with a probability of $P = 1/N$, the probability of the event not occurring and one of the other events occurring is $1 - 1/N = (N-1)/N$, which more closely approximates unity as N increases. In other words, it is highly likely that the specified event will not occur and very likely that one of the unspecified events will occur.

6.4. SPECIFIED COMPLEXITY

No matter the information-carrying capacity or complexity of a system, if all the outcomes are equally likely and significant, then it is not surprising when any one of the outcomes transpires. It takes a further constraint upon the system to impart significance to a particular outcome: a global context must be imposed upon the sample space, meaning that the event space must be specified. For example, if any combination of base pairs in the human DNA chain will do, then it is no surprise that human DNA evolved on Earth. However, if we assert that the only workable configuration is the one that exists today—a single outcome is the only viable outcome—then it becomes very improbable that the DNA chain formed by chance.

Specificity partitions the probabilistic sample space. Consider a standard deck of 52 playing cards again. If we ask what the probability of drawing any five-card hand is, the first thing we must note is that all five-card hands are equal in probability to be drawn, that probability being

$$(5/52)(4/51)(3/50)(2/49)(1/48) = 1/(2,598,960). \tag{6.1}$$

Although there is an exceedingly low probability of drawing any specific hand—an almost zero probability—games of Poker continue every weekend. However, in drawing any old five-card hand, nothing is surprising. Since the probability is very low, it takes a large amount of information to denote any specific hand as there are a lot of them. In other words, the information contained in any one-hand card combination must be capable of distinguishing it from the other 2,598,959 different hands. Again, this is the complexity part. In fact, this is even what complexity means in the ordinary sense: the measure of the amount of information to describe a given situation or circumstance.

Now, assume we're playing a game where the sequence $\{A\clubsuit\ K\clubsuit\ Q\clubsuit\ J\clubsuit\ 10\clubsuit\}$ is desired. Given that a context has been provided, mainly that we're playing Poker, and the desired outcome has been specified, we know the probability of drawing that hand is $1/(2,598,960)$. It would not be impressive if we were allowed to draw any old five-card hand. This can be seen by calculating the probability of drawing any five-card hand, which is found by multiplying the number of card combination possibilities by the probability of drawing any one of them, that is, $(2,598,960)(1/2,598,960) = 1$. Hence, despite the low probability of drawing any single hand, it is certainly possible to draw a hand at random.

So, the more specific the hand to be drawn, the less likely it is to be drawn. As another example, instead of drawing the five cards above of the same suit, if they can be from any suit, the probability is $4/(2,598,960)$ since there are four suits, which is greater than the odds of drawing the original hand. The complexity represented by drawing five cards of a specific suite is $I = -log_2(2,598,960) = 28.216$ and that of drawing the five cards of the same suite is $log_2(4/2,598,960) = 26.216$. As expected, the measured complexity decreased as the event space increased in size. Saying this another way, increasing the number of viable outcomes while leaving the number of possible outcomes constant reduces the system's measured complexity.

The ID claim is that specified complexity, given that the probability is sufficiently low, is a reliable indicator of design [55, 79, 80] under certain conditions that we will discuss in the next section. In other words, if the probability of an event occurring is sufficiently small due to its associated complexity, and that event is specified in some context, then that event only occurs through design. This is a strong statement, and it must be further explored.

Consider an avalanche of boulders and rocks rolling down a mountainside and collecting at the base. The resulting collection of rock is spatially complex, but it would not be classified as specified because its pattern of complexity was not pinned down in some way ahead of time. In other words, any arrangement of rubble is just as probable and meaningful as any other. There is no specification that prefers one or more configurations over the others. This is an example of the outcome of a chaotic, complex process not being specified. And it is a foreshadowing of where this section is headed. A specification is required to invoke specified complexity. But is this the case in the natural world, in particular, from in an evolutionary sense?

Of course, in terms of ID in nature, this concept is applied to the complex biological arrangements we see in nature, including large proteins, DNA, and living creatures. Intelligent design proponents claim the formation of biological arrangements like DNA is highly unlikely from a probabilistic standpoint. If one assumes that the entire human DNA chain of approximately three billion nucleotides is formed entirely by random, the odds for such an occurrence would be about $1/4^{3 \times 10^9}$. It is complex due to the low probability of the existing human DNA sequence being chosen by random, and it is specified because ID proponents seem to strongly imply that the extant structure of life on Earth is the only viable form that life on Earth could take. This is also a strong assumption, and a questionable one.

If only one of these sequences is allowable for life on Earth, then the odds of this combination occurring by chance is $1/4^{3 \times 10^9}$. The likelihood of a natural process ever hitting the mark by chance is minuscule, to say the least. As we calculated above, the complexity measure, in this case, is 6×10^9. But this is only true if there is *one and only one* life-giving combination.

We know that at least some DNA base pairs vary between humans, that we all do not share exactly the same DNA. The *1000 Genomes Project* in 2015 estimated that a typical human genome differs from the reference genome by up to 20 million base pairs [81]. This means that out of the $4^{3 \times 10^9}$ combinations within the human genome, there are at least $4^{20 \times 10^6}$ viable outcomes. In other words, the event space does not contain just one element but rather a large number of elements. However, in calculating the probability of one of these outcomes arising from pure chance, we get

$$P = \frac{4^{20 \times 10^6}}{4^{3 \times 10^9}} = 4^{20 \times 10^6 - 3 \times 10^9} = 1/4^{2.98 \times 10^9}, \tag{6.2}$$

which a rather small improvement in the odds, but it does demonstrate a couple

6.4. SPECIFIED COMPLEXITY

of things. The first is how the event space does not need to be just a single event in application to real-world problems. It can be and usually is a collection of events, which must be considered to give an accurate probability estimation. Secondly, approaching this problem like calculating the odds of dice rolls in a game of craps is completely wrong.

These additional elements of the event space are what is referred to in the ID literature as *probabilistic resources*. Two types of resources are defined. As is the case here, the broadening of the probabilistic event space is referred to as specificational resources since specification defines the events that one is looking for or trying to explain and ignores the remainder of possible outcomes. Therefore, increasing the number of viable outcomes increases the event space and increases the number of specificational resources available to the system at hand. (Incidentally, the other type is replicational resources. These represent repeated trials. For example, the number of tries one gets at obtaining ten tails in a row by flipping coins. The more attempts at such a feat, the higher the probability of success. The subject of probabilistic resources will come up again later.)

This section has summarized the term specified complexity as the ID community uses it. It is important to point out that there really is nothing new going on here, in either concept or application. It is simply applied probability theory. Complexity refers to the probability of an event containing a set of outcomes occurring. The larger the sample space with respect to the event space, the smaller the probability of success, which follows directly from the law of large numbers. Specification, on the other hand, is a definition of the event space, the event space being simply a subset of the total sample space.

However, there are a few items to point out before proceeding. The first is that a straightforward static framing of a dynamic system like evolutionary biology is not possible. It would be nice if we could formulate the problem in such a way that, starting at the time of the beginning of life on Earth, we could specify a goal or purpose of life, know that it was going to take exactly three-billion base pairs in human DNA, etc., and simply calculate probabilities in a static manner. But this is absolutely not the case for the evolution of life. Although shoehorning the evolution of life on Earth into an anthropomorphized, goal-oriented process makes probabilities easier to calculate, it results in an ill-formed solution. Life is not only dynamic in its functioning, it is also dynamic in its evolution and it never had a set goal. This contradicts the description of the specification as a pre-defined target [55]. There was never a pre-defined outcome, and as soon as this door is opened, as we have seen in Chapter 5 and will see in the rest of this book, the evolution of life on this planet not only seems possible but, in fact, inevitable.

Before we move on, let's examine further the meaning of the adjective specified. In a more rigorous definition in his book *The Design Inference* [79], Dembski defines a specification as an independent pattern that circumscribes a probabilistic event. Again, back to the example of drawing five cards from a standard deck. All hands are equally probable to be drawn, so what makes the royal flush so special? Alternately, when a person wins the lottery, what makes the number he or she chose so special? All numbers were equally likely to be chosen.

This is where the biggest and most subtle assumption of ID originates. What makes the royal flush special in poker is that someone said it is. The same is true with the winning lottery number. Both of these are independent specifications in the sense we have been discussing. The key word in the definition of a valid specification is that it is independent of the event it describes and the system that produces it. In other words, shooting an arrow at a wall and subsequently painting a target around the stuck arrow is not special precisely because the target is no longer independent of the event—in fact, it is directly dependent upon the arrow sticking event (i.e., that's where the target is to be painted). Now, whoever said that structure of life as we see it today on planet Earth was specified or predefined?

A bit of clarification is needed. I'm not referring to some divine predestination; quite the contrary. Another way of saying that the present form of life on Earth today was specified is to say that it exists in the only form it could. This is the big assumption that is hidden in the details surrounding the term specified complexity. I do not believe a single person in the scientific community today believes life in its present form was preordained or required. And without this assertion attaching the term specified to biological life becomes much more difficult. And I think that's precisely why the ID community does not go any further in applying this line of reasoning. I have yet to see a probability calculation, but it is conceivable that I missed it.

The biological landscape is not a prescribed structure that was brought about by planning. And it is not a coin trick of trying to land 100 tails in a row. It is a dynamic process across many dimensions, the morphology of its constituents being one of those dimensions. It is one of those open-ended word problems that every undergraduate dreads to see on a midterm exam.

When I first started as an undergraduate at the University of Illinois, I remember hearing the graduate students talking about the dynamics of circuits involving inductors, capacitors, and the like, and I wanted so much to understand the physical and mathematical intricacies of such devices. As it turned out, these were simple one-dimensional characterizations of real-life objects. And so I started there. The description of the behavior of the voltage across an ideal inductor in terms of the time-derivative of its current is $v_L(t) = L di_L/dt$, and that of an ideal capacitor's current as a function of the time-derivative of its voltage is $i_C(t) = C dv_C/dt$. I thought I had a good understanding of circuit components until I had to deal with what they really are.

In reality, these components are not simple elements that are functions of time only. Both of these devices rely on electromagnetic fields for their functioning, meaning their actual models are not only dependent on time but also three spatial dimensions. This more complex view of these elements comes into play when the wavelength of the exciting signals approaches the dimension of the circuit elements. Then the circuit model becomes a spatially distributed model.

6.4. SPECIFIED COMPLEXITY

In other words, all parts of the circuit elements must be taken into account, not just their terminals.

Life on Earth is not just a simple time-dependent model. It is a distributed model stretching across many different dimensions. And just like the electromagnetic components mentioned above, each organism within the ecology has boundary conditions relating to the environment, including temperature, air composition, sunlight, water availability, etc., and those relating to other organisms such as predators, prey, competitors, etc. This immense amalgamation of dynamic characteristics is in continual flux as it evolves. Rewind the clock back to three billion years ago and start again, and there is a definite probability that an entirely new set of fauna and plant life will emerge.

We think it had to be the way it is today because we are part of the resultant ecology, but it did not have to be this way. And that is the problem with the way ID theory formulates biological complexity as being specified. The specification condition is not met as the very biological complexity it is applied to is contingent not only on the environment in which life exists but also on random chance. This is such a stark fact of reality that Stuart Kauffman goes so far as to say that biology is beyond physics due to the indeterminacy associated with the final form of life as expressed through the evolutionary process, the complexity inherent in the laws that govern nature, and the random behavior built into reality. In other words, the outcome of the evolutionary process is so unpredictable that physics has no means of precisely determining it.

In the sense it is used in the ID framework, this notion of specification will need to be dropped to get anywhere in discovering how life came about. It poses the problem in the wrong direction. Life's development is not top-down from a goal, it is a bottom-up emergence, and there is no intent behind it.

The size of the state space required to represent the Universe is unimaginable. If we were to apply the same logic of specified complexity to even one instant in the time development of the Universe, we would have to conclude that the next instant of the Universe's time evolution is just too improbable to occur. This is Zeno's paradox on steroids. The only logical conclusion is that these things are not specified. Rather, they represent one of many possible outcomes. Biological evolution only becomes impossible and requires the aid of an intelligent designer if one believes from the start that it had only one possible outcome, in other words, that it was designed. (Refer back to Chapter 5 for more details.)

I agree that it takes intelligence to construct specified complexity. It is the only way that the concept of a specification can be understood. However, it has nothing to do with how life came about.

The last issue that must be dealt with in terms of specified complexity is the notion that complexity represents the improbability of such an event coming about. Just because something is complex does not mean it is improbable. Consider the Mandelbrot set we looked at in Chapter 2. The three-dimensional representation of the boundedness of the solutions of a simple second-order, complex equation is incredibly complicated. And yet, it is all derived from a single expression. Would one claim that the resulting Mandelbrot set is improbable? Certainly not, as it is entirely deterministic. This same complexity hidden in the functioning of systems with elementary rules is one of the drivers of biological complexity and diversification.

The same goes for the multidimensional optimizations given in Chapter 4. What appears to be a very difficult optimization problem can be solved by purely random excitation and a simple comparison rule. The complexity biological systems exhibit is a consequence of the information contained in environmental and species-species interaction. Anyone who has ever had a young child who loves building blocks has seen this principle in action. The child has a very simple set of assembly techniques centered around "stack things on top of other things". And the blocks have a straightforward set of practical rules: gravity enforces what can be done, and the list of options is minimal. Yet somehow, the result is always complex and extravagant.

Assuming that biological systems do not operate similarly is a huge and necessary assumption if one is to apply the concept of specified complexity to them legitimately. Random perturbation of the genome coupled with natural selection acting as a comparison operator results in a natural optimization method acting upon all biological systems. This does not imply that there is some global "optimal fitness"—an anthropomorphic idea—rather, the optimization should probably be thought more like optimizing the fitness of the boundary conditions between organisms and their environment, including other organisms. Additionally, the optimization does not occur for one organism in isolation; all organisms evolve in relation to their environment and each other. Therefore, although it is conceivable that an equilibrium state can be reached, there is no ultimate state being sought, no goal, no specification.

The evolutionary space does not have just one solution. In comparing the evolutionary process to a hiker in a national park wilderness, there is not just one hiking path on the side of the hill that visitors must follow. Hikers are free to take any path they want up the side of the hill. Evolution is like this: there is not just a single solution for life on this planet. Life was free to take any one of a myriad of paths. A hiker scaling the side of a wooded hill avoids trees, slippery patches, and holes in the ground and, in the process, produces an entirely new and likely jagged trail. This path is constrained by geographical obstacles and random choices and represents one of a million ways of trekking to the top of the hill.

The evolutionary path in our history was cut just like the adventurous hiker feeling out an appropriate path to the top of the hill. Evolution was not constrained to a single path that was specified ahead of time.

The point is that the evolutionary state we see today was not specified ahead of time. Instead of approaching it this way and calculating the odds of a very specific state being arrived at, it would make more sense to calculate the odds of an appropriate event containing *all* the possible ways that life could have shown up on Earth. But, again, as Kauffman points out, and rightfully so, there is no way to predict the particular outcomes using reductionist formulae. The only way to make this calculation would be to simulate all paths. Still, even then, there is no means of ensuring that the entire evolutionary space has been covered as random influences along the way potentially open up new areas.

Thinking that specified complexity can be used to describe evolution in the world around us and that a meaningful probability could be associated with it is, I think, wishful thinking.

6.5 The design inference

William Dembski, in his book *The Design Inference* [79], proposes a method of determining the presence of design by an intelligence. The main idea rests upon two central concepts: specified complexity and a probability bound, below which events are deemed not to occur.

We have already covered specified complexity, but it behooves us here to place it in a similar format as Dembski. The question at hand is, given an event E within sample space $\mathbf{E}$, can the outcome E be logically attributed to a chance hypothesis $\mathbf{H}$. For example, consider again selecting a five-card hand from a standard deck. We previously calculated the odds of selecting any five-card hand at $1/(2,598,960)$. If we are interested in the outcome of a royal flush of clubs, the probability is $P(RF_\clubsuit) = 1/(2,598,960)$. The event in question is the set $E = \{A\clubsuit\ K\clubsuit\ Q\clubsuit\ J\clubsuit\ 10\clubsuit\}$. The chance hypothesis, $\mathbf{H}$, is the assumption that these cards were drawn completely at random and with the probability given.

Now, assume there is a descriptive language set $\mathbf{D}$ from which descriptions D can be formulated that describe events E. And, finally, assume we have at our disposal some piece of information $\mathbf{I}$ we can use to formulate D. In other words, the information I is used to create a description D of the outcome E.

The crux of the matter is the claim that chance is a proper explanation of the event E so long as what is referred to as the *conditional independence condition* (CINDE), $P(E|H\&I) = P(E|H)$, is met (I have abridged this from Dembski's formulation and simply used I as information in general.) This makes intuitive sense: if an outcome is genuinely random, no amount of *a priori* information could predict it.

However, what we're after here is applying this logic to an outcome, specifically a sequence, after the fact. For example, could we justify the recorded sequence of fair-coin tosses as being a result of chance by using this methodology? The answer is, it depends. Essentially, two items must be satisfied. The first is a viable explanation of the event in question in terms of D and I. The second is that the probability of the event occurring due to random chance must be below a certain threshold. In this way, chance is ruled out, and acceptance of the alternative solution becomes the leading candidate.

There is one last condition of the design inference that must hold. The alternative explanation to random chance must be available and meet a specific *tractability condition*. Without this explanation, chance remains the strongest candidate explanation. Therefore, both the probability of the event occurring due to random chance alone *and* the difficulty in providing a viable alternative explanation must be below certain bounds. In this way, chance is ruled out, and the alternate explanation takes the stage.

To see how this works, let's assume that the dealer in a card game is unscrupulous and is dealing himself with hidden cards from his sleeve. He deals himself a royal flush in clubs in the particular round of the game we are considering. There is a chance hypothesis that the royal flush showed up simply through chance with the incredibly minute odds of $1/(2,598,960)$. On the other hand, the other players can certainly describe this event as being brought about by cards coming out of the dealer's sleeve. With this information, cheating becomes a viable explanation. And finally, given the extremely low probability of chance alone being the explanation, the logical conclusion is that the dealer cheated.

This is a brief and somewhat colloquial explanation of the design inference, which is based on favoring an alternative explanation for an outcome when the probability of that outcome occurring due to chance is sufficiently small.

Hence, a few requirements allow design to be inferred, and they have to do with the definition of specified complexity. The first characteristic of an object or sequence exhibiting specified complexity is that it be contingent. That is, the assumption is it cannot have been formed via unintelligent, prescribed rules as these can introduce no new information. Second, it must be complex: the object or sequence must have a low probability of being formed randomly. And finally, the sequence must represent a pattern or specification that distinguishes it from other, arbitrary sequences.

Now that we have a grasp of the design inference let's take a shot at applying it to the evolution of complex life by starting with the third requirement of specification.

Anyone who goes to the zoo and observes firsthand a portion of the diversity of life existing on this planet today can't help but be a little surprised at the complexity they observe. This is the fun part, and it is why people pay for tickets for entry in the first place. Obviously, the functional forms of all the animals seen on such a trip demonstrate a definite specificity, right? It certainly seems to fit the definition of specification as given above.

But how do we know this isn't just an assumption that we make? Remember the example of the landslide we talked about earlier? The rocks and dirt on the side of a hill, possibly due to an earthquake, come sliding down the slope and arrange themselves into an extremely complicated assembly on the roadway below the hill. All the atoms of that mass of rock and soil end up in a very complex arrangement that could potentially be delineated in three-dimensional space on a sufficiently sophisticated computer. The arrangement is not completely arbitrary. The rocks at the bottom support those at the top. The dirt rearranges itself to stabilize the overall amalgamation. The complete ensemble finds a minimum energy state. And all of this is done in a completely perfunctory manner—no intelligence is required.

6.5. THE DESIGN INFERENCE

How does this final configuration differ from the biological case? They both exhibit extreme complexity. Intelligent design argues that one exhibits a specification and the other does not. This means that the landslide ended in one of a myriad of possible combinations, all of them equally probable; thus, there is no specification. However, they contend, biological complexity *does* imply a specification. Working this implication backward means the event representing the formation of biological complexity comprises a small set of possible outcomes or only one outcome.

Lastly, the landslide is dictated solely by the physical laws of the natural world. On the contrary, however, the fact that ID attempts to infer design of biological complexity through the application of the design inference means there is the assumption that biological complexity is not constructed only by the physical laws, that it is contingent upon some other cause. An intelligent cause.

Also, as discussed above, two more ingredients are essential to the application of the design inference: the degree of complexity in the event of interest and the tractability in being able to assign an explanation other than chance to that event.

Let's go through how each of these requirements fares when applied to biological complexity.

There is no doubt that the organisms extant on Earth display an immense complexity in the Kolmogorov sense. For example, if one makes the naive assumption that the DNA of even a bacterium formed in one shot due to random chance, the resultant probability of such an event is vanishingly small. However, here is where the design inference applied to biological complexity starts breaking down.

There have been many times in the lab while working on very complex systems that I have observed completely unexpected results. Not one time did it ever cross my mind that a supernatural entity had caused the physics dictating the functioning of the device I was working on to go awry. Instead, *I assumed there was a mistake somewhere.* As will be discussed later, here is where ID runs afoul of the scientific community. Occam's razor rules out conclusions like ID in response to situations like this. If the probability of life's complexity makes it unreachable, then the reasonable conclusion is that there is something more going on, and further research is required. One does not conclude the existence of an intelligent designer as this proposition is more difficult to deal with than the initial problem. And it leads to nonsensical hypotheses such as causes that are not caused (this will become clear later).

It is clear that biological complexity is specific to the environment in which it originated here on Earth. In that sense, the number of possible forms it could have taken are constrained. That said, in light of the discussions in the previous chapter, the number of possibilities for the composition of life seems virtually endless. There are an estimated 16 million species on Earth today, which is a small percentage of the number of species that are estimated to have existed since life emerged on this planet. Attempting to portray the situation as though there was only one form life could have taken, and it is the one we see today, is more than a bit disingenuous.

This line of reasoning also suffers from the age-old argument from ignorance. Which is more likely, that there are more forms life could have assumed than we suspect, or that an unseen, supernatural, intelligent designer had a hand in bringing about the one possible outcome? It is not logical to assert that we know there is only one possibility or a few possibilities available and, since this seems statistically implausible, claim that a deity must be the ultimate cause. Again, this is why the scientific community has difficulty taking ID seriously.

In all scientific inquiry throughout recorded history, there has never been a verified occurrence of the physical laws being violated, and certainly not by a supernatural intelligence. Of course, many people throughout history have claimed this was the case, but science showed it was not the case at every turn. It is not even clear what it means to say that an outcome is not dictated by the physical laws of the Universe, regardless of whether we are aware of these laws. (This topic will be discussed in a later section.)

As mentioned above, there must be a tractable alternative to random chance available to accept the design inference. How exactly is the suggestion of an invisible, supernatural intelligence that intervenes in reality without violating the physical laws a tractable hypothesis? There is absolutely zero evidence that such a thing is even possible, let alone ever witnessed. This is the point that many theological philosophers miss. Take one of William Lane Craig's favorite claims that a god justifies the logical absolutes. Just because someone says a thing does not in any way make it true. In making such a claim, nothing has been solved. Fitting the operational parameters of a model of reality is not enough to justify or validate a hypothesis; it must actually be demonstrated. The story of a god who can do such things does not mean the problem has been solved. I'm not sure why this point is so hard to grasp by many people. Claiming an intelligent designer as a solution does not make that designer real or possible.

Finally, there is the requirement that the probability of the event of interest occurring by pure chance be sufficiently small. To this end, Dembski defines a *universal probability bound* in *The Design Inference* [79]. In the calculation of this bound, the assumption made is that any state transition in the known Universe requires at least one fundamental particle (electron, proton, neutron, etc.), of which there is an estimated 10^{80}. Furthermore, it is assumed that the Planck time only allows state transitions below the rate of 10^{45} Hz. And finally, the age of the Universe is well below 10^{25} seconds. Multiplying these three quantities together, we supposedly end up with the maximum number of state transitions that could occur in the age of the Universe, 10^{150}. Dembski then reasons that the probability bound that dictates a particular event cannot happen is $1/(0.5 \times 10^{150})$. The one-half is added because these are assumed to be binary states.

In order to more fully elucidate this probability bound, consider flipping a coin ten times, trying to get ten tails in a row. The probability of getting ten

6.5. THE DESIGN INFERENCE

tails in a row is $(1/2)^{10} = 1/1,024$. We would need to flip the coin $1,024/2 = 512$ times to get to a probability of $1/2$ of acquiring ten tails in a row. Tossing the coin less than 512 times means it is more likely we won't see this event, and with more than 512 tosses, we are more likely to see it. The number of times we repeat the coin tosses is what Dembski refers to as *probability resources*. In other words, the more attempts one has at achieving a particular outcome, the higher the probability of that event occurring. The point of interest is where the probability is 50%, the line between more and less likely. If the number of probability resources available to achieve a particular outcome is above this level, we can say the event is more likely to happen. If the number is below, we say it is less likely to happen.

The argument is that there are at most 10^{150} probability resources built into the Universe. Therefore, any event with probability less than $1/(0.5 \times 10^{150})$ is likely never to happen anywhere or at any time in the Universe. However, the implication in the ID community is assuming events with a probability below a specific probability bound *never happen*. This is an incorrect conclusion.

The notion of events not happening below a certain probability bound likely stems from the work of Émile Borel, a French mathematician from the first half of the 20th century [82,83]. Interestingly, he published the infinite monkey theorem in a book on probability. This thought experiment is familiar to popular culture and entertains the idea that a monkey randomly hitting keys on a typewriter for an infinite amount of time would undoubtedly type the works of Shakespeare by chance. He was one of the pioneers of measurement theory and one of the originators of the idea that events with a probability below a particular threshold can be assumed never to occur. However, he intended these to be simple guidelines and specific to the problem at hand. All engineers and scientists use this philosophy when making measurements. It is unfortunate that ID proponents grabbed on to the idea and used it almost like a law.

The probability $1/10^{150}$ is, of course, finite. I'm not saying that all highly improbable things happen—that is certain to be untrue. The issue here is using the notion of a probability bound below which things do not happen as a law, as a fact, when this is simply not the case. *It is possible for all events with a finite probability of occurring to happen.* And this is why the line of reasoning behind the design inference is destined to fail in scientific circles. It is intuitive, but it is proof of nothing. It has only the blunted teeth of a "probably".

That said, it is certainly worth taking note of if the probability of the spontaneous emergence of biological complexity is as small as this universal probability bound. However, the impetus such a calculation should engender is discovering the reason for the low estimate, not jumping to the conclusion that a supernatural force compensates for it. This is cheating on the midterm exam in plain sight. The virtue of science is that it does not permit this kind of kidding oneself or others. Designing an explanation in one's mind for a seemingly insurmountable problem does not mean that explanation is correct. Again, saying that god explains the logical absolutes does not mean that is the case or even possible. It is similar to an under-constrained mathematics problem. The equation $x+y=1$ has an infinity of solutions of the form $(x, y = 1 - x)$, but we need another constraining equation to find the actual solution that fits reality. And although a god may reside on the line $y = 1 - x$, it is only one of an infinite number of possible explanations and a highly improbable one.

6.6 The generation of new information

The main question of biology, and the one ID claims science can't answer, is concerned with the generation of new or novel information. In fact, ID outright claims that the only source of new information is from an intelligent entity. (I have not seen a definition of intelligence provided.)

It has already been shown in previous chapters that the new information found in biological complexity is directly linked to the randomness which drives evolution. Here, I want to address some claims from the ID community that new information can only be generated by an intelligence and cannot come from deterministic systems coupled with random excitation. The concept of intelligence will be considered in the following sections. Here we'll only be concerned with the notion that new information is impossible to generate via a random input.

It is almost self-evident that a deterministic system alone cannot generate new information. (I'm including deterministic systems with deterministic inputs as such inputs can be represented simply as outputs from additional deterministic components that, when coupled to the original system, result in a larger, single deterministic system.) The behavior of a deterministic system, by definition, is wholly defined *a priori*, and it cannot change.

A recursive equation is one of the simplest examples of a deterministic system. For example, the equation $x_{k+1} = 2x_k$, with $x_0 = 1$, generates the sequence $\{1, 2, 4, 8, 16, 32, \dots\}$. It generates this same sequence anytime it is iterated from the same initial condition. Plug in a different initial condition, and a similar sequence will result, but nothing new ever happens.

Another example is a computer program. It can have millions of lines of code, and yet it is still completely deterministic, and its behavior can be predicted from the beginning. No functionality that has not been programmed into the machine will be exhibited.

Even chaotic processes are bound by their determinism. Although it is interesting to see the time evolution of a chaotic system like the Lorenz differential equations (which we will talk about later), they are completely deterministic. So why is the weather impossible to predict beyond a month out? It is not necessarily because the atmospheric system is chaotic. Chaotic systems, if started with the same initial conditions and integrated, provide exactly the same trajectories in every run. What makes the weather unpredictable is the introduction of small, random disturbances, such as measurement error, into the system. The evolution of chaotic systems has an infinite sensitivity to initial conditions, so even an infinitesimal change in the initial conditions can result in an unbounded

6.6. THE GENERATION OF NEW INFORMATION

change in the system trajectories. This is the so-called butterfly effect. Even a change in the initial conditions as slight as what is caused by a butterfly flapping its wings in South America can result in a tornado in Kansas.

If the atmosphere was completely deterministic, the weather might get extremely boring. (Of course, it might not as, even though they are deterministic, chaotic systems exhibit a large degree of complexity.) Boring or not, it would always be predictable if we could simulate it to arbitrary precision. It is the introduction of randomness into the system that makes it interesting. And this certainly results in new information that is integrated into the overall system. This is easy to see by the fact that one would need to know what these random influences are in order to accurately predict the future of the weather.

A deterministic process is one that, by definition, cannot stray from its original content. All of its actions and state information are fixed from the beginning. Again, computers are good examples of this. If designed and programmed correctly, they do not make mistakes. If they did, we would not use them. That said, they do not produce new information either.

There are a couple of reasons why computers use binary arithmetic for processing and memory. The first is pragmatic: it is extremely easy to produce transistors on a silicon substrate using a lithographic process, particularly metal-oxide-semiconductor field-effect (MOSFET) transistors. The second is functional. In fending off noise within the digital circuitry, having two possible states very far apart from one another is very effective. In a microprocessor with a core voltage of, say, five volts, there are only two possible voltages any point within the circuit can take: zero volts or five volts. And these directly correspond to the binary values of zero and one.

It is certainly possible to build a computer with many states. For example, let's look at a decimal system with ten states. Here, the voltage levels that correspond to the numbers 0-9 might correspond to five volts divided into ten equally spaced values, $\{0.00, 0.55, 1.11\ldots, 5.00\}$. However, notice that now the different levels are only 0.55 volts apart and not 5 volts like they were in the binary example. Therefore, it only takes about one-tenth of the noise to interfere with the base-10 version compared to the original.

My point is that great lengths are taken to avoid indeterministic behavior in computers, and in any electronic or mechanical system for that matter. The same is true for writing, especially when it comes to mathematics. Anyone who has ever had to solve a mathematical problem that requires several pages of derivation will know what I'm talking about. We write out these derivations line-by-line to prevent making mistakes, as most of us can only go a few levels deep in our thinking before we are overwhelmed by the noise inherent within our minds.

Digital computers are very limited in some ways. There is no way I can compete with the speed and accuracy of Octave's symbolic toolbox, however, as we'll see in later sections, the fact that humans make mistakes due to random perturbations within our thinking process allows us to do things that computers cannot. Any piece of novel information within a deterministic system *is by definition a mistake*. And in the same vein, any mutation within a deterministic biological system can be considered a mistake as well. But it is these very mistakes that allow a system to acquire new information and behavior. In the same way, it is the introduction of mutations, or mistakes, that allow new information to be introduced into organisms as they evolve.

I agree with Dembski that neither deterministic systems alone nor purely random processes alone can create novel information. Still, I do not agree with his assertion that it cannot be accomplished with the combination of these two. In fact, I assert that is exactly how new information is created.

Intelligent design claims that the only source of novel information is intelligence. But what is intelligence? It certainly has the ability to make a choice. Etymologically, the word "intelligence" is borrowed from the Latin "intelligentia" meaning "the act of choosing between." The act of choosing in this context is non-contingent. In other words, this act of choosing is independent of any cause. The only non-contingent source of action we are aware of in this reality are random events such as radioactive decay. We have already established that novel information cannot originate from a deterministic system. Therefore, an agent that exhibits intelligence is necessarily indeterministic, but this is not the whole story.

In *The Blind Watchmaker* [47], biologist Richard Dawkins puts forth a simple simulation of an evolutionary process. This simulation is very similar to the bit-matching example provided in the chapter on optimization in this book. Dawkins's simulation starts with a random 28-character string, "WDL MNLT DTJBKWIRZREZLMQCO P". The job of the optimization routine is to convert this string to "METHINKS IT IS LIKE A WEASEL". The characters within this string are modified randomly to take on the characters A–Z and a space. The optimization rule requires that once the random process finds the correct character for a particular position in the string, that character doesn't change any more. Until the correct character is happened upon for a particular position within the string, that character continues to be randomly modified. The random modification of the characters is analogous to random genetic mutation and the sticking of the correct character, once it is happened upon, is akin to natural selection.

William Dembski, in his book *No Free Lunch*, says this about Dawkins's letter-optimization problem:

> Although Dawkins and fellow Darwinists use this example to illustrate the power of evolutionary algorithms, in fact it raises more problems than it solves. For one thing, choosing a prespecified target sequence as Dawkins does here is deeply teleological (the target here is set prior to running the evolutionary algorithm and the evolutionary algorithm here is explicitly programmed to end up at the target). This is a problem because evolutionary algorithms are supposed to be capable of solving complex problems without invoking teleology (indeed, most evolutionary algorithms in

6.6. THE GENERATION OF NEW INFORMATION

the literature are programmed to search a space of possible solutions to a problem until they find an answer—not, as Dawkins does here, by explicitly programming the answer into them in advance). For the sake of argument let us therefore assume that when it comes to biology, nature somehow selects targets without introducing teleology into the Darwinian mechanism, thus allowing us to set aside the teleological problem raised by Dawkins example.

I think that stating that preselecting an optimum is teleological is making much ado of nothing. What is being called the target here is actually the set of boundary conditions that constrain the optimization process. These would correspond to the functional constraints operating within the natural world that are ultimately generated from the environment.

Let's say we have a ball that is pushed off the hill of one side of a valley. There is fairly high grass on the slopes of the valley, which adds drag to the ball as it rolls along. The ball rolls down the hill and slightly up the other side before turning back, oscillating for a second, and then comes to a stop at the very bottom of the valley. The ball has acted as an optimization function and found a minimum energy state represented by the valley's lowest point. The slopes of the valley and the minimum at the bottom act as constraints on the ball's motion as gravity mindlessly acts upon it, driving the ball toward its minimum energy state—energy being the cost function—and solving this optimization problem with no intelligence at all. Natural laws involved include the ball's energy, $mgh + (1/2)mv^2$, where m is the mass of the ball, g is the gravitational constant near the Earth's surface, h is the ball's height above the valley's minimum point, and v is the ball's speed; the force pulling the ball toward the minimum $F = mg\cos(\theta)$, where θ is the angle of the force with respect to the slope of the valley; the ball's acceleration from Newton's second law, $F = ma$; and the drag force on the ball, $F_d = -k_d v^2$. (Of course, there are other secondary forces, but we have what we need here.)

Now surely, the objection that the valley's shape and gravity smuggled in extra information isn't going to be raised. Of course, they did. *That is where the context of the problem came from to begin with.* And this is why Dawkins's example is not cheating by smuggling in information. There seems to be some issue in the ID literature about deciding what problem is actually being solved. No, the optimization method did not create any complexity on its own. This information came from the constraints of the system. I'm not saying it is the case, but if each of the individual values of A, T, G, or C that make up a chain of DNA has a direct impact upon the fitness of a biological entity independent of the states of all other nucleotides within the chain, then the validity of Dawkins's example is recovered, with a corresponding probability of successful optimization much higher than if all states must be found in one shot. The increase in probability occurs simply because the states of the characters in the string, or the nucleotides within a DNA chain, are no longer solely dependent upon chance for determining their final values. I'm not going to argue yet whether this procedure produces specified complexity as defined in the ID literature. That's not relevant. What it does provide is a plausible explanation for the development of the complexity we see in nature.

Perry Marshall, in his book *Evolution 2.0*, references this same example of Dawkins, and misses the point by an even wider margin when he says:

> This was heralded as a success. However, Dawkins' software program was programmed to compare each new sentence to the *goal* sentence and either select it for continued "mutation" or reject it based on whether it more closely resembled the *goal* than the previous mutation. But his very own "1.0" Darwinian evolution explicitly forbids preprogrammed goals! So Dawkins' "Weasel" experiment had nothing to do with true Neo-Darwinism.
> *His program does vaguely resemble what cells do.* But don't forget—Dawkins has always insisted that evolution is blind and purposeless. His program is anything but blind and purposeless; its goal is precisely defined from the beginning! What Dawkins actually proved with this experiment was: If you want to evolve, you have to start with a goal.

And Douglas Axe, in his book *Undeniable*, misses the point entirely. Like many ID proponents, he mistakes the constraints on the problem, which is the phrase "METHINKS IT IS LIKE A WEASEL", for a "guiding" of the process. This is simply a complete misunderstanding of what is actually happening on Dawkins's computer or in nature. Again, the complexity is not "smuggled" into the simulations of these optimization problems. The complexity within nature is created by the constraints, the environment and competing species, and by the nonlinear, chaotic behavior of the biological systems themselves. There is no subterfuge going on here.

As we'll talk more about later, Dawkins's program is a simulation of evolution. It simulates *both* the evolving entity *and* the constraints on that entity. The phrase being modified represents the entity, and the target represents constraints on that entity. These constraints could be its environment, other competing entities, the amount of oxygen in the atmosphere, etc. For organisms in the natural world, the "preprogrammed goals" Marshall mentions are contained in the environmental cost function. For example, if an organism evolved to have larger lungs to increase its energy intake, surely no ID proponent would shout, "No fair!", the low oxygen density in the air *told* the entity how to evolve. This is a complete misunderstanding of how system optimization occurs.

Before going further, I must supply one more quote from *No Free Lunch*: "Complexity and probability therefore vary inversely—the greater the complexity, the smaller the probability. It follows that Dawkins's evolutionary algorithm, by vastly increasing the probability of getting the target sequence, vastly decreases the complexity inherent in that sequence. As the sole possibility that Dawkins's evolutionary algorithm can attain, the target sequence in fact has minimal complexity (i.e., the probability is 1 and the complexity, as measured by the usual information measure, is 0)." This statement seems to be a case of a confused author and is a telltale sign of expecting nature to be the derivative of one's work instead of the other way around. Let's say there was just one outcome

6.6. THE GENERATION OF NEW INFORMATION

that allowed sentient life on Earth, and that was the extant human genome. Are we not to call it complex because it was the only solution allowed? Of course not. The complexity comes in because, in this case, there is almost an infinite number of ways that life could have failed to materialize. A similar argument holds for Dawkins's example. In saying that there is no complexity associated with it because only one answer is ever generated by the simulation is a misunderstanding of how the simulation is intended to work. (I might also add that apparently, there is a certain and singular result when an intelligent designer acts to bring about a structure like human DNA, and yet it is detectable by its apparent specified complexity...)

Simulations are typically split into two isolated sections. One models the space upon which a function to be evaluated is acting, and the other section implements the function under test. This is how Dawkins's test is arranged and is how all the simulations in Chapter 4 are partitioned. There is no problem with Dawkins specifying an optimal configuration and then testing the performance of a candidate evolutionary method in finding that optimum. The same thing applies to the simulations present in this book. In both cases, the information of each section is isolated from the other section.

Let's consider an example that is a little more detailed. The way real-world simulations are performed needs to be understood to follow the discussions here. How would one go about, for example, simulating a new cruise controller for a car? In other words, a new controller has been engineered and needs to be tested. How could we simulate the controller with real-world conditions *without* actually building it and placing it in a vehicle?

The way this is done in practice is to partition the simulation into two pieces, as shown in Figure 6.8. The left half is the controller model that is to be tested, and the right half is a model of the vehicle and the real world. The two systems are connected only by the signals that would be available to each of them in an actual vehicle in the real world. In other words, there is no cheating by trading information between them. The only information traded is the same information that is interchanged in the real application.

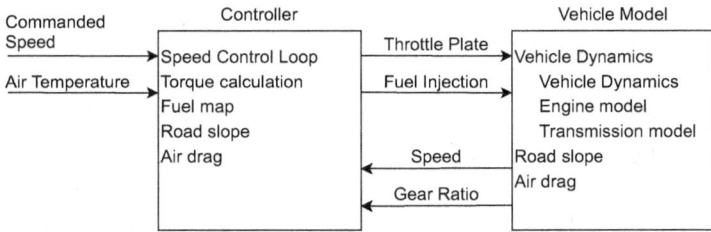

Figure 6.8: The structure of a simulation is divided into two parts. In this case, the left half is the controller that is being evaluated, and the right half simulates the real-world dynamics of a vehicle that is being controlled.

In the example, the engine controller is privy to variables like the speed commanded by the driver, outside air temperature, vehicle speed, and transmission gear ratio. Using these variables, the controller computes a throttle plate position and fuel-injector rate in such a way as to force the vehicle to the commanded speed. The computed throttle position and fuel-injector ratio are provided to the engine model on the right, which the engine model then uses to compute a torque at the transmission input. The transmission model subtracts any torque losses, scales the resultant torque, and then provides it to the vehicle's dynamic model. This model takes into account the vehicle inertia (using, of course, $F = ma$), air drag, the incline of the road, etc., to calculate the vehicle speed, which it provides back to the controller.

In a simulation of the evolution process, random mutation of the genome and fitness of an organism are determined by the real-world portion of the simulation. Phenotype generation from the mutated genome, and natural selection, are the evolutionary parts being tested. The mutation rate is determined by the environment and the phenotype (imagine different pigmentations in the skin, for example) is supplied by the evolutionary "controller." Fitness is also determined by the environment and competing species, and calculated by how well the phenotype fairs in competition. Of course, there are multiple ways to partition any problem, and that must be specified to understand the validity of the simulation.

The point of all this is that there is no information being "smuggled" in that doesn't belong. This idea can lead to confusion when it comes to the question of where the complexity comes from. The answer is from the real world, the environment, competing species, physical laws, etc. Including these items is not cheating. It is what makes the simulation valid.

So, where does any new complexity or information come from? We have shown that complexity in the formal sense can come about through global emergence from a set of very simple local rules. And it is known that the assemblages of proteins and other biological structures are built from large ensembles of amino acids and similar building blocks that adhere to a set of simple assembly rules based on, for instance, electrostatic attraction. We also know the inorganic world of physical laws and matter is capable of generating complexity. Since these are deterministic processes, we know they can't generate novel information on their own. Fortunately, reality is indeterministic. Combining these characteristics, the generation of novel complexity becomes a plausible outcome. However, it is only when this complexity is guided or constrained that we see it organize into the complex biological organisms we see today. This constraint, or guidance, is natural selection, and it acts to modulate the positive feedback implicit in self-replicating entities—in other words, it dictates where complexity will be concentrated or grow. That is a lot to take in, so let's attempt to break it down with an example.

These are difficult things to think about, and they only become more difficult when placed in the context of evolution. I suspect the most difficult part in understanding evolution is the presence of a replicator—a biological machine

6.6. THE GENERATION OF NEW INFORMATION 227

that duplicates itself on a generational basis. It would not do if we had a single master copy that continued to make copies of itself as nothing would change (this is why a carpenter always uses a single cutting template when duplications are required—he doesn't want changes to occur). However, if we make a slight modification and let the machine create a generation of new machines, and then those machines create a new generation, and so on, then the dynamics are much different. If the duplication process is subject to random error, then there is the possibility for the evolution of the process.

Think about building a structure with a child's building blocks—the kind of blocks that have protrusions and recesses that snap together. The game is to build the tallest stable structure possible—maybe so your child's action figure can climb it to reach the castle in the sky or something like that. You decide to use a moderately scientific method to build the best tower you can by constructing five slightly different towers and then shaking the table until all but one falls down. The one that survives is used as the template for the next round of towers. Each time, a single block is added to each tower in a random position. That is, a new block is placed randomly *in different positions* on each of the structures. Then the table is shaken again, and the process repeated.

Since the single block additions at each iteration are placed at random, this is an unintelligent process. The selection of which tower is to be the template for the next round is decided by simply shaking the table until four of the five towers have fallen down; there's no intelligence here either. In fact, there's no intelligence made use of in the entire endeavor. And yet, after 300 iterations, a tower stands taller than any you could have constructed by design. Complexity has emerged using a replicative process coupled with random modification and making use of a dumb selection method. How exactly is this possible if what the ID community claims is true?

In this example, the genome and phenotype are the same or are mapped to the same space. The random placing of the blocks is, of course, random mutation of the genome, and shaking the table is measuring the competitive fitness of the structures. The only one surviving being used as the template for the next iteration represents natural selection.

There are only so many examples someone can go through of dumb processes generating complexity. I believe the pertinent question is no longer whether complexity can be generated from a combination of deterministic complex systems and randomness, but, rather, if we actually see random mutation, complex systems akin to cellular automata, and natural selection via competition in nature. And, having considered all three of these in previous chapters, it seems the answer is, yes.

Does science have all the answers to all the details of this process? No. But it does have enough answers to qualify evolution as a valid theory and justify that evolutionary research continue. On the other hand, the aim of ID seems to entail sitting back and saying that since science doesn't have all the answers, the theory is not only in error but not worthy of pursuing further. Meanwhile, science just keeps moving forward and finding answers.

6.7 Intervention by an intelligent designer

The ID community claims that it must be the case that a supernatural designer is responsible for abiogenesis on planet Earth and the resulting diversity of life that we see today. The contention is that the odds are so against natural causes bringing these events about that they simply could not have happened naturally and must have had some kind of *ex nihilo* help. In this section, we're going to examine how such a designer might interact with reality to bring about the outcomes necessary for life as it exists on Earth today.

Generally, there are two ways in which a designer might interact with the physical world. The first is in a deistic fashion where the designer defines how the Universe's physical laws behave and then has no further interaction with it. This is the clock-work Universe: God winds it up, sets the rules, and then lets it run of its own accord. If someone believes that a deistic god dictates the functioning of the Universe, then even Newton's laws become evidence for a creator. Seriously, why would the complexity of life even need to come up in the discussion when holding a ball above the ground and dropping it is sufficient to demonstrate that God is active in reality? And such claims can't be refuted, of course. There is no way that a deistic god can be proven not to exist. However, this god is superfluous. If the physical laws dictate the behavior of the Universe completely, then there is no reason to add the assumption that a god exists. In the words of Laplace, such a hypothesis is not necessary and does not further explain the situation. Simply put, the Universe behaves the way it does, and that is all the explanation needed. Physics, although rejecting deism, assumes autonomous behavior of the Universe, and that there is no need for a god.

Our perception of the world around us is all we have, and if we don't embrace the above logic, then everything becomes a mass of confusion. I could assert one god and you another, each supposedly responsible for the physical laws. Neither can be proven not to exist. Science does not assert every explanation is on equal footing with every other. The characteristics that prioritize one particular hypothesis over others are evidence and simplicity. A hypothesis with no evidence is not pursued. Working in tandem is the rule typically referred to as Occam's razor that declares the simplest explanation the most likely to be true. These two concepts together rule out considering deistic gods. By definition, there can be no direct evidence of such gods, and a simpler explanation is that intrinsic rules govern the Universe. There is no need for such a god.

Most ID proponents will concede the above points, or at least they will not argue against them, because to do so would be futile. The Universe behaves in regular ways, observed and codified by physicists as natural laws, that can be tested and confirmed by anyone. Attributing these laws to a deistic god cannot be proven by definition. Therefore, we are all left with what we can observe and verify within the physical reality that we all share. In fact, that is what physics

6.7. INTERVENTION BY AN INTELLIGENT DESIGNER

is: a set of rules that describes how the Universe behaves that are discovered by repeated observation of the world around us.

The other option is that a god or designer actively intervenes regularly within reality. Not in a deistic fashion, but in an observable way. From a theistic vantage point, God stands outside the Universe as a creator, with the Universe as the creation, and, therefore, can act in such a way to modify the Universe by directly manipulating matter and energy. The direct interference by a theistic designer suffers, of course, from the problem that such intervention has never been observed.

There are plenty of stories of miracles (where I'm defining a miracle as a temporary suspension of the physical laws) occurring throughout history and around the world. Still, such a thing has never been confirmed. On the other hand, it has been confirmed countless times that reality does conform to the physical laws. Therefore, any logical understanding of reality must reject the notion that a supernatural entity forcefully modifies the physical world to effect design.

Hence, ID does not pursue claims of either a deistic or theistic designer, and there's a good reason for this. The former argues for a redundant and unneeded god, and the second isn't observed to happen in reality, thus rejecting the existence of a god. The solution needed by ID theorists must be one that has an impact on the physical world and yet does not physically interact with it in any detectable way. This sounds contradictory, and it is, but it is really the only way to shoehorn an effective but undetectable designer into reality.

To this end, ID suggests that a designer god might forego direct interaction with physical matter and instead modify the information that dictates how this matter behaves [55]. It is assumed this could happen by the designer actively modifying indeterminate processes. For example, it is suggested that modifying the probability distributions of quantum-mechanical variables would force particles, which behave according to these probability distributions, to move in the way a designer wishes, but—and this is important—without the designer needing to move them forcefully.

Existing at the macroscopic scale, humans expect the world to move in deterministic, predictable ways. For example, playing a game of pool requires the balls to move as independent objects that respond predictably to the forces acting upon them. In racquetball, the ball always bounces off the wall and back towards the players; it never goes through the wall. A car can be driven at any speed over a continuous spectrum; it does not prefer one or a set of discrete speeds. Although objects behave this way on the scale of human beings, this is not how the particles that make up these objects at the microscopic level behave.

At the quantum level, all particle locations, energies, speeds, and even instants in time, correspond to statistical distributions and not fixed values. For example, consider measuring the position of a given particle. As was discussed in Chapter 2, the position of a quantum variable corresponds to a probability distribution much like that of Figure 6.9 and, when measured, the particle can show up anywhere. However, it is most likely to be found at the mean of the distribution. In this way, quantum particles are not located at definite positions, but rather their probability distributions define *where they are likely to be found*.

The function shown in Figure 6.9 is what is called a *probability density function*, or PDF. It is a normalized histogram of the output of a process. An example will help demonstrate this better.

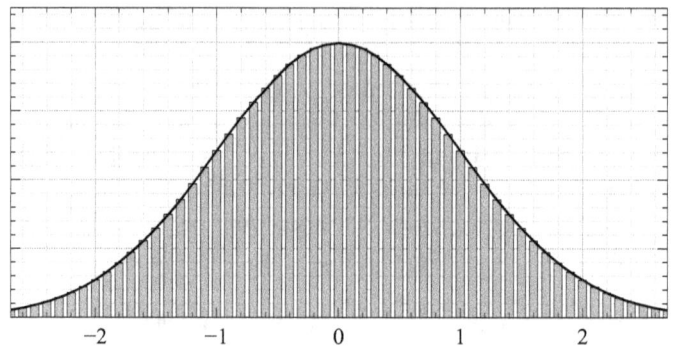

Figure 6.9: A typical normal probability density function.

Suppose there is a math function on your computer that, when called, returns a normally-distributed random number between positive and negative infinity, and let's say it has a mean of zero and a standard deviation of 1 (we'll talk about what these mean in a moment). Each time this function is called, it returns a random number. A PDF plot similar to that shown in the figure records the frequency of these numbers. That is, every time a number comes up between zero and one, it is added to a bin centered at 0.5, and every time a number between one and two comes up, it is added to a bin centered at 1.5, and so on. If we run this function thousands of times and add up the numbers in each of the bins spanning the horizontal axis, we get the plot in the figure. Note that the largest concentration occurs around zero; this is the *mean* of the distribution. Whenever the function is called, it is far more likely that the random number returned is near zero than far from zero. Also, note that the vast majority of numbers fall between ±1. This is called the *standard deviation* of the distribution. These two numbers—the mean and the standard deviation—parameterize the distribution. The mean measures where the center of the distribution is located, and the standard deviation tells us something about the width of the distribution.

If one were to measure the position of an electron in isolation repeatedly, the corresponding position histogram would look similar to the PDF in Figure 6.9. The electron is not guaranteed to be at any specific location; rather, the PDF associated with it provides only the probability of finding it at certain locations. The same is true of other electron properties, such as its speed, energy, or even its location in time.

6.7. INTERVENTION BY AN INTELLIGENT DESIGNER

In quantum mechanics, these PDFs are called wave functions because they behave like traveling waves. The mean of the wave function is centered on the particle, and the wave spreads out as time passes. In other words, the uncertainty in the electron's position increases with time as the wave function continues to spread. When a measurement of the electron's position is made, the wave function is said to collapse, and the electron is found to be at a specific position. This is the so-called wave-particle duality of quantum mechanics. All properties of particles are described by probabilistic wave functions until they are measured or interacted with and forced to choose a particular value within their associated PDFs. Another way of looking at this is that quantum objects are only well-defined within reality when they need to be.

Wave functions are not physical waves; they are probability waves, or waves of potentiality, that describe possible values of a given property of a particle. When one talks of the position of a particle like an electron, a reference is actually being made to the mean of the electron's position wave function as this is the most likely position to find it.

We don't see this behavior at the macroscopic scale because the matter we are used to dealing with is made up of large ensembles of particles. As these particles are in continual interaction with one another, their wave functions are continually being collapsed, which forces the positions of the particles to be localized. This is why we never see an object just diffuse away as the wave functions of its particles expand.

So how might a designer employ these aspects of quantum mechanics to affect reality without being detected? Dembski describes the notion below [55]:

> To be sure, an embodied designer could impart information by employing arbitrarily small amounts of energy. But an arbitrarily small amount of energy is still a positive amount of energy, and any designer employing positive amounts of energy to impart information is still in the business of moving particles. The question remains how an unembodied designer can influence the natural world without imparting any energy whatsoever. It is here that an indeterministic universe comes to the rescue.

Dembski goes on to explain how this indeterminism can be used to mask or hide an intelligent designer's actions in modifying the real world. In other words, since we can't predict the outcome of a random event, a designer could modify the outcome of that event any way he wanted without being detected. In addition, this modification is made with no energy expenditure. That is, modifying a probabilistic distribution of, for example, a quantum-mechanical variable does not require energy injection.

It is being suggested that an unembodied designer (an embodied designer, it is supposed, would be bound by the very indeterminism it intends to commandeer, rendering the argument moot) could, by accessing and altering the quantum mechanical wave functions of various particles, alter macroscopic outcomes, and this would require no energy to be injected into reality and, thus, would be undetectable. Let's assume that the PDF shown in Figure 6.9 corresponds to the wave function of a particular electron. The idea is that a designer could, in some way, shift the mean of this PDF from zero to some other location, say x_0, and thereby modify the natural world. However, since this entity did not physically move the electron, no transfer of energy was needed. When the wave function of the electron collapsed, it was simply more likely to choose a different location than it would have done otherwise.

With some imagination, it is not hard to envision how a supernatural designer could alter the wave functions of trillions of particles to bring about macroscopic effects, such as guiding the evolution of DNA on Earth. And that this could be done without using physical force and, thus, not introducing any additional energy to the system, hence being impossible to detect. The problem is that things do not work this way.

The properties of a quantum-mechanical system are not dictated by their associated wave functions; rather, the behaviors of the constituents of a quantum-mechanical system dictate the behavior of wave functions. Suggesting something like changing the quantum wave functions to induce alternate behavior is like saying that to change the potential energy of a mass at height h above the Earth's surface involves only changing the variable h that dictates its height. This is backward. The system dictates h, a measurement of the height of the object from the Earth's surface; the measurement does not dictate the object's position. The ridiculous nature of the latter example is just more obvious than the former, but they are of the same character.

Of course, the process by which the original idea is brought about is not completely clear. Maybe the intent was that the designer could use whatever magic to change the properties of particles, and the stochastic nature of these properties would hide this action. If the electron's position is a random process described by a wave function or PDF, then a designer could intervene in a wave function collapse and alter the location at which the electron appears as a particle. No one would be the wiser because its ultimate position was random anyway. (This does modify the associated PDF, however.)

Dembski provides an example in his book of a device that fires photons at a polarizer placed at an angle of $45°$ relative to the polarization of the photons. With this setup, quantum mechanics predicts that the photons have a fifty-percent chance of passing through the polarizer. A detector is placed after the polarizer that records a binary one if a photon is detected and a zero otherwise. Then he surmises what conclusion might be reached by the operator if the series of ones and zeros emanating from the detector formed an ASCII (a form of binary encoding for alphanumeric characters in today's computers) message that delineated the cure for cancer. In his own words from the book [55]:

> Given this setup, we have precluded that a designer imparted a positive amount of energy (however miniscule) to influence the output of the device. Nevertheless, there is no way to avoid the conclusion that a designer (presumably unembodied) influenced the output of the device despite imparting no energy to it. Note that there is no problem of counterfactual

6.7. INTERVENTION BY AN INTELLIGENT DESIGNER 233

substitution here. It is not that the designer expended any energy and therefore did something physically discernible to the device in question. Any bit when viewed in isolation is the result of an irreducibly chance-driven process. And yet the arrangement and coordination of the bits in sequence cannot reasonably be attributed to chance and in fact points unmistakably to an intelligent designer.

I assume this thought experiment is intended to be analogous to the formation of DNA or biological proteins with their respective complexities. As described above, this suggestion meets the criteria of not being detectable through physical interaction with the real world and yet being detectable via the improbable results of such interaction. That was the goal. Does this sound too good to be true, or too contradictory? It's very subtle. Let's see if we can sort this out.

Energy is not some tangible thing that can be carried around and used at will like it is portrayed in the movies. It is an accounting method that describes how reality behaves. What does it mean to say that an object suspended above the Earth's surface has potential energy? The energy ascribed to this object (and to the Earth as well) refers to moving the object over a distance within the gravitational field between the object and the Earth. When it comes down to it, all physical laws are based on information. They express predictable relationships between sets of information from differing sources. But don't get it confused and assume that physical laws dictate the behavior of the universe. Again, it is the other way around; the Universe's behavior dictates the physical laws. It's like one of those people who complain on a rainy day that the local weatherman said it wasn't supposed to rain. This is the confusion I'm trying to avoid. The forecast doesn't prescribe the weather; rather, the weather determines the forecast.

In other words, modifying the positions, or other properties, of quantum variables *is interacting with reality*. That's how physical interaction is defined in this reality. For example, if an experimental test set up with a detector behind a polarizer excited by photons delivered a cure for cancer, then it was (likely) interfered with. And a transferral of energy had to have taken place. It is not enough to assert that the bits acquired from the machine represent an ensemble of identically distributed, independent random variables and, therefore, since each outcome was random, that a designer is free to choose the outcomes behind the scenes and not be noticed.

This is an extremely subtle component of quantum mechanics. We are all aware of the conservation of energy in the macroscopic world. If you throw a baseball, electrochemical energy in your body is converted into mechanical energy that is then converted to kinetic energy in the ball. If one went down the chain of transference, the change in energy in the total system (you, the ball, the air, etc.) would be zero. Now, consider measuring the momentum of an electron initially at rest. The odds that the electron's speed and its kinetic energy after the measurement are still zero are very slim. It started with zero kinetic energy, and after the measurement, it has a finite kinetic energy. How did this happen? Actually, the energy differential derives from the measurement itself. Does this mean that conservation of energy doesn't hold at the quantum level? No.

Conservation of energy still holds in the average sense. Although they can be violated briefly, as can be seen through the Heisenberg uncertainty relations, the expected values (mean values) of particle properties still adhere to the conservation laws. Otherwise, it would be possible in some instances to violate these laws at macroscopic scales.

To further drive this point home, consider a 1 kg cube of gold centered at the coordinates $(0, 0, 1)$ one meter above the Earth's surface. The potential energy of this cube is given by $E = mhg = (1\text{kg})(1\text{m})g = g$, where g is the gravitation constant at Earth's surface $g = 9.81\text{m/s}^2$. Now, assume a supernatural designer intervenes and modifies the statistics of every particle of gold within this cube at a particular point in time such that the corresponding wave functions collapse exactly one meter higher than their original position, that is, at $(0, 0, 2)$ (it might take a series of successive collapses). The new potential energy of the cube is $E = (1\text{kg})(2\text{m})g = 2g$, twice its starting energy. In other words, changing the outcomes of random variables associated with the gold particles changes reality.

So, why go to all these lengths to show that an intelligent designer is involved in the world? Really, why not just say a designer created DNA? In fact, why not also claim that the physical laws are generated by God as well?

We're back to square one at this point. Even intelligent design advocates apparently do not regard the regularity within the Universe that can be described by physical laws as noteworthy evidence for a god or designer. This trend has been continuing over the past couple of hundred years. Any item that is subject to reductionism is no longer attributed to a god. The best example of this is the Universe itself. The Big Bang Theory, and its description of the systematic evolution of the Universe, has removed even creation as a valid argument for God (other than the initial cause). Lawrence Krauss' book *A Universe from Nothing* [84] provided one of the symbolic nails in the coffin of creationism by describing how the net energy of the Universe is zero, thus stymying one of the most hackneyed protests from creationists that something can't come from nothing. In fact, it doesn't have to do so. It turns out that nothing *can* come from nothing.

The success of science in using reductionism to explain the more mundane phenomena like gravity made these phenomena immune to convincing attribution to a supernatural entity by the ID community. However, less understood, more complex processes like evolution are still fair game. The assertion is made that there is no way nature alone could have constructed such complex structures, leading to the proposal that a supernatural designer was involved. Of course, no such involvement by a designer has ever been observed. Therefore, the next endeavor is to show that the designer acted in a way that was undetectable and, yet, is detectable. And, of course, science can't disprove such a being if that being does not interact with reality.

There is indeed no way to prove that a designer or god is not behind all of this. And that was never and can never be a goal of science. Science simply attempts to understand the physical reality we all experience. Adding a subjective god to the mix steps outside what we can collectively experience and, thus, classify as science. Therefore, the idea is rejected for being either unobserved in reality, simply unnecessary, or both.

6.8 Can intelligent design provide any answers?

There is overwhelming sentiment in the ID community that naturalistic evolution cannot provide a solution for abiogenesis nor answer the purported probabilistic impossibilities of the self-organization of highly complex biological systems. Furthermore, there is always the question of from whence came the new information within the DNA of all species of life. On the other hand, ID claims it can answer these questions. How so? Well, of course, by ID. This is one of the single most useless insights in all the history of intellectual enterprise. Let me just pop the proverbial bubble: now, instead of natural complexity, we need to explain an invisible intelligence which has never been sighted and, by definition, and much to Occam's dismay, must necessarily be more complicated than the original question. This is not scientific research; this is a universal roadblock with some pretty hefty baggage in the trunk. I wonder who this intelligent designer that stands outside the natural world might be? At this point, what might have been an interesting thought in passing, becomes a brickbat thrown, for the sake of the self-preservation of one's own beliefs, through the window of scientific objectivism.

I've tried to put this into words before and will try again here, but it is such an amorphous topic that it does not easily admit to forging in the furnace of the English language. *Just because you say a god or designer did it does not make it so, and it doesn't answer the question.* Explanatory power comes from explaining complex things in terms of simpler things. We attempt to explain new concepts with concepts we already understand. In explaining the origin or diversity of life, if one simply claims that a designer did it, then nothing has been explained. One set of words have just been exchanged for another. It would be different if the designer interacted with reality in a way we could detect—in that case we could investigate the designer. But this is not the case with ID.

Before going on, I want to give one more example of this type of "God did it" logical fallacy. In discussion with an atheist, Matt Slick, the founder of the Christian Apologetics and Research Ministry (CARM), provided this justification for the logical absolutes [85]:

> Since the logical absolutes are conceptual, they transcend all people at all times and are absolute in all circumstances...since they are absolute. Conceptual Absolutes cannot be accounted for in an atheistic worldview. But they can be accounted for in a theistic one. The Absolute God with an absolute mind has conceived of the logical absolutes. They are a reflection of His mind. At least I can offer an explanation for their existence where the atheist cannot.

This is simply an example of relabeling the explanation of the logical absolutes as "God did it" and has no explanatory power at all. The same effect could be realized by stating that magical fairies account for logic. Intelligent design is analogous in that it relabels the explanation for biological complexity as "an intelligent designer did it."

We talked previously about a single equation in two variables, for example, $y = 1 - x$. This equation has an infinite number of solutions that are all equally valid. They lie on a line in the xy–plane with a slope of negative one. The argument above is taking advantage of this same under-constrained indeterminacy. God is certainly a possible solution, but in no way is that solution guaranteed to be correct.

As a child, I lost a tooth once in a while and would participate in the common childhood game of placing the tooth under my pillow before I went to bed only to wake up the following day and find the tooth gone and a quarter under my pillow. I conjectured to my brother that it was mom putting the money there. Of course, the other, maybe more plausible explanation, given the information I had at the time, was that a tooth fairy, needing teeth for some reason or other, had made an honest exchange during the night. But let's be fair, this second theory certainly did explain the combination of missing teeth and money appearing from nowhere. And the science I possessed certainly couldn't explain it. What exactly would the statistical odds be that a tooth would spontaneously change into a quarter overnight every time I placed one beneath my pillow? Well, this supposed theory held water until I was awake one morning when my mother was making the exchange.

Creating an arbitrary account that explains the facts of a particular process, precisely by not explaining them, is not an explanation of that process. It's like fitting a tenth-order polynomial to ten data points; it tells you nothing and possesses no predictive power. Similarly, we have a mystery in the formation of life on this planet. Intelligent design provides the same mystery in a different form, but has no explanatory power either. An intelligent designer, or a god, can be and has been throughout the eons mutated to fit the data exactly. I could fit the data just as well, not approximately, just as well, by saying it was accomplished by magic. Or by appealing to the Greek or Norse gods.

A theory that is not sufficiently constrained, just like a single equation, cannot be shown to be false. For example, any point along the line $y = 1 - x$ is just as valid a solution as any other. How does one choose between ID, magic, Norse gods, or any other explanation for the emergence of biological complexity? They all can be made to explain this complexity by judiciously choosing appropriate values for the variables. On the other hand, scientists understand that the application of the evidence must sufficiently constrain an explanation and, due to these constraints, the theory is falsifiable. In other words, the real solution can be distinguished from competing theories.

Theories that are not falsifiable are essentially tautologies. The Christian religion is a particularly clear example of this. From a believer's point of view it

6.8. CAN INTELLIGENT DESIGN PROVIDE ANY ANSWERS? 237

is just simply true that a god created the Universe, and it doesn't matter at all that there is no evidence for it. Not surprisingly, ID is similar.

Not one time as a child, as I was learning about physics from my father, did I ever feel the need to attribute anything to magic. If there were any magic, it would soon admit to analysis. We have never run into any phenomenon within this reality that refused to yield under analysis. And this is not a bad thing. When seen at first glance, things like the characteristics of spacetime and light, love, and where we came from are mesmerizing and unexplainable. And they should stay this way for a while—this is half the reason we enjoy living in reality. Then, things change, and we go looking for explanations—this is the other half. This process repeats bringing us inexorably closer to the truth.

For ID to make any significant contributions, it would need to be accepted as a legitimate area of research by the scientific community. I'm not sure how the research of ID would occur and even if it would make sense. What exactly is something supernatural? We would need to start there. A first definition of the supernatural could be something that presently exists in nature of which we are not aware and, with the proper analysis, can be brought back into the category of the natural. I suspect that is not what the ID community refers to when they talk about the supernatural. Still, I think it is the only approach that could remotely be labeled as reasonable.

Nevertheless, I don't see how anything that affects the Universe could be in a category where it would remain classified as outside the natural. It is not as if there is a divine dictate that the natural is inferior to a higher, disconnected domain called the supernatural. As far as we can ascertain, this idea is a fiction created by a mind capable of co-opting concepts valid in one domain and attempting to apply them to an entirely different domain. For example, the knowledge that sound travels through the air as waves was used as a reason to assume that light, which is also a wave, should travel through a corresponding medium, originally called the aether. This is a perfect example of how our assumptions in co-opting ideas can be wrong: there is no medium that is required for light waves to propagate. This hypothesis was eventually rejected due to lack of evidence.

The notion that there is a supernatural undoubtedly comes from assuming that there must always be an up and a down, a light and a dark, a high and a low, and, therefore, a natural and a supernatural. However, just like the aether, just because it seems appropriate that such analogous concepts should exist doesn't mean that they actually do. In fact, the notion of a supernatural is even less convincing than an aether because it is self-contradictory.

Consider an engineer who is simulating an amplifier circuit on a workstation computer. It is a very simple circuit employing a negative feedback network to limit the gain. Specifically, this amplifier circuit has a feedback resistance of R_f and a shunt resistance of R_g, providing a voltage gain of $A = -R_f/R_g$. As the engineer repetitively reaches into the internals of the program to adjust the values of R_f and R_g to adjust the gain of the circuit, one might be inclined to conclude that from the simulation's point of view that the engineer's activities are supernatural. And that may well be a valid argument, but we *don't get to make that distinction just because it benefits our opinion.* The truth of the matter is there is a single reality: the simulation exists on a PC that is on the engineer's desk, both of which reside in the same overall reality. We don't get to claim that there are things *outside* this reality because as soon as that something is declared, it becomes part of this reality. The very action by which it affects our reality makes it part of our reality. Anything else is incoherent.

How does one determine whether an object is part of his or her reality? It is the very fact that the object interacts with reality. If it weren't participating in some way in reality, there would be no reason even to suspect it exists in the first place. The very notion that there might be a designer that affects this reality but is completely undetectable is to say that this designer is deistic in the sense of being a physical law. However, the designer concept is again superfluous. A slight digression is in order here.

Is the law of gravity governed by an intelligent entity outside our reality? The law of gravity simply states that an attractive force that propagates at the speed of light exists between massive objects (it's a little more complicated than this, but this is fine for our purposes). Now, if a being outside our detection provides this governance of the behavior of massive objects, how would anyone know? We would simply call it a physical law and go on, which is what we have done. So how is ID as proposed today a problem? If it genuinely describes a physical process we observe that always occurs, how could there be a problem?

The problem is that the framers of ID went far beyond the methods of Newton and Einstein. Intelligent design, as applied to the diversity of life on this planet, does not set down a fixed law from which predictions can be made. The predictive ability of a physical law is not important because it can be used in staged magic shows in predicting the future; rather, it is important to demonstrate that the rules do not change, and thereby what you think you've found is actually real. Each time the law is used successfully, the law is reaffirmed. Intelligent design is much different in this regard. It says that an entity introduced its own preferences into the formation of biological systems, which may change in the future. In other words, the preferences of a designer look a lot like randomness to us.

In addition, physical laws are not the same as the tricks used in staged magic shows precisely because *anyone* can wield the magic. I do not have to take Newton's account of gravity on his word. And why would I? If I took such a stance, the next thing I would be forced to believe is his notions that alchemy was true and that the Old Testament of the Bible contains secret knowledge. Newton was a genius, there is no doubt about that, but his opinions were and are no more valid than anyone else's. Only those things that can be verified, such as the observation that objects at the Earth's surface fall at 9.81 m/s^2, are reliable. What makes Newton's laws of gravitation actual laws is that they are verifiable by anyone anywhere and they do not change.

I don't want a misunderstanding here about non-changing laws, so I'm going

6.8. CAN INTELLIGENT DESIGN PROVIDE ANY ANSWERS?

to digress a bit further. Things can and do change over time. If this fact were not taken into account in engineering applications, there would be many dissatisfied customers. For example, the power loss in a power semiconductor switch like a MOSFET is proportional to the device's on-state resistance, which increases with temperature. If this is ignored, thermal runaway and destruction of the device are unavoidable in some instances. How is this taken into account? The mathematical description of the device's functioning must include a dependence upon the temperature of the device. As the temperature increases with time, the parameters are adjusted to maintain operation within the device's safe area of operation.

In the same vein, if Newton's gravitational constant was found to diminish by 1% per year from an initial value of 9.81 m/s^2, this characteristic would simply be integrated into the laws of gravitation as $g(t) = g_0 - (t-t_0)(0.01)g_0$. My point is that even a physical law which varies in time is still predictable so long this variation can be quantified. These things can be and are handled quite easily. And if there were an entity outside the Universe that was meddling with the DNA of organisms to achieve its own ends (whatever those might be), it would be detectable and would, without the supposition of a god, be codified into a natural law *if the meddling was predictable*. In fact, this is what science is doing right now, only leaving out the designer part because it makes no difference to the theory. It's the inclusion of this last bit in ID, that of a designer that makes no difference to the functioning of the theory, which is the primary objection from science.

If the inclusion of a designer makes no difference to the functioning of a theory, then the concept should be left out. In other words, if the designer is not detectable, one must assume that it is working through the natural laws to generate specific outcomes. Hence, from the point of view of science, nature is simply doing what it does, and the designer concept is superfluous. So, why does the ID community add a part to the story that isn't needed? The answer is, this is how baggage is smuggled in through the back door. And here is what appears to be the ultimate objective of ID: hidden in one of the smuggled suitcases is a god.

And to ward off any concerns that this conclusion might be ill-founded, here are the words from Judge John E. Jones III found in his 139-page decision from the 2005 *Kitzmiller v. Dover Area School District* trial:

> For the reasons that follow, we conclude that the religious nature of ID [intelligent design] would be readily apparent to an objective observer, adult or child.
>
> A significant aspect of the IDM [intelligent design movement] is that despite Defendants' protestations to the contrary, it describes ID as a religious argument. In that vein, the writings of leading ID proponents reveal that the designer postulated by their argument is the God of Christianity.

This subterfuge is likely the main reason the scientific community has no stomach for ID. Scientists are, by their nature, individuals who enjoy figuring things out. For them, nothing is better than a complicated mystery to solve, just for the sake of solving the problem. From this desire emerges an implicit guarding against cheating or deluding oneself, for if either is the case, the puzzle is not solved. I suspect that this plays a major role in scientists' rejection of ID. It is perceived as a cheat or that there are underlying motivations, and this just won't do.

Simply saying an unknowable intelligence is adjusting things behind the scenes knocks the game pieces off the board. The game is over. This is the reason why Dawkins states that a designer only leaves a bigger mystery to solve. Religious adherents may be content with no mystery to solve, but scientists are not. Even with an intelligent designer, the game would simply be pushed from one board to another. However, an undetectable designer is a game that can't be played. Scientific inquiry is over.

6.9 Is ID a valid scientific theory?

It seems over the past couple of decades, in particular, that the validity of ID is no longer measured on its own merits but, rather, what should have been an objective evaluation has turned into a winner-take-all debate between ID and evolutionary biology. Before we proceed, it is important to point out that the failure of either one of these two in no way indicates the success of the other, for that would be a logical fallacy as the two do not represent a true dichotomy. In reading the ID literature, one certainly gets the impression that ID would be validated by proving Darwinism incorrect.

In this section, I do not want to compare and contrast ID with evolutionary biology but, rather, examine how ID stands on its own as a potential explanation for the complexity of biological life on Earth. If it is a viable theory, it should provide its own explanation and not simply answer evolution's failures. It is very important that a theory be capable of standing on its own merit and not simply claim to answer the holes in a competing theory.

The primary question that ID attempts to answer is the formation of the biological complexity we see in the world around us. One example of such complexity is the structure of the very large number of proteins making up biological entities. This number is unknown at present, but we do know it is very large. Each of these protein molecules comprises a particular combination of amino acids. Although molecules exhibit varying degrees of tendency toward self-assembly, the vast number of them and their associated complexity is staggering. Another example of biological complexity that ID attempts to provide an answer for is, of course, DNA. The human genome is constructed from an assembly of approximately three billion base-4 pairs. This, again, is an overwhelming measure of complexity. Let's look at how ID proposes to explain these complex systems.

An intelligent designer did it.

Don't attempt to rewind the tape; you didn't miss anything. *That is the full answer.* It should be readily apparent even at this early juncture why scientists

6.9. IS ID A VALID SCIENTIFIC THEORY?

have difficulty taking the subject of ID seriously but, to be fair, we'll go on and see if there's more. Determining if a theoretical framework for ID can be constructed, it is necessary to consider supporting evidence.

First, can we observe this process in action? This is a very sticky question that we've discussed previously. There are two ways in which a designer could interact with reality: directly, where real physical forces are applied within reality to bring about particular ends, and indirectly, through what we might call divine creation where things are set up just so-so such that they unfold contingently through time. The latter option is a preloading of the initial conditions of the universe [55]. These are really the only two options available for an intelligent designer interacting with reality. The first case deals with what we typically think of as physical interaction within reality, and the second is concerned with the underlying structure of the Universe. Stated another way, the first deals with the use of physical forces, and the second the origin of those forces.

Although thoroughly embraced by theists via miracles in the Bible, the first alternative is not formally advocated for by the ID community. As I said previously, I assume this is because there is just simply no evidence for such occurrences. The only thing that is ever observed within the physical world is its strict adherence to a set of physical laws. Not once have these rules ever been observed to be violated. I can verify Ampere's law for magnetic fields anytime I want in my own basement (to some degree of accuracy), but I cannot observe Ampere's law being broken anytime. At all. If I could, then it would not be a law. If such a thing happens, it is rare, hidden from everyone, and thus will not be enough to refute Ampere's law. Speculating about how it might be broken by a deity is simply exercising the imagination.

No matter how enthusiastic theists (or possibly even deists) would like to assert that they know that our existence is defined by such-and-such, the fact is none of us really know that for sure. The only thing we can do is use our individual senses to, in some limited way, perceive the world around us and then verify that what we perceive is the same for others around us. And there is still no guarantee that solipsism is not the case. How would we ever verify such a thing? And so the best we can do is make a few assumptions about the world around us, one of the most important being that other people see the world similarly. Therefore, for a theory to be accepted, it must be testable by everyone and verifiable by them. Events that fall outside this criterion cannot be confirmed and, therefore, should not be accepted. Imagine if this were not the case; one would be forced to believe anything. Using our collective perceptions and understanding in this way is how science deciphers the functioning of reality logically.

So how do sciences like astronomy, geology, and well, for that matter, evolution, stand up to scrutiny if they can't be fully tested by everyone, or anyone for that matter, because of the time frames involved? This argument is ubiquitously used by ID advocates and creationists. Again, a theory is a consistent framework of supporting data and hypotheses; it is typically not just a single idea. For example, consider again the theory of gravity. Newton's universal law asserts that the attractive force between two massive bodies is proportional to their masses and inversely proportional to the square of the distance between them. This has been tested and verified repeatedly, and it represents the first piece in the puzzle. Second, we have Newton's three laws of motion (see Chapter 3), which define how objects move when acted upon by a gravitational force. The first law told us how much force is asserted by massive bodies, and Newton's laws of motion then tell us how these masses move under the influence of that force. But we're not done yet. Next comes Einstein's special relativity. This theory tells us mass and energy are equivalent—the two are flip sides of a coin. As the mass of a body increases, the force it exerts upon masses in its vicinity increases; however, this force also increases if the energy of the body increases. In addition, time slows and space contracts in the local region of a massive body. Einstein's general relativity reveals that mass actually dictates the shape of spacetime. All of these properties relating to gravity have been demonstrated and verified time and again. Each one of them contributes to the overall theory and reinforces all the others. If one of them falls, they all do. This is what a theoretical framework looks like and why it is so powerful.

Evolutionary theory has a similar framework that has been built up over the years. There is an explanatory portion that includes natural selection operating upon random mutation acting as the crux of theory. However, corroborating claims and information are stacked upon this foundation, including paleontological discoveries and morphological mappings, the arrangement of the fossil record and stratification records, the history of the genetic record imprinted within the genomes of living entities, etc. The fossil record displays a consistent evolution of ancient, simpler species compared to those around us today. Stratification of these records within the rock layers places them in chronological order. An abundance of dating methods confirms the timeline of the fossil record and evolution on Earth. Further confirmation comes from the decoding of the genomes of many plants and animals, including humans, that makes clear that all lifeforms on Earth are related and descended from a common ancestor. Just as with the theory of gravity, all of these lines of evidence and more come together to support the theory of evolution. All of them depend upon one another, and all of them fall if one link in the chain fails. So far, they are consistent.

Notice that both of these theories comprise assertive statements and evidence. The theories are predictive frameworks built upon supporting evidence. They are not built using negative statements or negations. On the other hand, there are no assertive statements for ID, only negations. The primary statement is that evolutionary biology cannot explain biological complexity and, therefore, it must have been introduced by a supernatural designer. Additional arguments for ID are similar. There is no possible way the biological complexity exhibited in nature can come about by random chance in combination with natural selection. The probability of such an occurrence, although not zero, is too small; therefore, said complexity must have come about via a designer. Radiometric and carbon-14

6.9. IS ID A VALID SCIENTIFIC THEORY?

dating are fraught with errors, leading to gross age estimation errors in the fossil record; the world is actually much younger than science claims. There's no way a functioning eye or bacterial flagellum could have formed through evolutionary processes due to both being irreducibly complex. There are no transitional fossils in the record; therefore, it did not happen that way. Although evolution is readily observed on short time scales, the variants of Covid-19 being an all too close to home example, there's no way it happens cumulatively across millions of years. This is just a sampling of the arguments used for ID, but it should be readily apparent that they are characteristically different from those of scientific theories; vis-á-vis they are all negative statements.

Despite claims to the contrary by the ID community, the entire paragraph above, and additional statements I left out, represent the combination of an argument from ignorance fallacy and a false dichotomy. Claiming that something that has not been proven false must be true is an argument from ignorance fallacy. That ID is true if evolution is false is a false dichotomy. Intelligent design is not a theory but rather an attempt to negate an existing theory.

The design inference is based entirely on the idea that if the odds for a certain event are below a specific non-zero probability, then the event could have only happened by design [79]. Let's say for argument's sake that the formation of DNA is impossible through Darwinian evolution; this does not imply that it had to have been designed. This is silly. Perhaps there is a law of the Universe that we have not discovered yet that tends, under certain circumstances, to generate complex molecule chains (I suspect that law is Darwinian evolution...). Perhaps panspermia is correct, and life was started on Earth by extraterrestrial entities who themselves were formed by some unknown process. I'm not advocating either one of these alternatives, but I want to make it clear that the contention between ID and evolutionary theory does not represent a proper dichotomy: proving one false does not prove the other to be true. There are plenty of other possibilities that shouldn't be accepted either until there is sufficient evidence for them. And this is exactly why ID needs its own assertive framework and not just a refutation of Darwinism.

Let's assume there is a force in the Universe that manipulates physical reality, specifically biological entities, to produce complex information over time, and let's say that the DNA of a particular organism initially has the sequence $x = $ AGTCCCTTAGAGTT .... The designer does not change this sequence all at once—this we know, because if it did, we would see an entire organism appear at once, and that we don't observe. Instead, small changes are made to the genome at regular intervals until it reaches the intended sequence. Also, the changes do not occur sequentially in an arbitrary manner. If they did, the organism would likely die partway through its metamorphosis. No, the changes need to be made in such a way that a life-sustaining continuity is maintained until the final sequence is reached. I suspect that even the reader most devoid of a sense of humor knows I'm being a bit facetious at this point. If random positions were changed in the sequence, with the constraint that each must be viable, in transition to a desired sequence, *this looks very much like evolutionary theory.* Although the changes would appear random to an outsider, a viable trajectory would be traced through the genetic phase space, and a final, more optimized form would be arrived at.

This brings us to the second way a supernatural designer could intervene in reality. If this designer took a behind-the-scenes, deistic approach to manipulate reality, we would never know it. As in the example just given, there would be no way to tell if this was a designer-directed activity or the functioning of evolutionary biology. In either case, rationality would dictate that one go with the latter conclusion that it was caused by natural means. Occam's razor rules out a deistic designer.

Intelligent design embraces neither of these approaches for two reasons. The first is, as has just been explained, neither approach is useful for ID from a practical standpoint: the first has never been observed, and the second, by definition, can't be observed. So the approach that ID has taken is a hybrid of both. That this is the case was best illustrated by the suggestion that the designer might work through the indeterminism of quantum mechanics to address the former point. It must be understood that this point is not suggested as a matter of explanatory discourse but rather as a means of getting around the fact that we do not see the laws of physics being broken on a regular basis. In other words, since the designer obviously doesn't interfere in the world, there must be a different explanation, and the designer interacting through nondeterministic processes is a possible candidate as this alternative can never be observed or proven to exist. As Dembski says in *No Free Lunch* [80], "What are the candidates here for something in nature that is nonetheless beyond nature?"

Again this argument won't do to prove a designer exists. A designer that works in the background, hidden from everyone, cannot be proven to exist. Of course, and this is key, it can't be proven not to exist either. Therefore, there must be additional evidence, and this comes from the work in the form of specified complexity, which claims that the complexity observed in nature, for example, within DNA, not only cannot be accounted for by natural causes alone but that it originated by design. All of ID, with regard to biological complexity, hinges upon this claim.

Specified complexity, as was explained earlier, is the claim that specified events with a sufficiently low probability of occurrence must be the products of design. The ID advocate asserts this claim is falsifiable, as any good theory requires, by biologists simply showing, for example, that Darwinian evolution is capable of explaining irreducibly complex systems and specified complexity. But this is not true. Aside from erroneously pitting ID against biological evolution again, the point is that if a supernatural designer works through nondeterministic processes to effect guided evolution, even if Darwinian evolution was shown to be capable of generating the necessary complexity, it would not negate the notion of an intelligent designer that operates deistically within reality. The argument for a designer that works unseen in the world could continue. It is true, as Dembski

6.9. IS ID A VALID SCIENTIFIC THEORY?

states in *The Design Revolution* [55], that with the demonstrated success of evolutionary theory, Occam's razor should finish ID off. But this doesn't seem likely. I suspect the explanation of an unobservable, unembodied designer that operates through nondeterministic processes within nature would continue to be advocated by the ID community. It would be a weaker argument, yes, but it would not go away.

Scientific theories add to our knowledge of reality in an epistemological manner by partitioning our knowledge and understanding of the world into non-inclusive sets. A theory like that of a deistic god, that can explain anything, provides absolutely no partition, but rather encompasses the entire set. Explanatory statements are, by definition, limited. If they're not, then they apply to everything, and the result is no explanation. It is unclear how an ID theory could be tested any more than a deistic god manipulating Newton's clockwork universe from behind the scenes. From a foundational understanding, such theories cannot be tested because they simply cannot be contrasted with anything. They are not contingent upon anything. It is only through the differentiation between partitions of knowledge that proofs provide insight into other regions of our understanding. Intelligent design does not offer any toe hold to assess its functioning in terms of other known concepts, making it impossible to investigate or use in a meaningful way.

It is insightful that supernatural entities have never been shown to play a part within the functioning of reality. Of course, as was mentioned above, there is no way to disprove the presence of a deistic god working in the background to create biological complexity and even physical laws such as gravity. The rules of physics function the same no matter their claim of origin. Intelligent design takes this one step further and says that although the designer is acting in the background and is unobservable in its actions, its handiwork is nonetheless recognizable in the resultant complexity of biology. It seems that a far more reasonable conclusion would be that this complexity is a result of natural processes and that we simply have not put all the pieces in place yet to understand fully how it came about. Leaping to the alternative explanation that it was brought about through divine intervention is not a good choice.

The arguments for specified complexity being a reliable indicator of design appear not only erroneous but also misapplied. The first issue is trying to define an absolute probability bound below which we proclaim the impossibility of an event occurring. This is not a hard bound, and it cannot guarantee that such an event is impossible. In fact, it says just the opposite: although the event is improbable, it is certainly possible. The second problem is with the definition of the value of the universal probability bound, which is calculated from the approximately 10^{80} fundamental particles in the known universe, the maximum frequency at which these particles can transition states, estimated from the Planck time to be 10^{45} Hz, and a generous estimate of the age of the universe at 10^{25} seconds. Multiplying these three quantities leads to a minimum probability for any realizable event of 10^{-150} (see earlier discussion).

Calculations are then made in the book *No Free Lunch* that estimate the probability of random assembly of a bacterial flagellum. The probabilities arrived at are well below the universal bound just given, resulting in the conclusion that the formation of the flagellum was by design and not chance. The most likely explanation, however, is that both of these alternatives are wrong. Calculating the probability of the assembly of a biomechanical structure like the bacterial flagellum based solely on chance is erroneous and misleading. The proteins necessary for the construction of the flagellum are not simply created randomly from DNA templates and then randomly find their way to the proper locations. No biologist believes that cells assemble this way. Nor did the processes that accomplish these tasks come about entirely randomly.

Intelligent design advocates often use this same misleading approach to characterize the origin of the complexity within DNA. It wasn't that just one day, completely by random chance, a bunch of molecules of the appropriate varieties just happened to land in the proper locations and form the appropriate bonds to create a long string of nucleotides. This process cannot be modeled as a series of independent random variables arriving at a predetermined outcome. Virtually everyone in the scientific community would agree that can't happen. And, although the evolutionary paths can't be fully tracked at this particular time, that's okay. That is science—it continues to progress. We have only been looking at these subjects from a rational perspective for at most a thousand years—the blink of an eye in geological time.

The last issue with the approach of specified complexity is the specified part. Dembski claims this part is usually implicitly satisfied for biological systems because they perform some clear function [80]. And to be fair, he attempts to be conservative in his probability calculations in taking into account the degree of specificity observed in, for example, the bacterial flagellum. Regardless, this is a very difficult parameter to pin down. Consider visiting another planet where an entirely different form of life has evolved; how difficult would it be for any of us to point at the subsystems of an alien life form and declare that they are constructed for a clear purpose. My point is that the biological subsystems that were resident within the ancestors of today's life might well appear alien when compared with the derivatives of those subsystems today. In both cases, the subsystems served some functional purpose, but they could be vastly different from one another. We don't really have a good idea how versatile such variants are, that is, how large the specification space actually is for biological systems. Behe's irreducible complexity certainly cannot be proven as the utility space of variants is so large as to be unsearchable.

The claim that the specified complexity of biological structures is an indicator of design rests on the probability of this complexity forming naturally being so small as to declare the occurrence impossible. Note that the probability calculated in this case, even by ID advocates, is not zero. It is finite and, therefore, admits that such an occurrence is possible in principle. Now, let's also ask, what is the probability of a supernatural designer existing that is capable of inducing

6.9. IS ID A VALID SCIENTIFIC THEORY?

the formation of biological complexity [86]? Here is where ID and science diverge. How do we calculate the probability of a designer existing based on the specified complexity of that designer? It turns out that we can't perform the calculation because we *can't even formulate the problem*. This argument is along the lines of those who claim everything needs a cause *except* a god. Using the same logic applied to the formation of life, it seems that the appearance of a supernatural designer would be exceedingly rare as it is more complex than the life it is supposedly guiding into existence. Since the designer is more complicated than the complexity found in life, and since life requires a designer based on its level of complexity, the designer must also need a designer.

At this juncture, the ID community will drop the "complexity requires a designer" requirement and fall back on a special pleading fallacy, claiming that their complex intelligent designer is exceptional and does not require an explanation. This ill-formed logic goes against everything scientific. One does not get to stop the chain of investigation just because one does not like where the evidence leads. And one is not justified in applying a particular line of reasoning in one instance and then asserting it does not apply in another instance because magical reasons say so.

As has already been made clear, the evolutionary landscape is dynamic, tightly coupled, and has no goal. It is simply a large system driving toward a dynamically changing equilibrium that is dictated not only by the environment but also by the interaction between existing life forms. Even the concept of specific functionality within this environment is difficult to make sense of, let alone pinning down a particular example of specified complexity.

However, evolutionary biology provides a consistent framework to carry on this investigation partly because its claims are not as broad as ID. A simple optimizing framework driven via persistent excitation is provided to explain biological dynamics. Intelligent design goes much further, removes the need to explain complexity by shoving the work onto an invisible designer, and refuses even to touch the subject. One of the most illuminating passages in *No Free Lunch* is on page 313 and has a list of key problems that an ID research program should endeavor to investigate. The last four of these include questions about the morality of the design, the beauty of the design, the intention of the designer, and, lastly, discovering who or what the designer is. Dembski is quick to add in the next paragraph, "To be sure, the last four questions are not questions of science, [...]." I have no additional comment as these stand for themselves.

This is the beginning of the problem of introducing concepts like ID into popular scientific discourse. It opens the door for hypotheses of every kind that can't be justified and with no evidence to back them up. Dark matter is one of the foremost problems in physics at present. It appears that up to 85% of the matter within the known universe is, so far, invisible to our investigations. Dark matter is inferred by scientists from many sources of observation within the Universe, including the rotation of galaxies not adhering to Kepler's second law, gravitational lens distortion around galaxy clusters, anomalous perturbations within the cosmic background radiation, and more [87]. The question is, given that it is unknown what this additional mass might be, do we keep searching for the source of the gravitational attraction holding galaxies together and making the universe flat, or do we also attribute it to an invisible, supernatural presence? This question is not a joke. If we follow the lead from ID, it certainly makes sense to categorize dark matter as a supernatural phenomenon.

At this point, I'm just going to cut to the chase and say what is on everyone's mind: ID is a lead-in for the existence of a god, specifically the Christian God. The use of the concept of an unembodied designer, although maybe unintentional, is very close to the conventional concept of a god [55, 80]. There's a reason that concepts like this are rejected in science. Simply put, the supernatural has never been shown to exist. Let me explain again. If it exists, an observable supernatural interaction within the natural world, either its direct influence or the result of such interaction, appears to us to be part of reality. Newton's clockwork universe is true for all we know, but we cannot test for it. This is the trick that is being attempted with ID: an unobservable interaction with reality by a designer that ends up being detectable after the fact through specified biological complexity.

Again, let's use the same logic with gravity. A god is acting behind the scenes to make massive objects attract one another. All we see is a ball falling toward Earth, which we claim is proof that the god is acting behind the scenes. We can't see the god push the ball toward the Earth, but we see the result of it doing so. This argument is just as valid as that for a designer creating complex biological molecules, the interaction being invisible to us, and inferring a god did it by examining the result. Even if this is true and the rules of physics as we know them can't explain the result, we will be forced to attribute the spontaneous formation of complexity to natural law. Could it be any different? Humans are at an extreme disadvantage in that if such deities exist, we are not privy to them, only the effects of their actions. Whether derived naturally or by a god, we call these universal behaviors observed in nature physical laws. The regularly occurring, spontaneous generation of biological complexity would be no different.

A new means of slipping ID into the sciences has now been introduced which says, if the instances of intervention by a supernatural designer are not regular, but only occur at random times, presumably at the whim of the designer, then from our perspective, it would appear as if miracles are occurring: the laws of nature being suspended at random points in time, observable by us, but exhibiting no apparent cause. *And who knows? They might be miracles.* This approach takes advantage of the fact that natural laws are generated from regularity. Now, it would be very difficult to say that gravity has been tampered with by a divine entity because we have never observed an instance of rocks not falling when dropped. However, the generation of biological complexity cannot be brought about on demand like dropping a rock, therefore, it is tempting for the ID community to attempt to argue that it is not driven by natural law. This is very misleading. No one in the scientific community is claiming that complexity spontaneously occurs like gravity pulling on a rock, rather, this complexity

6.9. IS ID A VALID SCIENTIFIC THEORY?

is built upon a simpler set of natural laws *that are repeatable and consistent*.

But their argument gets even more convoluted when randomness is introduced. The claim is that it would be possible for stochastic processes to hide miracles (see Section 6.7 of this chapter). The designer could be using the inherent indeterminism of nature to hide his miraculous works. Very convenient. This kind of behavior of the designer is very similar to his making of the fossil and DNA records appear just as if evolution did occur. But, if we assume that the designer's intent is once again to hide his actions, how do we approach this problem from a scientific perspective?

As described previously in this chapter, a reverse-biased PN junction appears as an insurmountable energy barrier to current flow. Again, it's like the wall in a game of racquetball that the ball can't go through. If the ball possessed sufficient energy, perhaps by traveling at 100,000 m/s, it could breach the wall, but the fact is that a human player cannot approach these energy levels even with the best swing. In the same way, the electric field generated across the junction of the diode possesses an energy far above that of any electron trying to cross the junction. It is just as impossible for the electron to cross the barrier as it is for the ball to go through the wall—*and yet the electrons do just that on occasion*. Particle tunneling through energy barriers is very common in the microscopic world. The important thing here is how this phenomenon was assimilated into physics. These events were not regarded as miracles and were brought under the umbrella of physical law using statistical techniques. On the contrary, ID claims biological complexity is the result of a miracle.

The quantum-mechanical process describing tunneling is not deterministic, but that's okay. It is simply characterized as a stochastic process. I argue that the same would occur for the spontaneous formation of biological complexity if it showed up randomly and could not be explained deterministically. The other option would be to take the route that ID has and regard events for which we do not yet have a full explanation as miraculous. The problem is, at that point, scientific inquiry will stop. Divine answers are final answers by definition.

In this section, I did not intend to systematically go through all the standard characteristics of a good scientific theory and assert that ID has not lived up to them. That is boring, and I don't believe things are ever that clear-cut. Instead, I have tried in a more colloquial manner to show how ID fails any logical means of seeking the truth in this reality. It does so for two primary reasons. The first is that it is a kind of giving up: since we can't yet explain what is happening in biology, let's just say a god or supernatural designer did it and move on. Every single time this kind of thinking has been put forth in the past, it has failed, and science has picked up the baton and solved the problem. The second reason is that ID favors the Christian God as the explanation. It is impossible not to pick up on this by reading the literature on the subject. If one cannot filter their preconceptions and preferences out during the course of one's work, then it is inevitable that the work will become tainted, and it seems, intentional or not, this has been the case with the development of ID.

Science, by definition, only involves the investigation of the reality which we all share. It cannot do otherwise. If there have been prior examples of science successfully applied to so-called supernatural subjects, it is only because science moved these supposed supernatural subjects into the natural realm. If a designer god is to be investigated within the realm of science, that god will necessarily need to come under the umbrella of the natural world. If one doesn't like the word "natural", then a different term can be chosen. The term "reality" might be a better choice. Regardless, as soon as this god is extant within reality, science will begin to characterize it. It will become a part of reality, and its machinations will be assimilated into physical law.

This is the nature of investigation. A designer who stands outside of space and time doesn't exist to us until that designer makes an incursion into space and time. It could not be otherwise. We simply can't infer the existence of something that doesn't exist in this reality. This topic is confusing because it is a non sequitur. If something doesn't exist in reality, then it doesn't exist. I can't make this more clear than I have.

6.10 The Anthropic Principle and fine-tuning

The first time I read about the anthropic principle and fine-tuning, I was probably about fifteen years old, and I found the concept quite intriguing. It was not until much later—ten years or so—that I would hear about fine-tuning again, this time in a completely different context. It would be commandeered by ID proponents and religious advocates to make claims that the Universe is fine-tuned by a designer for the existence of human life. This notion is entirely contrary to the original idea of the anthropic principle. Since it has been and still is such a popular concept in ID arguments, I wanted to cover it here with the intent of dispelling the idea that it is an argument for the Universe having been designed.

Carter's original form of the anthropic principle was stated in two parts as follows [88]:

1. We must be prepared to take account of the fact that our location in the Universe is necessarily privileged to the extent of being compatible with our existence as observers.

2. The Universe (and hence the fundamental parameters on which it depends) must be as to admit the creation of observers within it at some stage.

The first statement is called the *weak anthropic principle* and basically says that we observe the Universe as it is because we are here to observe it. Conversely, if the Universe were not conducive to the development of intelligent life, we would not be here to take note of it. This statement has a number of implications, but that list does not include the inference that the Universe was designed explicitly for the existence of intelligent life. Human life formed in the Universe simply

6.10. THE ANTHROPIC PRINCIPLE AND FINE-TUNING

because it could, not because the Universe was designed for it. This is a subtle distinction.

The second statement is called the *strong anthropic principle*. It makes the stronger claim that the Universe must have properties that allow the emergence of intelligent observers at some point in its existence. Whereas the weak version only claims that if life shows up in the Universe, then the Universe must be such that this can occur, the strong version asserts the requirement that the Universe, at some point in its existence, is conducive to the development of intelligent life. The weak version is so apparent that no argument is needed. However, the strong version generates a great deal of controversy because there appears to be no evidence for an absolute requirement that the Universe be suitable for life as we know it or any other version of life for that matter.

The weak anthropic principle can actually be useful to science as it is a statement of fact: the particular universe in which we find ourselves must possess properties that allow life such as ours to exist. For example, the fact that we are carbon-based lifeforms means that the Universe must be capable of producing carbon. This point led theoretical physicist Fred Hoyle to make the audacious prediction in 1954 that since the Universe must create carbon-12 for life as we know it to exist, there must be an undiscovered variety of carbon with an energy of 7.65 MeV that is formed within dying stars [89]. It is only through this exotic state of carbon that stable carbon-12 can be formed, along with many of the other lighter elements such as oxygen, nitrogen, and neon. Five years after Hoyle's prediction, the emission lines for this strange form of carbon were found in starlight. Hoyle's prediction of the stellar process that leads to stable carbon-12 is a perfect example of using the weak anthropic principle to make testable claims about the Universe.

In contrast, the strong version makes no testable predictions. How would we ever verify that the Universe had to evolve to allow the development of intelligent life at least once during its existence? It is difficult to fathom the reasoning behind this claim, although several prominent scientists have tried. The strong anthropic principle will not be considered any further here, and the term anthropic principle will be synonymous with the weak version from this point forward.

If the Universe had different parameters, it seems not too far-fetched that life far different than we are familiar with could have evolved to fit these parameters. This does not mean that life could form within any old kind of universe; the parameters of the Universe still had to satisfy certain constraints. There are, it seems, limits on how different the Universe could have been and still allowed life of any kind to form. For example, one might conjecture that a universe in which matter does not form would not support life.

The various mathematical models of the Universe employed by physicists are parameterized by many physical constants such as the speed of light, cosmological constant, the masses of electrons and protons, gravitational constant, etc. As many physicists, including Paul Davies, Steven Weinberg, and Stephen Hawking, have pointed out, the measured values of these constants appear crucial for the formation of life such as ours in the Universe. Even minor deviations in any one of them would result in a Universe that would be devoid of life as we know it. The notion that the fundamental physical constants conspire to high degrees of fidelity to allow for human life to exist is called the *fine-tuning problem*. It argues that if any one of the nearly two dozen constants deviated even slightly from their present values, the result would be a universe very inhospitable to life.

Some changes would only rule out carbon-based life, such as human beings. Other changes would rule out the formation of stars and planets and, hence, would seemingly deny life altogether. Of course, there is no way that we, in our nascent understanding of life and reality, can answer this question with any concrete conviction, but the idea seems plausible to many scientists.

The argument for fine-tuning of the Universe usually begins by taking one of the almost two-dozen physical constants, independently adjusting its value, and estimating the effect upon the Universe at large using a mathematical model. For example, making the gravitation constant smaller might mean that the particles within the cores of stars would be less compressed, resulting in the heavier elements never being created. Conversely, if the gravitation constant were a bit larger, the stars would burn much hotter due to the increase in central pressure, resulting in stellar lifetimes too short for the emergence of intelligent life. That doing things this way results in an accurate picture of reality is a very big "if." The independence of the model parameters must be assumed from the outset, which may or may not be the case.

We need to enter some very subtle waters here to get to the heart of this matter. For instance, consider the claim made by Brandon Carter, as presented in Robin Collins's paper *The fine-tuning design argument* [90], that if the force of gravity had been stronger or weaker by one part in 10^{40}, stars like the Sun would never have been possible. The assumption here is that the force of gravity can change independently of all other universal parameters. Furthermore, before we can ever get to the question of the independence of parameters, the question of what it even means to change the fundamental physical constants must be addressed.

Engineers often make use of dynamic simulation to verify or optimize designs. The simulation of a dynamic system using mathematical software such as Octave or Scilab is accomplished by integrating the differential equations describing the system using computer code. These equations are discretized and then iterated in time to arrive at the system's dynamic behavior. It is important to note that the system equations are descriptive and reductionist in their nature and include the likes of Newton's laws of motion, Maxwell's equations, semiconductor models, thermal equations, etc. They are descriptive in the sense that they are derived from observations of the functioning of the natural world. For example, Newton's second law of motion relating acceleration and force may be used to describe the motion of the shaft of an electric motor. Keep in mind that we do not know what causes Newton's second law to hold true. It is difficult even to ask

6.10. THE ANTHROPIC PRINCIPLE AND FINE-TUNING

the question. We just know that it accurately describes the behavior of any mass with a force applied to it. Similarly, we do not understand why the universal gravitational constant has the value it does. We just know from observation that it has a particular value.

The physical laws from which the differential equations of a simulation are built are reductionist in that they describe overall, average behaviors. When calculating the force of a massive body upon a second massive body, some strong assumptions are made such that the calculations are tractable. For example, one uses point masses and takes advantage of symmetry to coax the problem into a solvable form. Simulations such as that described above certainly do not consider the interactions between the individual particles themselves; instead, they rely on the emergent behavior of particle ensembles.

The models of the Universe are much like those used to simulate a dynamic system on a PC. They are constructed in precisely the same way. (The models need not be constructed exclusively of differential equations.) Moreover, just as it is easy to tweak parameters within the electric motor simulation, it seems it would be tempting to adjust the parameters of the Universe to infer how its behavior could have been different. The fine-tuning argument is based on the assumption that this tweaking of parameters is meaningful. Whether it actually is meaningful is the question at hand.

Consider a dynamic system described by the state equations in (6.3). This system has three equations parameterized by the constants a_1, a_2, and a_3. Given that the constants are specified, we can integrate this system with respect to time and inspect its dynamics. We can repeatedly modify the constants while running the simulation to observe the impact of changes in the constants on the system dynamics. This is an entirely legitimate effort as the constants are independent of one another.

$$\begin{aligned} \dot{x}_1 &= f_1(\mathbf{x}, a_1) \\ \dot{x}_2 &= f_2(\mathbf{x}, a_2) \\ \dot{x}_3 &= f_3(\mathbf{x}, a_3) \end{aligned} \qquad (6.3)$$

Now, assume the constants are not independent and are constrained as in equation (6.4). In this case, we cannot arbitrarily set all three constants to any value we desire. If we change the value of a_1, the values of the remaining constants shift with it as dictated by the constraint equation.

$$g_1(a_1) + g_2(a_2) + g_3(a_3) = k \qquad (6.4)$$

This situation arises all the time in the simulation of dynamic systems and is the point I am trying to make about the fine-tuning argument: we have no idea whether the fundamental physical constants are independent of one another. How could we? There is no way we can modify any of the constants and check. When we modify them in our theoretical models, we are simply observing the model's behavior, not physical reality.

Remember, the physical laws and constants are descriptive in their form: they are determined empirically through observation. The Universe is not a simulation running within reality; *it is reality*. Changing the parameters of a model of reality and expecting a valid result would imply that these parameters are not descriptive. This would mean that we know where they come from and could explain them, which we cannot—that is why they are in the equations to begin with, because we cannot thoroughly define the fundamental, underlying processes of reality. This is the formalism of reduction. The physical constants that emerge in a model are indicators that the model does not truly represent reality; that our understanding is lacking. As the model evolves to become a more accurate representation of reality, the constants go away or are reduced in number. Therefore, it is difficult to understand what it even means to say that any one of these constants could be independently tweaked.

As a crude example, consider a car engine that provides a torque T_e to the wheels of radius r, which are turning with an angular speed of ω. The diagram of the wheel is shown in Figure 6.10 (only one wheel is used for simplicity). The force applied to the road is, of course, the torque supplied to the wheel divided by its radius, $F_w = T_e/r$. The wheel's angular speed is ω, meaning the vehicle speed is $v_w = r\omega$. Hence, we have the two equations

$$F_r = \frac{T_e}{r}$$
$$v_w = r\omega. \tag{6.5}$$

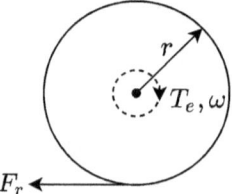

Figure 6.10: A wheel of radius r with torque T_e applied to the shaft and spinning at an angular speed of ω. A force of F_r is applied to the road.

Let's rewrite these as follows as if we did not know how the torque and speed scaled with r:

6.10. THE ANTHROPIC PRINCIPLE AND FINE-TUNING

$$F_r = k_1 T_e$$
$$v_w = k_2 \omega. \tag{6.6}$$

We now have two fundamental constants to describe the wheel's motion. The first, k_1, translates the wheel torque to the force applied by the wheel on the road, and the second, k_2, relates the wheel's angular speed to the linear velocity of the car. If we assume the two constants are independent, we might also assume that we can double the power (power is torque multiplied by angular speed) by doubling the torque constant k_1. This seems perfectly legitimate when examining equations (6.6). How would we know any better if all we had were these equations? However, in this case, we are privy to the underlying structure of the problem.

In examining the original equations in (6.5) it is easy to see that $k_1 = 1/k_2$; one cannot be changed without changing the other. Doubling the torque constant k_1 will not double the power delivered to the wheel because the wheel speed will be cut in half due to k_2 being halved.

Now, I am not saying that the cosmological models are this simple. I am certain they are not. But I am saying that a reduced model with parametric coefficients does not tell the whole story, and there is no guarantee that the coefficients are independent. Consequently, when someone adjusts the gravitational constant in an insular way and asserts that the Universe could have actually formed that way with the attending ramifications, it seems the validity of the claim is potentially questionable.

The next step in the ID fine-tuning argument is to calculate realizable bounds for the physical constants. For example, Collins says that if the strength of the initial expansion rate of the big bang had differed by as little as one part in 10^{60}, the Universe would have either collapsed back on itself or expanded too quickly for the formation of stars [90]. This single number tells us nothing about the probability of the expansion rate of a newly forming Universe falling within the acceptable range. Perhaps the possible variation was only 0.5 parts in 10^{60}, then the odds of the Universe being within the required bound is 100%. There is simply not enough information to calculate probabilities here. Despite this fact, ID proponents continually cite astronomically small numbers and claim they indicate a marksman designer of consummate skill.

For example, consider an excerpt of a rebuttal of the weak anthropic principle by William Lane Craig from his Reasonable Faith website [91]:

> Let me use one that John Leslie, a Canadian philosopher, has given who has specialized in these fine-tuning arguments. He imagines your being dragged in front of a firing squad of one hundred trained marksman all with rifles aimed at your heart to be executed. You hear the command given: ready, aim, fire! And you hear the deafening roar the guns. Then you observe that you're still alive—that all of the one hundred marksmen missed! While

> you would be perfectly justified in saying, I shouldn't be surprised that I don't observe that I'm dead, because after all, if you were dead you couldn't observe it; you should nevertheless be extremely surprised that you do observe that you are alive in view of the enormous improbability of the one hundred marksmen all missing! In exactly the same way with respect to the universe, we shouldn't be surprised that we don't observe conditions of the universe incompatible with our existence because if they were incompatible we couldn't be here to observe it. But it doesn't in any way follow from that we shouldn't be surprised that we do observe initial conditions of the universe which are compatible with our existence in view of the enormous improbability involved. So this argument that is given by these Anthropic theorizers is simply a fallacious argument.

This is a false analogy. My only thought on the matter is that I wish he had calculated the "enormous improbability" involved in getting the Universe's initial conditions and/or physical constants correct. Going back to the strong nuclear force example: it was asserted that this force is allowed to vary by up to 5% and still be acceptable for the formation of life. But over what range can it actually vary? Infinity? If so, the probability of hitting the desired range is zero—0.05 divided by infinity—and indeed, it must have been divinely tweaked. But the point here is that we don't know over what range it could vary. Making assertions about probabilities in this way is what Sam Harris would call playing tennis without the net.

For the sake of argument, let's assume that the formation of a universe with these particular constants is very improbable. We can also grant the ID design hypothesis. What then is the probability of a designer existing to tune said parameters? I have not seen this calculation either. Based on what we can observe and intuit from the physical world in which we reside, the likelihood of the existence of such a designer is very slim indeed, for not a shred of evidence of a divine designer has ever been observed. With sufficient data, it might be easy to accept that the configuration of our universe is highly improbable. Still, this conclusion would in no way make a god any more likely.

Suppose we accept that the Universe's configuration is highly unlikely to form naturally. In that case, we are left with an improbable Universe on the one hand and a very improbable, magically designed Universe on the other. We have reached a point where this entire argument is just silly. We have two competing probabilities, neither of which we can even begin to compute, thus making the entire discussion moot. I will only say that we have seen vast evidence for a Universe forming naturally, but we have seen zero physical evidence for a god. Although I do not believe there is enough evidence on either side to answer the fine-tuning question—which is precisely the problem—I know which side my money is on.

6.11 Is there a supernatural?

I have talked a great deal about what a theory actually says about physical reality. As a change of pace, let's consider what a theory doesn't tell us.

Back millions of years ago, before modern humans came on the scene, our ancestors undoubtedly used the properties of nature, including gravity, forces, inertia, speed, position, etc., to hunt their prey and live their lives in general. There may indeed have been no understanding at all about the physical meaning of these quantities. For example, they may have used observations of their proximity to prey or predators to respond appropriately. However, these were likely just simple, reactive actions based on instinct and very simple analysis—massive objects thrown that impacted prey usually led to the prey's demise. Sharp objects, when applied with force to predator or prey, typically resulted in death. Falling off a cliff was bad. This knowledge was likely used every day.

It was probably not until later that we came to separate these phenomena from ourselves and observe them in their own right. Rocks have positions associated with them. If we see a rock on a hill today, and it is not disturbed, it will still be there when we come back tomorrow, leading to the notion that external things have their own physical reality. If a rock rolls down a hill, its motion is determined by some force independent of the person watching it. The ability of human beings to remove themselves from the picture is the precursor to objectivism and science.

Once this step has been taken, people can observe the Universe and start noticing regularities or patterns within it. Again, there have been many independent measurements and confirmations of the square law of gravity, so many that the descriptive behavior of massive objects became a physical law. A virtually uncountable number of these patterns have been codified as theories or laws over the centuries.

All science is built upon observable patterns in the Universe. It is important to understand that there is no meaning ascribed to the cause of these regularities. Theories can be, in a sense, superseded by other theories. For example, Newton's laws of gravity and motion eventually became a special case or approximation of Einstein's theories of relativity. However, there is no reason for the Universe to act in these ways; it just does. For instance, if you have a habit of driving to the local theater every Friday night after work to catch the latest film, associated with this behavior would be both a pattern of driving to the cinema and a reason for this behavior (that is because you enjoy watching movies). Physical patterns in nature only have the former.

We take the patterns we see in nature and record them as laws. These laws are very useful for predictions of how reality will behave under all kinds of circumstances. However, the laws do not tell us why the Universe behaves the way it does. Sometimes scientists slip up and use colloquial language. For instance, a professor of physics might say that the reason massive objects fall to the Earth's surface is gravity. Gravity is not the reason. Gravity is the description of massive objects' tendencies to be accelerated toward one another.

Whenever we discover reality exhibiting a new behavior that we were not previously aware of, we want to characterize it, experiment with it, and form a theoretical description of it. All of our laws were derived in this way. And, if these laws vary, we don't immediately jump to the conclusion that there is something fishy going on with reality. Rather, we assume that we have not seen the entire picture of what is going on and attempt to dig deeper to discover what we missed. For example, when Ernest Rutherford, working with radioactivity in the early 19th century, noted that the intensity of particles emitted by a radioactive material decreased with time, the assumption was not that physics was broken. The actual reason is that once an atom decays, it can't decay a second time. Therefore, there are fewer particles left to decay, leading to the rate of atoms decaying decreasing with time.

It appears that all patterns of phenomena are encapsulated in nature and waiting to be discovered. So, what would it mean to discover something supernatural? And how would we recognize a supernatural thing?

Suppose a new force or phenomenon that goes against what we already know about reality comes to our attention. In that case, we do not (or most of us don't, I would hope) jump to the conclusion that something supernatural is happening. For example, in the latter half of the 90s, a team of astronomers made up of Saul Perlmutter, Brian Schmidt, and Adam Riess discovered, contrary to general belief at the time, that the Universe's expansion from the Big Bang is not slowing down but, rather, is speeding up. This discovery earned them a Nobel Prize thirteen years later.

This discovery seemed outlandish. How could the Universe's expansion be accelerating against its own gravitational pull? What force would be driving it to do so? Well, even today, there is no answer to that question. The unknown culprit has come to be known as *dark energy*. (The adjective "dark" is not used to suggest some evil intent. Instead, it means unseen or undiscovered.) Apparently, there is a so far unobservable energy permeating the Universe that leads to its inflation. Now, and here is one of the main questions of this section: should the reason for the accelerating expansion be regarded as supernatural? Things get a bit subtle here.

Let's jump back to gravity for just a moment where we may get a bit of help in sorting this issue out. Do we know the cause of things falling to the Earth? Not really. We just know that they do. We even know at what rate they fall. However, we do not have a reason as to why they fall. The natural laws are descriptive, not prescriptive; although one could say that relativity explains Newtonian physics, there is no explanation for relativity.

Any phenomenon we observe in nature, be it gravity or dark energy, is destined to be incorporated into the laws of physics. There will be no meaning behind the laws—they will simply be descriptions of the functioning of reality.

What then differentiates the natural from the supernatural? It seems the supernatural must be defined as an action, rule, or cause that lies outside the

6.11. IS THERE A SUPERNATURAL?

reality we are aware of. It is difficult to conceive of an example of the supernatural where this would not be the case. If the cause exists in the reality we observe, we will search for the origin of the cause. If there is a rule that prescribes in some way how this Universe behaves, we will observe its consequences and formulate a description of it. If an action actually has an effect in this reality, we will measure that effect and work backward toward the source. All of these things we will assign to the category of natural. Therefore, the supernatural must lie outside this set of phenomena.

Even if there were a magical being that somehow resided outside the time and space of our Universe, we would have two problems to solve. The first is what exactly it means to reside outside of space and time. I suppose this is possible despite the fact that we can't easily reconcile it. I can certainly write down a set of equations that do not involve space or time, although the context they reside in is necessarily contrived. The second problem, and it is unavoidable, is how to describe the magical being and find its origin. Immediately we would lump it into the natural and look for its explanation, as we rightly should.

If aliens that had a billion years of evolution on us visited Earth, what would our perception of them be? As Arthur C. Clarke said in his third law, "Any sufficiently advanced technology is indistinguishable from magic." Would we call them gods and proclaim that they stand outside our space-time reality? I would hope not. Any scientist who has been studying reality long enough knows that for every question she answers, there are a dozen more generated. In other words, reality tends to remain a mystery. Hopefully, we have learned this lesson to the extent that we favor the conclusion of "I don't know" in lieu of "it must be the work of a god".

In every case throughout history where we have assumed the explanation was supernatural or a god, science has demonstrated this not to be the case. And it has done so persuasively enough that even religions come to believe the results. Despite Galileo's troubles, the entirety of the modern world believes that the Earth revolves around the Sun, and no one thinks that believing so is disrespectful to a god. Every time some new mystery comes up, there is a tendency for those like the ID movement to claim that this time the situation is different and that maybe we really need a supernatural explanation. It never turns out that way, however.

All observable phenomena will come to be attributed to the natural world, rendering the supernatural nonexistent and nonsensical. But there is one avenue that is still open, and theological philosophers and ID proponents often take it. It makes use of the claim that the supernatural is undetectable as we've already discussed. At first glance, this may seem to have some merit, but it only seems so to people who have not considered the concept in any detail.

First, let's pin down the characteristics of such a supernatural entity, and I will bring in descriptions from the theological side to do so. This seems appropriate as they are the majority who make these kinds of claims.

An entity (or god) like this would firstly be immaterial. It would not be composed of matter or energy. Second, it would exist outside of space and time. These are the characteristics of something that does not exist. If I were to ask you the attributes of something that does not exist, your first thought would probably be along the lines of nothing: no matter, no energy, no volume, and no change. I'm not trying to be facetious. Rather, I'm pointing out something subtle, hoping it won't be missed. The definitions used for the supernatural, at least regarding gods or divine designers, are descriptions of nothing. Now, why would this be the case?

It certainly would be possible for someone to say that the real reason massive objects fall to the Earth when dropped is that a deity pushes them down with his hand. Of course, no one can see or detect him, and he is omnipresent and is always there to enforce the rule of gravity. How would this change our views of reality? It wouldn't. The reason is that there is absolutely no evidence, by definition, for this deity. In fact, everyone could make up their own version of a deity or god that governs the physical laws but is not detectable. And, well, that is exactly what has happened.

How could there possibly be thousands of versions of religion worldwide with all of them in conflict with one another? There should be some way to sort this situation out, but there isn't because these gods are completely unobservable, and their supposed influence upon the world is entirely indistinguishable from nature.

The supernatural, by definition, does not exist. If a phenomenon of reality is found and measured, it is automatically lumped into the category of the natural. In fact, the primary characteristic of a natural phenomenon is that it manifests in reality and is observable. Supernatural events, by definition, do not affect reality and are not observable and, hence, for all intents and purposes, for we entities in this reality, such events do not exist.

6.12 Is intelligent design unfairly rejected?

Imagine you are driving down the road one night heading home after taking in a movie or going out to eat with friends, and suddenly you hear a loud *thump-thump-thump* emanating from the left front wheel of the car, and the vehicle becomes difficult to steer. You quickly slow down and fight the steering to get the vehicle to the edge of the road. You turn on the hazard lights and get out to look at the wheel. Nothing immediately jumps out as a cause of the noise and the steering problem. However, you notice that the wheel is turned hard to the left. You quickly walk around and look at the passenger-side wheel, and it is pointing straight ahead. Reaching beside the cockeyed wheel, you find what you are looking for: a broken tie rod (the tie rod links the wheel steering assembly to the steering actuator). Since the wheel was not held straight without a tie rod, the tire was free to wobble any way it wanted.

What could have caused this mechanical defect? You call a mechanic, and he shows up in about 20 minutes with a tow truck and quickly agrees that it is

6.12. IS INTELLIGENT DESIGN UNFAIRLY REJECTED?

a busted tie rod. You inquire what would have caused the tie rod to break, and the mechanic is quick to reply, "Gremlins are likely the cause."

You step a little closer and ensure that you do not smell alcohol on his breath and then laugh because you realize he's joking. However, he just stands and soberly looks you in the eye, prompting you to ask if he's joking.

"Nope. Gremlins are my guess," he answers.

Incredulous, you ask, "Wouldn't fatigue and rust of the metal be more likely?"

"A lot of people think that," the mechanic replies. "And we do know how oxidation of the steel occurs, leading to a reduction in the tensile strength. The guys down at the shop agree that the covalent bonding of oxygen with the metal is a good first candidate for this weakening. However, we can't quite pin down the physics underlying the fundamental particles that take part in the chemical interactions. It would be simplistic and barbaric to talk about charged electrons and protons without knowing exactly the fundamental structure of the quarks that make them up."

You're done talking at this point and simply climb into the passenger seat of the tow truck to get a ride into town.

This is a silly story, but it illustrates the nature of the ID argument as a whole.

There is a reason why the scientific community does not buy the notions of gremlins, demons, gods, angels, prophecy, the significance of astrology, good luck charms, the effectiveness of prayer, the supernatural, the existence of Santa Claus, or that perpetual motion machines exist and the guy in that Youtube video could really provide the world with free power using only a battery, an induction motor, and that thing in the little black box if the power company would only not shut him down. And science certainly doesn't entertain theories about processes that are not detectable, such as non-intervening gods. If one can put away all presuppositional concepts for a moment and just absorb this notion, its ridiculousness will sink in. *Science does not investigate things that do not exist.* And it's not me saying that an undetectable intelligence doesn't exist—it's right there in the description of a supernatural designer from the ID community: an intelligence whose actions are indistinguishable from chance or the perfunctory machinations of the natural world.

In his book *The Design Revolution*, William Dembski states, "Skepticism has a standard bag of tricks for keeping a gullible public in line. I want to lay out here why that bag of tricks is unlikely to succeed against intelligent design." I was a little surprised when I read this line. It seems difficult to believe that scientists, in general, have an ulterior motive to lead the general public to disbelieve accurate information. It seems utterly ludicrous that a group of scientists know that life was really created magically, and they deny it by making up stories *that actually do make sense.*

There are multiple reasons the scientific community rejects ID. First, the entirety of its argument is that evolution can't explain biological complexity. Although proponents will claim that ID is more comprehensive than this, it is not. There are no other claims made, especially none that argue positively for ID. The fact is that no positive arguments can be made. How does one argue for a mechanism that can't be found, that, by definition, is indistinguishable from natural physical laws? As we talked about previously, ID is intentionally framed this way. If there were evidence in reality for it, we could examine said evidence, but there is none. Therefore, ID enters through the back door by claiming that it acts through an undetectable means such as skewing quantum probabilistic distributions—modifying reality via information—to softly and quietly (and indiscernibly) bring about change.

We don't buy this kind of logic in other aspects of life or science; why would we buy it here? We don't make such silly arguments like magical fairies actually lift planes off the ground; you cannot see them, and there's no way to detect them, and they do the same thing that aerodynamic lift does, but they are there. Occam's razor is intended to eviscerate just these types of claims. And it's easy to see why. One shudders to think what else would be allowed if we entertained such nonsense. It amounts to opening the door to religious war and policymaking in the name of a divine being who can't be seen, can't be detected, and has no observable interaction with reality. One does not have to stare too far over the horizon to envisage this outcome. After all, it wouldn't be the first time it has happened.

There is much lamentation in the ID community about mainstream science totally refusing to accept anything outside material reality. It blatantly rejects the supernatural. Many people who argue for a supernatural element seem to have not really thought much about what that would mean. It is a completely confusing concept because it is self-contradictory. Anything that interacts with reality is, by definition, part of the natural. Splitting reality apart and calling one portion natural and another supernatural is not only redundant but also confusing. On the other hand, something outside of reality that does not interact with the natural world does not exist for all intents and purposes. We cannot, by definition, tell the difference between its insularity and its nonexistence.

The scientific community is not prejudicial to ID due to some conspiratorial dogma. Science just doesn't investigate things that do not exist.

Let's assume for argument's sake that there was some *measurable* evidence that nature was using what we would call intelligence to guide the generation of biological structures. Assume that cells had what appeared to be a serial number on their sides (this is an alternate version of one of Dembski's thought experiments), a series of prime numbers were encoded in the so-called junk DNA of the cells, and it was explicitly shown that some complex feature of the cell was unreachable through the evolutionary process (irreducibly complex). Would science then concede an intelligent designer of the type the ID movement is considering? I suspect and hope the answer would be a resounding "no."

The first conclusion would be and should be that the cause for this display of intelligence resides within the natural. Concluding that the reason is something supernatural leads to a dead end and, as argued above, to complete confusion.

Much to the consternation of ID proponents, *panspermia* (panspermia is the theory that life was seeded on Earth from extraterrestrial sources) or something similar could be accepted as a reasonable explanation. The reason is that it is a positive, testable hypothesis that resides within this reality. Maybe the extraterrestrials that delivered life to Earth are many light-years away, and perhaps we never develop faster-than-light travel enabling us to test the theory, but at least in concept, it is testable.

The last point that doesn't score any relationship points for ID from the scientific community is that there really does seem to be an ulterior motive behind the ID movement. The Dover trial provided direct evidence of this kind of subterfuge. It is very clear that separating ID from the beliefs of Christianity is virtually impossible. It is also quite apparent from the ID arguments there is a belief that science has a dogma and a desire to maintain the status quo such that ID proponents believe they, at least to some degree, are in an argument of opinions with the scientific community. This is very telling. I think the scientific community respects anyone's personal beliefs on faith; I just don't think they care about such *opinions* when it comes to actually doing science. This is a good thing.

I believe if ID presented just one cogent, coherent and positive argument for the existence of an intelligent designer influencing the emergence of biological complexity, they would get their proverbial foot in the door of science. But to do this, the arguments would need to admit to scrutiny and test. And if that were the case, I don't think it would take long for science to sort things out and reject them with an explanation. I believe this is the real problem and why ID stands on the periphery of science.

6.13 Conclusion

It is one of the hallmarks of scientific research demonstrated throughout the years that nature always follows an elegant path. Nature is not half-assed in its functioning. Of course, there are many times, when first coming to understand how a particular part of nature functions, that one wonders why it would ever be that way. Later, as the bigger picture develops in the mind, the beauty and elegance become apparent. I'm not saying that an aesthetic argument like this holds any water—it doesn't—but there is no way that I believe that nature would admit an explanation like ID. It is far more imaginative than that. And this has become readily apparent as science learns more about how life formed on Earth.

The main problem with ID is that it proposes an unfalsifiable hypothesis with no supporting evidence. One might argue that the low probability of the unguided formation of the genome or the unintelligent construction of complex biological machinery is evidence that these components could not have formed naturally. However, there is no way to calculate these probabilities without fully understanding the underlying processes by which they do form. If we were to simply calculate the probability of a snowflake forming by random-chance positioning of the molecules that make up the snowflake, we would assume snowflakes never form. However, add in the simple, mindless rules that dictate the motion of these particles, and the formation of snowflakes becomes almost inevitable. Biologists who work on the problem of biological complexity for a living do not know how to calculate these probabilities, and the ID community certainly doesn't know either.

Discovering what appear to be impossibilities in nature is the impetus for further investigation. It is the fun part: we don't know, so we inquire. No answer is ever known before it is investigated. This is the very definition of inquiry—answers must be sought or they will not be found. The ID community is confused about this fact because they believe they already have the answers. They would shut down the objective consensus that science demands of us as a group, and they would provide the answers to all of us. This is a dangerous road.

Contrary to the sentiment of ID proponents, conclusively proving an evolutionary pathway to biological complexity would not falsify ID. Intelligent design and evolutionary theory do not comprise a true dichotomy. In fact, I would put money on the prediction that when evolutionary theory is shown to be absolutely the case, ID proponents will fall back on divinely guided evolution. This is the problem: ID is unfalsifiable. One can just as easily argue that an undetectable divine authority mediates gravity and the other physical forces, and this will also be unfalsifiable.

The requirement that a theory is falsifiable is not just so the theory can be sorted out from competing theories. It is more fundamental than that. If a theory is unfalsifiable, then that theory does not have, cannot have, any explanatory power. It doesn't say anything new. This should be obvious: a theory that can't be tested in reality does not affect reality. If it did, it could be tested. And this is exactly what ID has done, create a theory that doesn't do anything. That a theory is not falsifiable is a sure sign someone is trying to prove the existence of an immaterial idea or concept instead of something concrete in reality. A theory must be falsifiable in order to correspond to something real.

So far, ID has provided no positive demonstration of anything useful in describing the origins of life or its diversity on Earth. In *The Design Revolution*, Dembski states that explanations such as these are not included in ID's burden of proof, but they are required to be explained in detail by evolutionary theory. That is very convenient and, again, is precisely why ID is not taken seriously by the scientific community.

I hope it is not the case, but when considering the coupling of the ridiculous nature of this logic and the "Wedge" document described in the opening of this chapter, one is led to suspect that intelligent individuals like those working on ID, would only go down that path for ulterior motives. That's too bad if it's the case.

I understand how it can be scary not to believe in something. I have been there. But science has to be divorced from fear or the desire for reality to be a certain way. The imaginary leads to conflict because everyone has their own

6.13. CONCLUSION

imagination, and they are all different. Somehow we have to get past placing our beliefs in front of objective reality. There was probably an evolutionary imperative at some point for us to have that ability, but that time is past. If we do not rely on rationality instead of our subjective fears and beliefs, things will not get better in this world. The ID movement believes materialism is the cause for a breakdown in morality (and it is certainly questionable whether a moral breakdown even exists), but I contend that violence will undoubtedly increase exponentially if we leave behind objectivism and fall back to relying on unfounded belief. We cannot just make up imaginary gods and supernatural designers that have no impact in reality because everyone will tend to make up their own versions, and if history has taught us anything, these versions will not get along.

It was very encouraging to see the parents in the Kitzmiller case hold the school board accountable and prevent the degradation of science in public schools. It is clear that the ID movement will not stop with the introduction of a supernatural designer into science. It is also clear that designer is the Christian God. And it is clear that, if they can, the ID movement will reintroduce religion into every aspect of our lives.

Chapter 7

The energy basis of life

> "Nature does not hurry, yet everything is accomplished.", Lao Tzu.

When many non-technical people think of energy, they have images in their minds of science fiction starships firing bright, neon energy bolts in a galactic battle, or superheroes developing devastating blasts of light from their hands or eyes. Energy has a very definite meaning in physics, commonly expressed as a force acting over a distance. The force could be one of several, including gravity. When a ball is held above the ground and dropped, as the ball falls toward the ground because the force of gravity is pulling it down, this is a force acting over a distance. But don't get confused. The ball already possessed the energy in the form of potential energy before it was dropped. As it accelerates toward the ground, this potential energy is converted to kinetic energy. There is a direct tradeoff between the two, but energy is conserved. (So where did the ball get its potential energy in the first place if energy is conserved? From you when you lifted it above the ground. The ball's potential energy increased, and the amount of chemical energy in your muscle stores decreased.)

The conservation of energy is probably the most important law in physics, and what it says can be both blatant and incredibly subtle. The Universe contains the same amount of energy today as it did at the Big Bang. Now, this might seem like an oddity to some readers. In an elastic collision between two ideal billiard balls, both momentum and energy are conserved. (This is the definition of an elastic collision.) However, in an inelastic collision—say between two tennis balls—momentum is still conserved, but energy is not. Or, take another example of a block sliding down a ramp. If the ramp is frictionless, the block has, when it gets to the bottom, exactly the same kinetic energy as it had potential energy before it was let go at the top. But, if the ramp is not frictionless, then the block loses energy as it slides down the ramp—its acceleration is slowed. So, does conservation of energy really hold all the time?

Yes, it does. These kinds of problems are of the type one sees in early physics classes. For the inelastic tennis ball collision, the initial energy of both is equal to their ending energies *plus* any energy lost to the environment in any form (primarily heat). The same is true for the block sliding down a ramp with friction.

When we talk about energy not being conserved in an inelastic collision, we really mean that the *usable* energy is not conserved. When the ball is moving along the table in single direction, we can make use of that energy. Perhaps we use it to knock over our buddy's cup as a joke. Or maybe we use the block sliding down the hill to generate electricity with an attached string and a generator. Whatever the case may be, the energy in both situations can be recaptured and used. That is not true for the energy lost to the environment (or to the ball itself) as heat from the collision. There is no way to "line this energy back up" where we can make use of it. Nevertheless, that energy was not destroyed.

Before going on, I want to clear up one item that may come to the reader's mind. Although not confirmed as of yet, it appears that the net energy of the Universe is zero. So why are we talking about it at all if it's zero? Despite the total energy being zero, this does not imply that there is no energy acting in the Universe. Locally, the energy content is not necessarily zero. For example, a ball above Earth's surface has a net potential energy and, if it is dropped, it will gain a net kinetic energy. So why doesn't this change the net energy of the Universe? Because of symmetry. As you stand on Earth and lift the ball, there is a reaction force on the Earth's surface that pushes it away as the ball is lifted, if only by a infinitesimal distance. And when the ball is dropped, you release the force on the ball and Earth holding them apart, and they move back to their original positions. As far as the Universe is concerned, nothing happened. I'm assuming that this experiment is being performed far away from any objects that would have a direct effect on either the Earth or the ball—that is, break the symmetry of the situation.

So, what is energy? That is a very difficult question. When most textbooks answer this question, they answer what energy does, not what it is. The simplest answer is that energy is just a number that is derived from more fundamental physical quantities. For example, the kinetic energy of a mass m moving at a speed v is $KE = (1/2)mv^2$. Or, the potential energy of a mass m at a height h above the Earth's surface is $PE = mgh$, where, again, g is the usual gravitational constant. Everything is made of energy including, of course, mass, but this is not a very satisfying answer as it doesn't tell us anything. This is simply renaming everything "energy".

In a more fundamental sense, energy is a measure of things happening in the Universe or the potential for things to happen. When injected at a point into a system, it will propagate throughout, making as many things happen as possible. Once this process is kicked off, it won't stop, like water ripples emanating outward from a stone thrown in a lake. The ripples get smaller, but they never stop, and they will touch everything on the lakes surface.

We will discuss it later, but this is entropy increasing. Entropy is a measure of the number of combinations of which energy can take. Think of it in terms of money. Banks print money on demand when a loan is made. This money is then injected into the system. When it first enters the system, money is worth, say, an amount of V. However, as that money keeps changing hands as it diffuses through society, a "macroscopic" effect called inflation takes place. By the time the money reaches everyone in the economy, the value of money has dropped exactly such that this new money makes no difference. Energy and entropy are very much like this.

As energy diffuses through a system and into the environment, it continually excites as many states as it can. Entropy is a measure of the number of these possible states. Whereas at the start, the energy might have been in a form that could be directed and used, such as providing torque from the gasoline exploding in an automobile engine cylinder, in the end it is spread over so many directions that it is completely unusable.

As energy propagates through a system, it evinces the dynamics of that system. It is important to note that the energy does not imbue the system with a particular dynamic behavior; that behavior is innate to the system. As an example, consider a simple electronic circuit. Such a circuit will have a particular set of dynamics associated with it. If an impulse is applied to the circuit's input, it will excite these dynamics, and then possibly interesting things happen. (An impulse is a very short pulse that is used to place energy in a circuit's storage elements such as capacitors and inductors.) The dynamics of these systems can be quite complex, and they will all come alive when the system is excited by an energetic impulse.

However, from the system at hand, the energy will propagate into the environment, exciting every molecule it can find; we call this heat loss, and it represents irrecoverable energy. Any machine, or life form, must compensate in some way for this process. If a gasoline engine didn't have access to a continual supply of energy—in the form of the gasoline—it would stop and simply cool down by releasing its heat energy into the atmosphere. Life is the same way. Both of these processes "live" within an energy gradient: ordered energy enters and unusable energy exits. If this were not the case, it would not take long for both to come to thermodynamic equilibrium with the environment, meaning the car is out of gas and the life form is dead.

The main point is that life is not a simple thermodynamic process to analyze as it operates outside of thermodynamic equilibrium. We will see that entropy, a term that is bandied about in ID circles, is completely misunderstood most of the time, and the increase of which is actually a requirement for life's existence. Entropy really means nothing more than the space of possibility is increasing.

Another possible result of energy propagating through a system is the spontaneous generation of complexity. Snowflakes, river beds, lightning, and many other phenomena, all occur as a result of this property of energy diffusion. For example, the fractal shape of river bed geometry is not just by accident.

Adrian Bejan described the tendency of nature to generate global organization from local rules in his book (with J. Peder Zane) *Design in Nature* [57]. In this book, he calls the observation that nature finds its own course via the expression of energy *the constructal law*. His observation that energy, as it propagates, tends to generate complexity is in the reverse direction of the way we usually think about physical processes. We generally think that complexity must exist first and it guides energy flow. This is not the case. The complexity is self-generating and emerges spontaneously *due to energy flow*.

Energy flows in the most probabilistic directions and in doing so, it carves out complexity. We will look at that process in this chapter. What emerges is a picture of life existing in change. In equilibrium, nothing interesting really happens. It is only in the gradient, or the point of change, that the most interesting things happen.

Life is not order. And it is not disorder. It emerges in the nexus of these two extremes. It rides the wave from order to disorder. This makes a classification very difficult—it is something entirely outside the bounds of these labels.

I need to point out one last thing. The intent of this chapter is not to go through an in-depth study of biology. Again, this was never meant to be a biology textbook—those are easily found elsewhere. Also, if the reader is expecting a solution to life's emergence on Earth, that won't be found here either. The intent here is to show that some aspects of physics, which the ID community claims bar the natural evolution of life, are, on the contrary, absolutely essential for life.

7.1 Feedback systems

A dynamic system is one for which the relationships between its moving parts can be described by mathematical equations in a geometric space. For instance, a car propelled by the torque from its engine represents a dynamic system. The car's position, speed, and acceleration are all related through Newton's equations, for example, in a Cartesian-coordinate system. These variables, when concatenated, form what is called the *state vector* of the system. The state vector is a list of numbers that completely defines the state of the particular dynamic system.

The behavior of dynamic systems can be expressed in a number of ways. The most common framework into which these systems are cast is a feedback structure. I think it is clearly understood that these systems have inputs and outputs. Something is supplied to the system, and something else is provided by the system. For example, a vehicle's accelerator pedal can be considered an input, and the speed of the vehicle an output. The more the pedal is depressed, the faster the car goes.

So, how do we get the car to go at a speed of 55 mph so we don't get a speeding ticket? We use feedback. In this case, the speed of the car—through the driver watching the speedometer—is fed back through the driver's brain in order to adjust the pedal to the proper position to hit 55 mph. If the speed is above this level, the driver sees this and lets up on the pedal slightly. If the

7.1. FEEDBACK SYSTEMS

speed is below this level, the driver presses harder on the pedal. This is a classic dynamic system with feedback.

The driving skill of people differ wildly. Some can bring the vehicle right to the correct speed and hold it there. On the other hand, there are some who can't do it at all and the car keeps surging and slowing down. These are the most annoying drivers for me. It is clear that not all feedback loops are created equal. It turns out that by adjusting the feedback gain, we can have a situation where the speed just keeps increasing, keeps decreasing, or oscillates.

Consider a standard swing-set swing that experiences no air friction or drag in the chain pivots from which it hangs. You sit down in the swing and push backward off the ground with your feet. The swing rises into the air and is subsequently pulled back down by Earth's gravity. From that point on, the swing simply oscillates for eternity (remember, there is no friction). This is an example of what is termed a marginally stable system. It is not stable, and it is not unstable. It is called marginally stable because its operating point sits right on the boundary between stability and instability. The amplitude of its oscillatory motion will not increase over time (instability) and it will not decrease (stability).

Of course, the swing system can be made stable by adding a little bit of friction. The friction resists the change in the swing's position. In other words, it produces a drag force that is proportional to the swing's speed. Everyone knows what this feels like when we have stuck our hand out a car window while the car speeds up. At low speeds, the force of the air is hardly felt, but as the car increases its speed, the force becomes much more apparent. The same goes for the friction felt by the chainrings around the bolts at the swing's top pivot point. Since the friction force is always directed against the velocity of the swing, it will eventually rob all of the swing's energy and the swing will settle to a stop. The friction is acting as *negative feedback*. The word negative here does not mean bad; it means that the direction of the frictional force opposes that of the swing's movement. Systems with negative feedback are stable as they always approach a state of equilibrium, a minimum energy state.

If negative feedback results in stability, then positive feedback drives a system toward instability. As an illustration of positive feedback, go back to the swing example again. This time instead of applying a force that opposes the swing's motion, apply one that is in phase with it. A simple example is a friend who stands behind you, and each time you start swinging forward, this friend pushes you. In response, the height the swing reaches continues to increase. Every kid remembers this scenario from grade school. Instead of dying out, the oscillation amplitude of the swing continues to build as the individual pushing continues to add energy.

Another example of a feedback process occurs while driving a car. If you have ever pushed down on the accelerator pedal and, as the car lurched forward, had the seat catch come loose, allowing it to slide all the way back, you have experienced negative feedback. The harder the acceleration forward by the car, the faster the seat accelerates backward in relation to it, causing the pressure on the accelerator pedal to be released and the car's acceleration to decrease. On the other hand, pressing hard on the brake pedal can result in positive feedback if you slide forward in your seat. The hard punch to the brake causes you to slide forward in your seat, thus applying even more pressure to the brake pedal, causing the car to decelerate faster, and so on. It is easy to screech the tires like this accidentally.

Feedback systems are incredibly important in engineering and biology. They are used in virtually every technological device in existence including vehicle cruise control, household heating and cooling systems, power supplies in almost every electronic device in existence, battery-charging systems, missile guidance controls, and on and on. From a biological standpoint feedback structures are used in metabolic pathways, gene circuits, the blood-clotting cascade, neurotransmitter levels in the brain, and many more. In light of their ubiquity in both biological and technological systems, an overview of feedback systems is given here.

The incandescent light is a very good example of a system with negative feedback. The current that flows through a resistor is given by Ohm's law, $V = RI$, where V is the voltage across the resistor, I is the current flowing through it, and R is its resistance to current flow. (The resistance of a resistor is a parameter determined by its composition and geometry.) For readers with no background in electronics, this situation can be related to the flow of water through a pipe. The pressure of the water is analogous to voltage, the rate at which the water flows is similar to electric current, and one divided by the pipe's diameter is akin to resistance (the smaller the pipe, the greater the resistance to flow).

The problem with the light bulb is that the filament, through which current flows causing the bulb to glow, is metal and, hence, a good conductor with low resistance. If the resistance of a cold filament is one ohm and the light is connected across the 120 volts in a common household, by Ohm's law, the resulting current flow is (120V)/(1 ohm) = 120 amps. That is the level of current that flows in an arc welder. What saves the day (or the night) is that the resistance of the tungsten filament increases as the filament's temperature increases. Consequently, although a very large current may flow for a brief instant when a light switch is turned on, due to its minute mass, the filament heats rapidly, causing its resistance to quickly increase. At about the temperature where the filament radiates white-hot, the resistance has increased such that the current is just sufficient to hold it at that temperature. To summarize, the heat induced in the filament by the current flowing through it causes its resistance to increase, providing negative feedback to the current. At the point of equilibrium, where the bulb operates, the resistance of the filament has climbed to the precise value such that the power dissipated is exactly what is needed to keep the temperature at this point.

As a second example, when I was a kid, my dad helped me build—well, he did most of it—a *regenerative radio*. We built it around a 6U8 tube. I still miss the

7.1. FEEDBACK SYSTEMS

sound of the background hiss from the electron emission in the heated cathode of an old-time tube amplifier. Those days are gone as the semiconductor junctions used in contemporary amplifiers do not exhibit that kind of noise.

The regenerative receiver was invented in 1912 by an electrical engineer named Edwin Armstrong at Columbia University [92]. When electronic amplifiers came on the scene around one hundred years ago, they had a severely limited signal gain, unlike the operational amplifiers in common use today. One means of increasing the gain of these amplifiers, as discovered by Armstrong, was to use some positive feedback. As we have seen, positive feedback is a means of feeding energy back into the system that adds up over time. The regenerative receiver feeds a small portion of its output back to the input that is *in phase* with the incoming signal, so the result is additive. In other words, a small percentage of the output is fed back around to reinforce the input signal—an ingenious idea.

Of course, nothing comes for free, and the regenerative receiver suffered from a couple of serious drawbacks. The first being that the feedback had to be very fast compared to the signal being amplified. For example, if a high note came into the receiver's front-end followed immediately by a low note, the feedback needs to be fast enough that the feedback signal, which helps the high note, does not then degrade the low note. Not meeting this requirement results in distortion of the audio. The second problem is that positive feedback needs to be curtailed somehow, or it will run away entirely and result in complete instability. In the old radios, there was a knob that was turned to adjust the amount of positive feedback, and this adjustment could be very touchy—going too high in gain would result in an annoying screech in one's earphones.

A generic, single-integrator feedback system with a feedback gain of k is shown in Figure 7.1. The input to the system is u and the output is x. The error signal e is constructed from the sum of u and the output scaled by k, that is, $e = u + kx$. The error signal is then integrated to provide the output x.

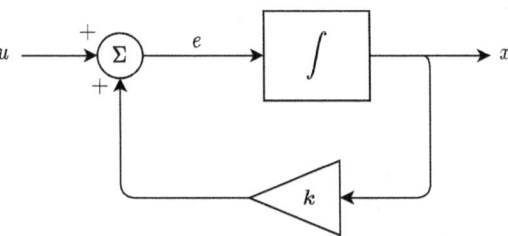

Figure 7.1: A single-integrator system with feedback gain k.

Mathematical integration, represented by the symbol in the figure that looks like a large "S", is an operation that sums its input over time. It is the area under the signal e with respect to time. This is shown explicitly in Figure 7.2 for an example signal $e(t)$. The integral of $e(t)$ from $t = 0$ to $t = t_0$ is simply the area under the curve from $t = 0$ to $t = t_0$ as shown in the figure. Similarly, the integral of $e(t)$ at t_1 is just the total area under the curve $e(t)$ on the interval $0 \le t \le t_1$. The integrator in Figure 7.1 continuously adds up the area under the curve of $e(t)$ and provides that value as the output, x.

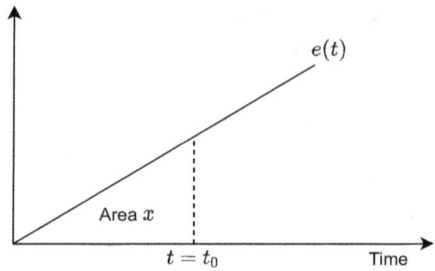

Figure 7.2: An example of integration. The integral of $e(t)$ from $t = 0$ to $t = t_0$ is simply the area under this curve from time $t = 0$ to $t = t_0$.

Integrators are ubiquitous in dynamic systems. The speed at which a ball falls to Earth is the integral of its acceleration. Likewise, the ball's position is the integral of its speed. The volume of water in a bucket is the integral of inflow of water into the bucket. The temperature of an object is proportional to the power it absorbs over time. The population of a species is the integral of its reproduction rate minus its death rate.

We can easily solve for the differential equation that describes the behavior of the system in Figure 7.1. The signal e can be written as $e = u + kx$, and since x is the integral of e, then the derivative of x must be equal to e, $\dot{x} = e$. (Again, the dot on top signifies differentiation with respect to time.) Substituting the first equation into the second gives us the differential equation we are seeking: $\dot{x} = u + kx$. Let's look at the case with zero input. Setting $u = 0$ and rearranging this equation, we have

$$\frac{dx}{dt} = kx. \qquad (7.1)$$

In this case, we start with a positive initial condition on x, therefore, $x(t)$ is always positive as it can't cross through zero since, if x is ever zero, then its rate of change is also zero. It is clear that if $k < 0$ then $x(t)$ decays toward zero; and if $k > 0$ then $x(t)$ increases. We can integrate to solve for $x(t)$ as follows:

$$\int_{x_0}^{x(t)} \frac{x}{dx} = k \int_0^{t_0} dt$$

$$\ln \frac{x(t)}{x_0} = kt$$

$$x(t) = x_0 e^{kt}. \qquad (7.2)$$

7.1. FEEDBACK SYSTEMS

It is clear from this last equation that the response climbs exponentially when $k > 0$ (positive feedback) and decays when $k < 0$ (negative feedback). A simulation of this system is shown in Figure 7.3. The initial condition is $x(0) = 1$ and the feedback gain is $+5$ (positive feedback) for $0 \leq t \leq 1$ and -5 (negative feedback) for $1 < t \leq 3$.

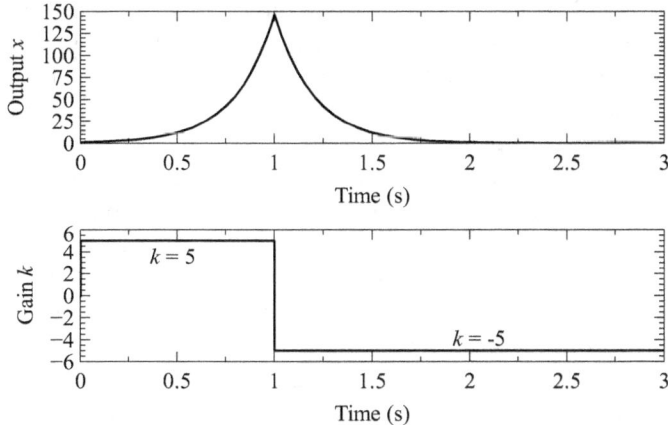

Figure 7.3: The output x (top plot) and gain k (bottom plot) of the feedback system of Figure 7.1. When the gain is positive, the response climbs exponentially, and when it is negative, the response falls exponentially.

This same differential equation can be used as the simplified model for the population of a particular species. The gain k, in this case, would represent the fitness of the species to reproduce. If the reproduction rate is greater than the mortality rate then the positive feedback is greater than the negative feedback and the population increases exponentially. On the other hand, if the mortality rate is higher than the rate of reproduction we have negative feedback and the population decreases exponentially. If on the odd chance the mortality rate exactly equals the reproduction rate, the population stays constant.

From an evolutionary theory standpoint, the fitness of an organism, as dictated by its genetics, modulates the feedback gain. When this leads to a positive gain, the population of the species grows exponentially, which also causes the genetic material these organisms carry to grow exponentially as it is passed on to successive generations. This is one of the primary mechanisms that drives evolution. Beneficial mutations within the genome of species are those that result directly in the population of the species increasing, that is; they make the feedback gain more positive.

We'll use the concept of feedback again in later sections.

7.2 Entropy

Something emerging from nothing notwithstanding, entropy and the second law of thermodynamics are probably the most often cited reasons by ID theory that science can't possibly explain physical reality and, therefore, a god or other supernatural being is necessary. We've already touched on the something-from-nothing problem. We take up the thermodynamics concerns in this section. As will be shown, entropy is not a magical quantity, and the second law of thermodynamics is not an unbreachable rampart preventing a natural explanation of biological complexity. Neither concept inhibits life. On the contrary, they are necessary for life's development.

Many books that take up the subject describe entropy as the increase of disorder from order. Although not entirely incorrect, equivocating entropy with disorder inevitably leads to confusion. Stuart Wachowicz provides a definition of entropy in one of his *Tomorrow's World Viewpoint* videos "As the degree of disorder of a system" and that the second law of thermodynamics states "...in a closed system disorder increases with time, unless acted upon by an organizing force." [93] He then goes on to say that "This law presents an insurmountable obstacle to the concept that life on Earth was somehow generated from nonlife, as living tissue requires matter to organize to a very high degree of order." The implication is that since the second law of thermodynamics requires the disorder, or entropy, of a closed system to increase, it is not possible for life, which is obviously ordered, to have arisen spontaneously from disordered nonliving matter. Determining if he is correct will require first that we understand the second law of thermodynamics and what entropy really is.

Consider the grid with four distinct squares shown in Figure 7.4. The ball in the upper left square can position itself in any one of the four squares. The overall grid is the *system* under consideration. The four positions that the ball can assume are called *microstates* of the system. Microstate is just a name that was arbitrarily chosen to indicate a configuration of the system. It undoubtedly stems from statistical mechanics where, for example, something like temperature is a composition of interactions on a microscopic scale. But, today, it could just as easily represent the number of combinations that can be formed from a standard deck of cards.

That the ball is present in the grid will be called a *macrostate* of the system. A macrostate in statistical mechanics is a parameter that describes the overall behavior of the system. For example, the temperature of a gas is a scaled version of the average kinetic energy of the molecules in that gas. Temperature is a macrostate that describes the collective behavior of an ensemble of microstates.

The *entropy* of the system is found by counting the number of microstates that correspond to the macrostate, labeled W in this case, and then applying Boltzmann's relationship $S = k_B \ln W$. Here, S is the system's entropy, k_B is the *Boltzmann constant*, equal to 1.38×10^{-23} J/K, and the function is the natural logarithm. There are four microstates that all correspond to the same macrostate "a ball on the grid," therefore, the entropy is calculated as $k_B \ln 4 = 1.386 k_B$.

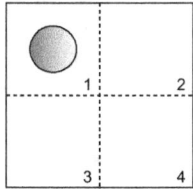

Figure 7.4: A ball in a 2x2 grid. The ball can be moved to any one of the four positions, or microstates, of the system.

In this definition of entropy originally conceived by Boltzmann, the number of possible configurations or microstates of the system, which correspond to the same macrostate, determines the entropy. If there is only one microstate then the entropy is zero (the logarithm of one is zero). On the other hand, as the number of microstates increases, the entropy increases as the logarithm of this number. (Note that the logarithm is used only to compress the size of the entropy figure. When one is dealing with molecules, as is often the case when speaking of entropy, the numbers can get very large. The logarithm reduces the size of these numbers while still maintaining their relative sizes.) One can now see that, although increasing entropy is akin to increasing disorder, it is not exactly the same thing.

The same grid of four squares is shown in Figure 7.5 with an additional four squares added. In (a) there is a wall between the two grids, only allowing the ball to reach the original four squares on the left. Hence, the entropy of the system in (a) is the same as was calculated above. However, if the wall is removed so the ball can reach all eight squares as in (b), the new entropy calculation is $S = k_B \ln 8 = 2.080 k_B$, which is larger than before. As the number of microstates increased, the entropy of the system also increased.

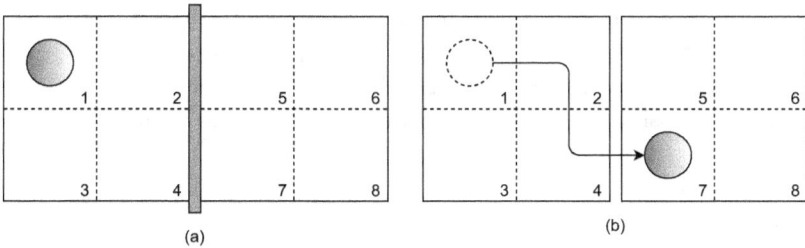

Figure 7.5: The system on the left has a lower entropy than the system on the right because the ball can only assume four positions in (a) while it can assume eight in (b).

Figure 7.6 shows a box on the left in (a) that has a divider down the middle.

There are N particles—they could be gas molecules—that are contained on the left of the box. The right half of the box is empty. Let the number of states, positions in this case (I'm simplifying the problem to only positional states for now), that each particle can assume be denoted by m. So, there are m states when considering a single particle. There are m^2 states for two particles. Think about it like this: if particle 1 is in state 1, then particle 2 can be in m different states. The same is true if particle 1 is in state 2, and again if it is in state 3, up to m. If we add up all these combinations the result is m^2.

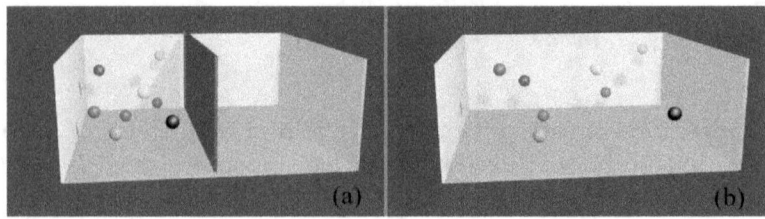

Figure 7.6: A box with a volume of ideal gas molecules trapped on the left in (a). When the divider is removed, these molecules are allowed to travel to the right side of the box in (b).

Similarly, for three particles, there are m^3 combinations. And for N particles, there a grand total of m^N different combinations or microstates. The entropy for this system is then $S_h = k_B \ln m^N = N k_B \ln m$.

Now, if we remove the divider as shown on the right in the figure, the particles will randomly bounce into the remainder of the box. At that point, the volume in which the particles reside doubles, thus doubling the number of states each particle can occupy. Hence, the entropy of the box without the divider is $S = k_B \ln(2m)^N = N k_B \ln 2m = N k_B (\ln m + \ln 2)$. The entropy increases by an amount of $N k_B \ln 2$.

We now know what entropy is, but how does it behave? The *second law of thermodynamics* declares that the entropy of an isolated system—a system that has no energy or mass transfer into or out of it—always increases or remains the same. Remember, entropy is not exactly disorder. The second law says that the number of possible configurations or microstates of an isolated system must increase or remain constant.

One could start with the particle arrangement of Figure 7.6 with all the particles arranged in a line on the left side of the box and given some time, they would once again diffuse into a disordered state throughout the entirety of the box. This doesn't happen because there is some force that drives toward disorder. It is simply because there are far more disordered microstates than there are those that are ordered, therefore, the probability of the particles randomly selecting a disordered state is astronomically higher than for selecting an ordered state. The reason, for example, why the particles, after being mixed throughout the box, do not by themselves go back to the left side of the box, is because there is a far

7.2. ENTROPY

larger number of arrangements where they are strewn throughout the box. This is a very subtle and important aspect of the second law: the tendency toward higher disorder is driven entirely by the probability being higher of landing in a disordered state than arriving in an ordered state.

The second law being based on the probabilities of microstates of the system means that the law can be deduced analytically, unlike many other physical laws such as gravity, which must be deduced through observation. Consider taking a handful of ten coins and tossing them up in the air. It is possible that they will land on the floor all heads up or all tails up. The probability of any outcome is equally likely. However, if either one of those outcomes occurred, we would consider it odd. One reason for this feeling of oddity is that we easily recognize those two outcomes, whereas the others, we don't. But there is another reason, and it is embodied by the concept of a macrostate.

The number of combinatorial ways that k coins can come up tails in a toss of ten coins is shown in Figure 7.7 and is calculated by (this is the standard N choose k expression):

$$C_k(N) = \binom{N}{k} = \frac{N!}{k!(N-k)!}. \tag{7.3}$$

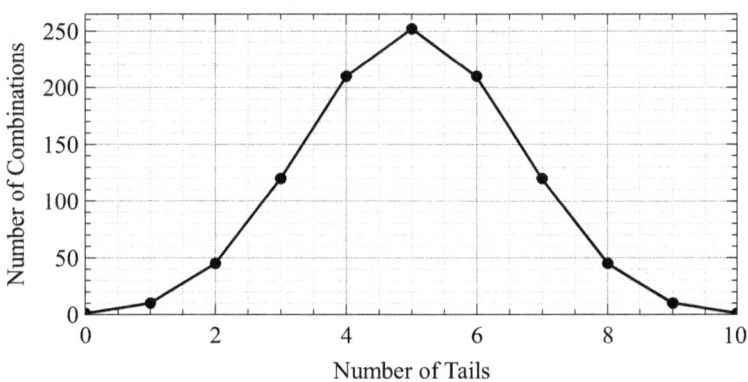

Figure 7.7: The number of combinations that a ten-coin toss can land with k tails.

Despite their probability of occurrence being equal with any other outcome, as seen in the figure, there is only one way to get ten tails (or ten heads). Similarly, the number of ways to get nine tails (or nine heads) is ten. However, there 252 ways to get five tails and five heads. We would call nine or ten tails far more ordered than five tails and five heads. Therefore, in any toss of the ten coins, it is far more likely that the coins will land in a disordered array than landing all tails or heads. This is akin to the increasing entropy case above. There is no force that makes the coins prefer a disordered outcome; rather, a disordered outcome is more likely simply because there are far more ways to produce a disordered outcome as compared to a highly-ordered outcome.

One of the biggest arguments from ID against the natural development of life is that the second law of thermodynamics precludes the emergence of order from disorder. Let's look at a couple of examples that demonstrate this is not the case. The *fundamental thermodynamic relation* $dU = TdS - pdV$ must be understood to see how order is brought about. If the pressure and volume are held relatively constant, then the internal energy U of a system varies proportionately to its entropy with temperature T as the proportionality constant.

An example of the tradeoff between energy and entropy is demonstrated by the common lab experiment comprised of mixing together the two crystalline chemicals ammonium nitrate (NH_4O_3) and barium hydroxide octahydrate ($Ba(OH)_2 \cdot 8\,H_2O$). Although originally solids, when mixed together, these two chemicals form an aqueous solution. The reaction is given by:

$$Ba(OH)_2 \cdot 8\,H_2O + NH_4NO_3 \longrightarrow Ba(NO_3)_2 + 2\,NH_3 + 10\,H_2O. \qquad (7.4)$$

The *Gibbs free energy* for a reaction is given by $\Delta G = \Delta H - T\Delta S$. If the Gibbs free energy for a reaction is negative, this indicates that the reaction is favorable in that it releases energy that can make the reaction self-sustaining. A positive Gibbs energy implies the reaction needs energy added to proceed. The *enthalpy* change, ΔH, is the amount of energy the reaction produces (negative enthalpy change) or the energy it requires (positive enthalpy change). The enthalpy of a substance is its internal energy. The reaction temperature is denoted by T, and the change in entropy by ΔS.

In the case of mixing ammonium nitrate with barium hydroxide octahydrate, the enthalpy change is positive, meaning the reaction is not favorable from an enthalpy standpoint. However, the change in entropy is larger and leads to the Gibbs free energy being negative, indicating the reaction is spontaneous. If these two chemicals are mixed together, they will form an aqueous solution from the original constituents, and the temperature of the solution will drop substantially—the reaction is *endergonic*; that is, it pulls heat energy from the surroundings. It is not the energy released that continues to drive the reaction as the enthalpy change is positive. Rather, it is the change in entropy. In other words, the tendency of the constituents to proceed toward a more disordered state is strong enough to drive the reaction.

As the reaction proceeds, the entropy of the mixture increases drastically as the order inherent in the original crystalline components gives way to the disorder of the liquid form of the resulting mixture. From the fundamental thermodynamic relation, the extreme increase in entropy requires an accompanying temperature drop, and this is exactly what one sees as the reaction proceeds. In this case, the entropy of the system is increased by decreasing the entropy of its surroundings.

7.2. ENTROPY

I don't want to miss this point: it is astonishing that a reaction can be favored simply because of the tendency for the molecules involved to fall into disorder. Yet, in another way, this should not be surprising at all. The amount of information required to describe the molecule mix increased, requiring a compensation in energy. This reaction is irreversible—the aqueous solution cannot be again separated into the original crystalline reactants without the introduction of additional energy. In one sense, this extra energy is required to undo the creation of additional information introduced by the increased entropy of the mixed molecules [94].

The snowflake that emerges from a water droplet in cold air is an example of system entropy being decreased at the expense of increasing the entropy of the surroundings. The water freezes into ice crystals as heat flows out of the water and into the air surrounding it. The order emerges from a commensurate drop in entropy, leading to crystallization. This is an example of order spontaneously appearing from disorder. Liquid water has a far higher entropy figure than does ice because the molecules of the water can move in far more ways than when they are locked in crystals of ice. The disorder, in effect, is pushed out of the snowflake and into the atmosphere.

These examples demonstrate a couple of important points. The local entropy of a system can be increased or decreased via the exchange of energy with the surroundings. Even if the Earth is a closed system, this does not mean that every piece of it experiences an increase in entropy. Overall, yes, if it is closed, then entropy must increase or remain constant. But local regions can decrease their entropy at the expense of increasing entropy elsewhere in the system, and this is exactly what life does.

But, the Earth is not a closed system. The Sun provides energy to the Earth at the cost of increasing its own entropy. This energy is used by life to reduce its own entropy. As Erwin Schrödinger said in his book *What is Life?* [95], "What an organism feeds upon is negative entropy. Or, to put it less paradoxically, the essential thing in metabolism is that the organism succeeds in freeing itself from all the entropy it cannot help producing while alive." As we'll see later, although this is not exactly a correct way of putting things in the modern vernacular, it is accurate in its essence.

Concerning the driving force of life, Schrödinger was well aware that the obvious answer was the wrong answer. He noted that the consuming of matter in eating and drinking cannot be the source of what makes life special as "Any atom of nitrogen, oxygen, sulphur, etc., is as good as any other of its kind." He came to the same conclusion regarding energy. It is neither matter nor energy that fundamentally drives life. There is something else going on. We will discuss what this might be in a later section of this chapter.

As we discussed previously, entropy is an expression of the increase in the information of a system. In this case, it shows up as randomness—information or complexity without a specific context—but it still must be paid for in terms of energy. It is interesting that the generation of novel information is not free;

it requires effort to create. That effort shows up as heat losses in a system and results in processes being irreversible [94]. The generation of the additional information that corresponds to the increase in complexity required to describe the increased number of degrees of freedom of a system represents a loss in the usable energy of the Universe. Undoing this information—that is, reducing the number of degrees of freedom of the system—also requires an expenditure of energy.

It is the indeterminism in this reality that leads to system irreversibility and sets the direction of the arrow of time. It is not that objects anthropomorphically desire to go toward disorder; it is that there are far more disordered states than there are ordered states. It's like playing the lottery: there are many losing numbers and one winning number, meaning what you're likely to find in your pocket is a losing ticket. But you don't lose because the Universe thought you should; you lost simply because the odds were against you. Entropy is simply the Universe following the rules of probability.

7.3 Self-assembling systems

The functions of life are extremely complicated and involve many complex processes beyond just the idea of evolution. We have previously looked at the topic of cellular automata and the complexity it is capable of generating by using just a basic set of simple rules. It should come as no surprise that life also makes use of unguided self-assembly of complex systems. The assembly is subject to a very limited set of rules that are dictated by a few physical laws.

Self-assembling systems are just what the name sounds like; they are disordered systems that spontaneously organize themselves into ordered structures using only local interactions. There is no overarching guidance that brings about the order. Just like cellular automata, the only information that any one part of the system has is that of its immediate neighbors. And out of these local rule interactions emerges a global, organized structure.

An assembly process like this seems to go against the second law that says disorder should reign. This misunderstanding is exactly why it was noted in the last section that claiming that the second law always tends toward disorder is a bit misleading. The reason the systems we're going to discuss here self-assemble is because the resulting structures, although seemingly more ordered, are the most probable configurations of the overall system.

There are many examples of self-organizing structures in nature, including riverbeds, snowflakes, trees, lightning, molecules, etc. They all have one property in common: they are built from the bottom up using only local information. Consider the magnetic ball bearing assemblies shown in Figure 7.8. These metallic spheres are magnetized across their diameters, meaning that there is a south pole diametrically across from a north pole. They make good games for kids because these spheres will self-assemble when thrown together. The shapes in the figure had help from a person in choosing those particular structures. However, these

7.3. SELF-ASSEMBLING SYSTEMS

spheres in the presence of some mechanical vibration, akin to thermal noise in the self-assembly of molecules, will assemble into chains, cylinders, triangles, and many other shapes without outside assistance [96]. In generating these complex structures, the spheres rely only on the local information of the magnetic fields from their immediate neighbors. Obviously, there is no intelligence involved.

Figure 7.8: Self-assembling magnetic spheres. (Photos by Meredin and Jules_88 from Pixabay.)

A small arrangement of the magnetic spheres is shown in Figure 7.9. Each sphere is permanently magnetized across its diameter. The magnetic field lines are shown in black and the intensity of the field is color-coded (the darker colors are more intense). At the contact points, the field becomes stronger as indicated by the higher density of field lines and the darker shading. This arrangement is stable in that the magnetic field of each sphere is attractive with the four spheres surrounding it. In other words, these simple spheres are capable of forming into extended sheets. The sheets can be of varying thickness as the spheres will also attract other spheres from above and below (the directions into and out of the page).

If many of these spheres are deposited on a flat surface that is vibrated to give them some random motion, they will self-assemble into a sheet. Similar to the individual cells in the cellular automaton examples, each of the spheres have essentially two local rules: south poles are attracted to north poles and, conversely, like poles repel. If, for example, the south poles of two spheres are facing one another, there will be a torque generated on both such that they spin until they have opposite poles facing one another and, at that point, they snap together. (If two spheres were perfectly aligned with like poles facing one another, they would repel away from one another. This is unlikely, but if it does occur, the vibration will induce random motion of the spheres, and they will later get another chance to attach together.)

This type of self-assembly can be generalized to include shapes that are not spherical. The triangles and squares in Figure 7.10 have just one behavioral rule: edges of the same number are attracted to one another. If a handful of the triangular pieces are randomly placed on a flat surface and then vibrated, edges of the same number will find one another and stick together, building a hexagonal structure. The same is true for the rectangles, leading to their forming

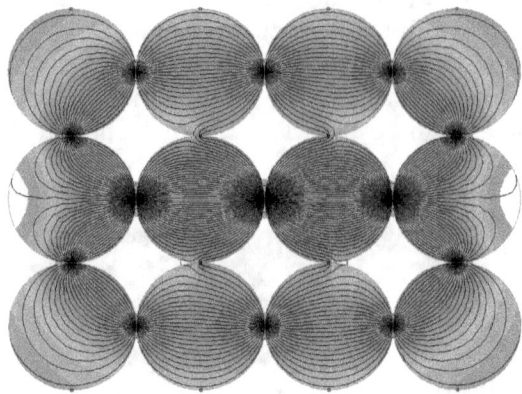

Figure 7.9: Plot of magnetic field lines for self-assembling magnetic spheres. Darker colors are higher intensity.

a rectangular structure.

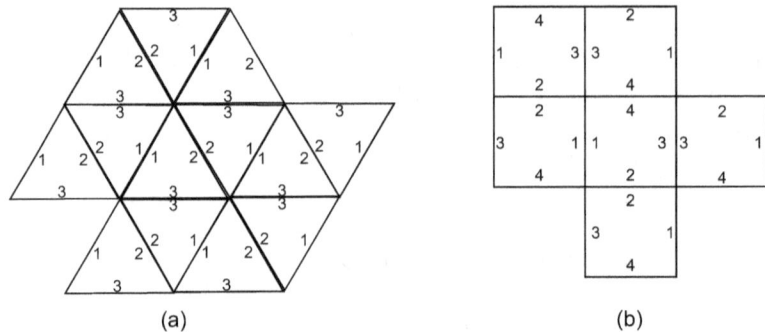

(a)   (b)

Figure 7.10: Self-assembling triangles and squares. The only rule is that each number is attracted to itself.

The manner in which each of these shapes bind to one another is determined by a set of very simple local rules that can be coded just like the cellular automata rules. As an example, the assembly of triangular pieces is shown again in Figure 7.11 with a color rule added: one color for a three on top, a second color for a three on bottom. With the addition of shading, the emergent order is readily apparent.

Again, it is important to note that the assembly of these structures is from the bottom up, not the top down like most of us are familiar with. In building a house, construction workers use a blueprint, whether it is written on paper or

7.3. SELF-ASSEMBLING SYSTEMS

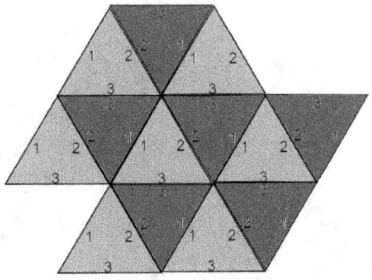

Figure 7.11: Self-assembling triangles have been colored to distinguish their orientation.

simply in someone's mind. This blueprint provides a top-down description of how the house is to be built. What this means is not that the top of the house is to be physically constructed before the bottom but, rather, that the conceptual design of the house starts with the big picture: how big the house will be, will it have a basement, how many rooms will it have, etc., and proceeds to the finer details such as where the floor joists will lie, the window positioning, the geometry of the front steps, etc. In this way, the outer shell of the house is determined and then the geometry of internal components. Larger structures determine the design rules for smaller components that make up the larger pieces. The top, overall picture dictates the lower, smaller constituents.

Whereas the rules for the cellular automata we looked at previously were programmed as bit sequences that determined the state of a cell based on a boolean match of the three immediate cells in the prior row, in nature the rules are derived from the physical laws such as ionic and covalent bonds, Van der Waals forces, dispersion forces between nonpolar molecules, dipole-dipole forces, etc. These forces only act in very specific and simple ways, determining how objects that possess them interact with one another.

Water is a good example of a system that assembles itself. Water is formed from the elements hydrogen and oxygen. In isolation, these two elements form the covalently bonded diatomic molecules H_2 and O_2. However, if a mixture of these gasses is infused with sufficient activation energy to break the bonds, the molecules will split into hydrogen and oxygen atoms. The hydrogen and oxygen atoms will then, while releasing substantial heat, join together to form water or H_2O as shown in Figure 7.12.

These two types of atoms join together with covalent bonds in such a way as to satisfy the requirement that their outermost electron shells be filled. That is what covalent bonding means. One may think of the electrons of an atom as orbiting in shells around the nucleus. The number of electrons that each shell can contain is $2n^2$, where n is the shell number. Hence, the first shell can possess $2(1)^2 = 2$ electrons, the second $2(2)^2 = 8$, the third $2(3)^2 = 18$,

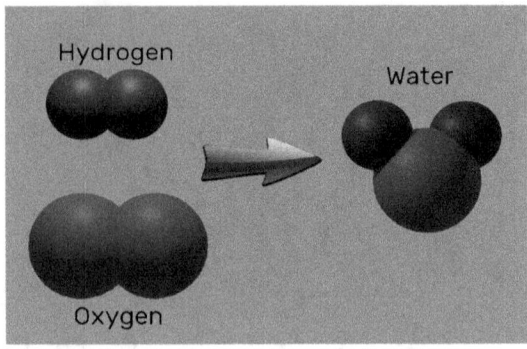

Figure 7.12: The two darker spheres on the left represent hydrogen (H_2), and the lighter, larger spheres are oxygen (O_2). The molecule on the right is water (H_2O). The reaction that goes from the left to the right is $2H_2 + O_2 \rightarrow 2H_2O$.

and so on. Hydrogen only has one electron, therefore, its first shell, which can hold two electrons, requires one more electron to be full. Oxygen on the other hand has eight total electrons, meaning that its first shell has two electrons and its second shell has six. Thus, oxygen needs two additional electrons to fill its second shell with eight electrons. To satisfy the requirements of both atoms, two hydrogen atoms join with one oxygen atom, and the three share two electrons: each hydrogen atom gains one electron in the sharing, and the oxygen atom fills its second shell with these two shared electrons. At that point, the two hydrogen atoms are covalently bonded with the oxygen atom.

Again, the atoms are not intelligently directed to bond in the way they do. The electron shell filling order holds with some variation for all atoms and represents a very simple set of rules that govern molecular construction from the ground up. Water is a very simple molecule. However, extremely complex molecules can be formed using the same set of local rules.

Once water is formed, if the reaction occurs at room temperature, the result is not a gas but rather a liquid. The phase change is due a property of the water molecules. The water molecule, as shown in Figure 7.12 is not symmetric; the two hydrogen atoms are on one end and the oxygen atom is at the other. Effectively, the electrons spend more time on the oxygen-end of the water molecule, resulting in an electric dipole with the hydrogen end positive and the oxygen end negative. When many water molecules are placed together, the opposite polarities of the dipoles tend to attract one another, bringing the molecules much closer to one another than they would normally be in a gas, resulting in liquid water. It is the close proximity of the water molecules, due to their dipoles, that makes liquid water behave as it does.

Sodium chloride, or table salt, is a molecule formed from the elements sodium (Na) and chlorine (Cl). The sodium atom has 11 electrons total, meaning it has one in its outermost shell. The chlorine atom has 17 electrons total, meaning it

7.3. SELF-ASSEMBLING SYSTEMS

has seven in its third shell. When a sodium atom comes in close proximity to a chlorine atom, the electron in the sodium atom's outermost electron shell is added to the seven atoms in the chlorine atom's outermost shell. This leaves both atoms in the stable configurations of having their outer shells full. (The reason we say that the third shell of the chlorine atom is full without holding its maximum 18 electrons is because the fill order is altered due to energy constraints. Once the first eight electrons in the third shell are filled, the next two electrons added go to the fourth shell, and then the rest of the third shell is filled. Therefore, once the third shell has eight electrons, it is energetically stable. These rules can be found in any undergraduate chemistry textbook.)

After the exchange of the electron, both atoms become ions. *Ions* are atoms having unequal numbers of protons and electrons, hence they possess a net electrical charge (normally atoms are electrically neutral). The sodium ion, Na^+, has lost an electron and therefore has a net positive charge (electrons have a negative charge and protons a positive charge). The chlorine ion, Cl^-, has gained an electron and carries a net negative charge.

The opposite charges of these ions pull them together into a cubic lattice as pictured in Figure 7.13. The result is common crystalline table salt. The only rules needed for the construction of salt from the sodium and chlorine ions is that opposite charges attract and like charges repel (very similar to the magnetic spheres earlier). Using these local rules, the ions self-assemble into cubic crystals of salt. For example, as most people have seen, as water evaporates from a saltwater solution, crystals of sodium-chloride will appear.

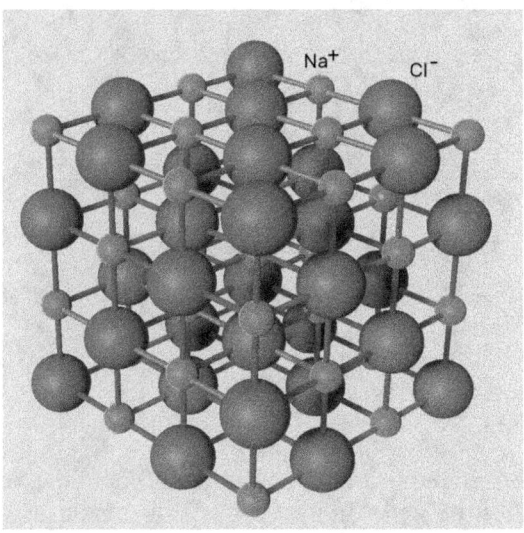

Figure 7.13: A salt crystal constructed from sodium (Na^+) and chlorine (Cl^-) ions. The ionic charges of the atoms lock them together into a cubic structure.

288   CHAPTER 7. THE ENERGY BASIS OF LIFE

Electrostatic forces are not the only ones that bring about self-assembly of biological components. When brought together, many substances will mix thoroughly because of the tendency for a system to move toward maximum entropy. For example, if you mix red and blue paint together, the two will diffuse into one another and produce purple paint after a time and maybe with a bit of agitation. However, some mixtures do just the opposite and separate, for example, oil and water. Many of these materials do not mix because energetically, they more favorably interact with themselves rather than with the other component of the mixture. Nevertheless, some materials do not mix for an entirely different reason.

As we have already discussed, water forms a polar molecule of one oxygen and two hydrogen atoms. The two hydrogen atoms are covalently bonded to the oxygen with an angle of approximately 104.5° between them. Water molecules are electrically polarized with the oxygen end being more negative than the hydrogen end. As a result, in liquid water where the molecules are free to move yet not overly agitated by thermal noise, the molecules tend to arrange themselves into tetrahedral structures [97]. This is depicted for one water molecule in Figure 7.14. The water molecule is at the center of the figure and clearly has a larger oxygen and two smaller hydrogen atoms.

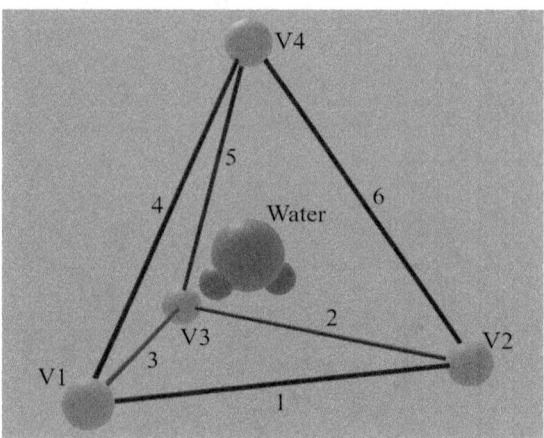

Figure 7.14: Tetrahedral arrangement of water molecules. This shape is created due to the polar characteristics of the molecules.

The four vertices of the tetrahedron represent couplings to other water molecules. In the figure, the water molecule in the center is aligned such that it is hydrogen bonded with vertices V1 and V2. In other words, the two hydrogen atoms are aligned with edge 1 of the tetrahedron. All of the molecules line up in a similar fashion to form large groups within the liquid water.

One important thing to note about the configuration shown in the figure is that the water molecule is free to align with any of the other edges of the

7.3. SELF-ASSEMBLING SYSTEMS

tetrahedron as well. And water molecules do just that, continually changing rapidly in liquid water. This means each molecule has six degrees of freedom it can access at any time. These six alignment possibilities have a direct implication for the maximum entropy of a mass of water. The molecules have some order enforced upon them simply because they are polarized, leading them align in an orderly fashion. However, the ability to dynamically move between these six degrees of freedom (the edges) increases the overall entropy of the system.

All this leads to the question of what happens when a second substance is added to the water that is not polarized. The molecules of such a material will not be able to bond with the water molecules. For example, if one of these nonpolar molecules resides at one of the vertices of the tetrahedron structure that water tries to form, three of the edges of the tetrahedron will effectively be removed, reducing the number of degrees of freedom of the central water molecule from six to only three. This is a substantial reduction in the entropy of the overall system, which is not a favorable outcome. This problem is mitigated by the water segregating the nonpolar molecules.

The nonpolar molecules will tend to form large groups with water surrounding them. This reduces the contact surface area between the water and the molecules of the secondary substance, thus reducing the number of water molecules that have their motion restricted by a missing edge of the usual tetrahedral structure. Grouping the nonpolar molecules together in a large volume is optimal for minimizing contact with the water molecules because volume increases as the cube of distance and surface area increases much slower, as the square. The result is that polar molecules like water shun nonpolar molecules.

Nonpolar molecules that fit in this category are called *hydrophobic*, meaning afraid of water. On the other hand, polar molecules that can bond with water are called *hydrophilic*, meaning to love water. Hydrophilic molecules will arrange themselves in the matrix of water molecules without disrupting it.

The spontaneous segregation of hydrophobic molecules by water is the first hint that such a process could be used in self-assembling structures, guided and enforced only by the second law's requirement of entropy maximization. Phospholipids are a group of large molecules made up of two fatty acid chains on one end and a phosphate group at the other end. These molecules are *amphipathic*, meaning they possess both hydrophobic and hydrophilic parts. The fatty acid end of a phospholipid is hydrophobic, and the phosphate end is hydrophilic. When phospholipids are immersed in water, the segregation process described above causes them to line up in a bilayer (a double layer) with the hydrophobic ends mating together and hydrophilic ends facing the water as shown in Figure 7.15. The result is a bilayer membrane that can wall off water. These bilayers are commonly used in nature for the construction of biological membranes.

Bilayer membrane construction is a good example of the second law of thermodynamics not being an impediment to life but, rather, an asset. The construction of biological membranes using the polar and nonpolar properties of certain molecules is performed by exploiting the fact that the overall entropy of

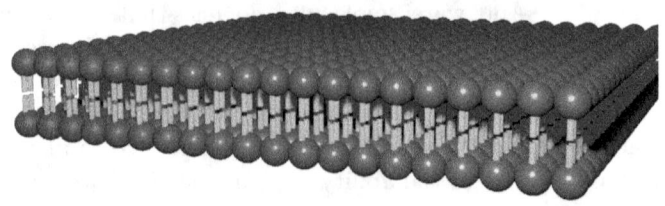

Figure 7.15: Self-assembling phospholipid bilayer membrane. In water, the hydrophobic ends of the molecules line up facing one another with the hydrophilic ends facing the outside of the membrane.

a system is always maximized. The layers of phospholipids increase their order at the expense of an even larger increase in the disorder of the water molecules. A local decrease in entropy is allowed by the second law so long as the overall entropy of the environment increases.

There are many structures in the human body that are self-assembling. Most of them are not well understood yet, but progress is always being made in that direction. Before we leave this section, let's look at some of the processes that life employs. Although they resemble those of human-designed technology, they are fundamentally different in that the control signals are actually the inputs and outputs themselves. This is not the case with electronic controllers. There, the control signals are constructed representations of real-world variables. In addition, researchers are beginning to understand how self-assembling processes are created.

Life is a process that takes in energy and matter and outputs less ordered energy and matter. From these gradients, it extracts the energy and materials necessary to bring about ordered structure in itself. But it is not as if once enough energy and mass have been taken in to build an ordered structure, an organism no longer needs to take in any more mass or energy. No, the gradients that enable life must be sustained indefinitely; if they stop for even a short time, the internal processes of the organism will descend into irreversible disorder and life will cease.

It is not energy and mass that are the primary requisites of life. Rather, it is the energy and matter gradients or flows that make life possible. Like a surfer riding a wave, life continually rides these gradients. This is a huge clue to the essence of what life really is. We will come back to this point later, but for now, a short look at the nature of these processes is necessary.

In engineering, feedback and feed-forward loops and many other topologies are ubiquitous, especially in the design of analog electronics. Biological processes operate in a similar manner but the techniques for their analysis are not fully developed yet because the problem is so much more difficult, in particular, the degree of interconnectedness. Generally, engineers design systems in a modular

7.3. SELF-ASSEMBLING SYSTEMS

fashion to separate the overall complexity into more manageable pieces. Biological systems do not do this as a matter of course because they are not designed. This is not to say that biological processes do not admit analysis, however. They do.

A very simple system is shown in Figure 7.16 with a single input $A1$ and a single output $B1$. These two signals may represent particular molecular streams, $A1$ being the input raw materials, and $B1$ the completed molecular product. The input $C1$ might be a catalyzing protein that modulates the reaction from $A1$ to $B1$. The catalyst does not get used up or destroyed in the process, rather, it modulates the speed of the reaction.

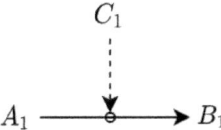

Figure 7.16: Simple system with one input, one output and a single constraint.

It might be that the isolated reaction from $A1$ to $B1$ takes place with a time constant of τ_B. That is, the reaction is governed by a single-order differential equation with the solution $B1 = kA1(1 - e^{-\tau_B/t})$, where t is time and k is a constant. The response starts at zero and over the course of time, as e^{-t/τ_B} goes to zero, approaches the value $kA1$. The time constant τ_B determines the speed with which the final value is approached. If $C1$ acts as a catalyst, it modulates the time response of the reaction, causing it to speed up or slow down.

In the literature, $C1$ might be called a *constraint* as it constrains the reaction from $A1$ to $B1$, in this case modulating its speed. There can be, of course, other types of constraints, but here we'll only consider catalyzing constraints.

Even though a catalyst is ideally considered to remain inert over much longer time scales than the reaction it catalyzes (it can change on time scales of the reaction, but on average it remains the same), it will eventually degrade and its concentration will fall. This might seem like a bad thing from one perspective as the reaction would eventually slow down to an unusable speed. However, if the catalyst is supplied by a second process as shown in Figure 7.17, then the speed of the reaction can be modulated by another part of the system.

In the figure, there are two inputs $A1$ and $A2$, two outputs $B1$ and $C2$, and two constraints $C1$ and $C2$. The important thing to note is that the constraint $C2$ is produced by a first reaction involving $A1$ and used to constrain a second reaction involving $A2$ and $B1$. There are multiple reasons why this topology might be useful. The first is to achieve gain. If we assume that the production of $C2$ is difficult for a metabolic chain in its current configuration, it would be good to add a secondary process stage that provides $C2$ as the output. The module containing $A1$ and $C1$, and outputting $C2$, can be considered a gain stage or,

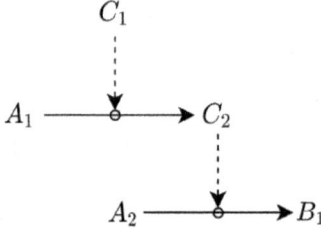

Figure 7.17: A system with a dependent constraint.

from the engineering world, an impedance match. Since the required quantity of $C2$ was not produced implicitly in the system, a second process with a higher yield of $C2$ was introduced.

These network configurations are not there just to unnecessarily add complexity. In the engineering world, they represent functionality required for a circuit to operate in a specific way and I would venture the guess that they are included in biological networks for the same reason.

As soon as network components like these are available, some amazing functions can be achieved by introducing feedback as depicted in Figure 7.18. There are two processes present, each governed by one of the constraints $C1$ or $C2$. The constraint $C1$ is an output of the second process (this second process has two outputs, $C1$ and $B1$) and is fed back to act as a constraint governing the first process. If $C1$ increases the efficiency of $A1 \rightarrow C2$, and $C2$ increases the efficiency of $A2 \rightarrow C1$, then the loop has positive feedback. If one of the two constraints reduce the efficiency of one of the reactions, the feedback is negative.

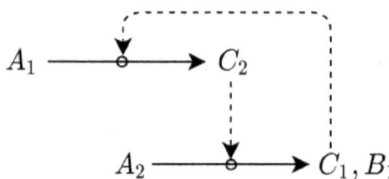

Figure 7.18: A system with a dependent feedback constraint.

A feedback loop like this is called in the biology literature a *closure* for obvious reasons. The existence of feedback networks like this in biological structures provides organisms the tools necessary for complex process control, in particular, the regulation of metabolic pathways. Keep in mind that these are not just control signals that control functional processes—this is what is done commonly in engineering. There is not necessarily a distinction between control and functional pathways in biological systems; they can be one and the same. In other words, the desired output product can also act as a control variable.

I don't want to get too caught up in describing the structure of biological control systems. I introduce it here because it will be beneficial later in this chapter. For a more complete introduction to the topic of constraint closure, see Montévil and Mossio [98].

As has been introduced here, nature does not need to rely on intelligent design for the development of self-assembling systems. Assembly is driven by the mathematical interaction of natural forces to construct biological structures from the bottom up. Life is an expression of how energy and information flow within reality and the degradation of these two quantities does not imply that disorder is the inevitable outcome in all locations. The progression to disorder in one area can be the catalyst for order in another. Life only exists outside of equilibrium, in the gradient flows of energy and increasing entropy.

The interesting thing about a system exhibiting constraint closure is that it produces the very constraint needed for its own production. This represents a positive feedback network where the signals are the very products themselves. Such a thing is unusual in human designs, but ubiquitous in biological structures. And it is a definite hint that biological control loops are not designed.

7.4 Self-replicating systems

A *self-replicating system* is a system that can reproduce or duplicate itself. We are very used to the idea of cars and TVs being produced in factories by automated equipment, but that is not self replication. For a device to be self-replicating it must make use of raw materials to produce a copy of itself without relying on outside assistance. Although cars and trucks carry employees to work and deliver raw materials for producing vehicles, they do not produce themselves; rather, the equipment in the manufacturing plants do that job while being overseen by humans.

Some ambiguity exists as to where the boundary on a self-replicating entity resides. For example, as we discussed previously, if a collection of diffused, diatomic oxygen and hydrogen is ignited, the molecules will spontaneously form water. But this would not be called self-replication because the water molecules do not produce additional water molecules. The molecules form due to the affinity between the hydrogen and oxygen atoms. Conversely, organisms take in the necessary raw materials, modify them, and assemble copies of themselves using their own DNA instruction sets. This is very different from what we have been discussing as self-assembly.

John von Neumann was the first person to study self-replicating machines in a scientific way. He designed in the 1940s what he called a *universal constructor*, which was a machine that operated as a cellular automaton [99]. His machine consisted of three subsystems: a description or design plan of itself, a universal constructor that was capable of making copies of the machine, and a copy mechanism whose job it was to copy the design plan to be passed on to the replicated machine. In his conception, the copier could introduce random errors allowing evolution to proceed and for the machines to grow in complexity.

So, what does a simple self-replicator look like? One with which we're all familiar is a computer virus; a piece of code that can spread copies of itself. Although viruses use the hardware of a computer platform to accomplish their tasks, from a firmware perspective, they are self-replicating as they use their code to produce exact copies of themselves, for example, as follows:

```
    n   : for k = 0 to 2
n + 1 :    copy(A(n + k), A(n + k + 3))
n + 2 : end
```

The first and third lines of this code form a loop that repeats three times, with k being equal to 0, 1, and 2. As it does so, its body, line 2, is repeated as well. The numbers to the left of the colon are memory locations storing the program. The first time through the loop, k = 0, and the copy command copies the code at address n plus the offset 0 to address n plus offset 0 plus an offset of 3, which means it is placed in location n + 3. The second time through the loop, the second line copies the line of code at address n plus offset 1 to address n plus offset 1 plus an offset of 3, meaning the second line copies itself to location n + 4. This process repeats until the block of code is copied to a location starting right after its ending location.

This process is shown diagrammatically in Figure 7.19. The code block loops three times, copying its three lines to the memory location following itself. Then this second block executes and copies itself to the three locations immediately following itself. And this sequence repeats indefinitely or until the computer running the code is out of memory. This is a very simple computer virus. Once executed on a particular computer, it writes itself over the entire memory. (Of course, a practical virus is more complicated than this simple example.)

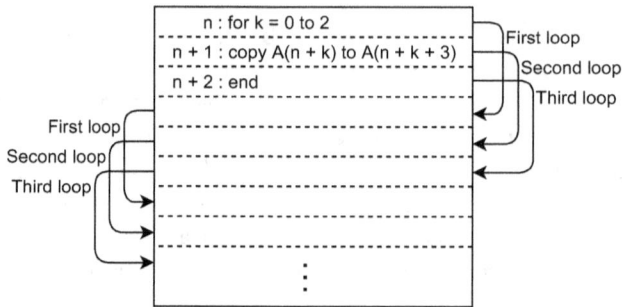

Figure 7.19: An example of a self-replicating program. The three lines of the program simply copy themselves to the next three lines, and then these new lines of code execute. This process repeats indefinitely.

If a virus like this is run on a multi-threaded machine, multiple copies could

7.4. SELF-REPLICATING SYSTEMS

be running simultaneously, meaning that each block of code would be executing independently, creating replicates of itself. Whereas in the previous case, the multiplication of the virus code was linear, here it would be exponential. The first block would copy itself, meaning there would now be two copies. Each of them would copy themselves again, making four copies, then eight, 16, 32, etc.

Note that in the processing world of a computer, our simple virus possesses all three components of the von Neumann machine. However, in this case, they all refer to the same thing: the code itself. The code is the machine to be replicated. It is also the replicator. The code is also the design plan.

There are many computer simulations today that attempt to mimic life and its ability to replicate and evolve including Avida, Biogenesis, and Conway's Game of Life [26, 100]. Many of them are cellular automaton simulations, but some rely on neural networks and other means. All have the common denominator of replication.

Replication is an absolute must for the production of complexity in biology (and everywhere else as well). It is what mediates the dynamics of change of organisms, and therefore could be called the most important aspect of life. I'll just mention a few aspects of what has been learned about how biological replication is accomplished, but as there is no general consensus yet, it will be very brief.

There is every indication today that life started in what has come to be called an *RNA world*. (Ribonucleic acid (RNA) is one-half of a DNA chain.) This term references the prehistory of the Earth before there were any organisms as we know them today, and all that existed were prebiotic molecules such as amino acids and simple proteins. These molecular precursors to biotic chemistry showed up spontaneously on the ancient Earth, and were possibly even brought here by extraterrestrial means, such as the Murchison meteorite [101]. It is interesting that the RNA molecule can not only carry genetic information, but it can also act as a catalyst for chemical reactions similar to enzymes and proteins.

The most important property of RNA molecules, however, is their ability to self-replicate. They not only carry information but also possess the ability to auto-catalyze copies of themselves. Of course, the copy process is not perfect, allowing mutations induced by UV radiation from the Sun or chance occurrence by other means, and once this door was opened, evolution was an inevitable consequence. Those RNA chains that were better at replicating themselves became the most prevalent. It was not life, but the underlying process of replication and the foundation of evolution had been born [102]. Once replication had been achieved, the real game began.

Before evolution could fully take hold, the competing entities needed spatial separation. The hypothesized "primordial soup" was simply a conglomerate of assorted replicating molecules relying on their immediate surroundings for the raw materials necessary for replication. (Again, I'm walking on thin anthropomorphic ice here—these molecules replicated simply because the chemistry was there, not because of any desire on their part.)

How the biological cell became the basis for life on Earth is still unknown today, but there are some glimpses of the origins of the cell emerging through recent research. One of the primary things that must be kept in one's mind when considering these subjects is that life had no goal. Only with that perspective can the proper course be discovered. There was never a plan behind the first biological cell, so how then did it come about?

There have been many theories as to how the first protocells formed. One idea is that compartmentalization in the first cells was provided by *coacervates* [103,104]. Coacervate is a liquid infused with macromolecules such as proteins or nucleic acids. It will separate spontaneously to form dense coacervate droplets within a dilute phase. These droplets can serve as stable boundaries for isolation of material without an actual membrane.

The leading contender for protocell formation, however, is still that of a spontaneously generated phospholipid bilayer, which was discussed in the previous section. It is clear that somewhere along the way, groups of replicating molecules, capable of metabolic processes, became encapsulated within a membrane. As has already been discussed, phospholipid membranes form spontaneously due to entropic effects. The speculation about the formation of protocells involves these membranes closing upon organic molecules that were already in solute, and thereby providing spatial isolation. Of course, once a primitive cell had formed, the next question is how did it divide to bring about reproduction?

There has also been recent work indicating that fission of these cells could have been brought about through the increase in internal temperature stemming from metabolic behavior [105]. Basically, the temperature change causes the phospholipid membrane to fold in on itself and the cell to then split.

How life replicates itself along with all the componentry of the cell, and how this process originally started (abiogenesis), is one of the least understood aspects of life. Research has been ongoing for many decades but the problem is extremely complex. This naturally leads ID proponents and creationists to interpret biologists' inability to clearly articulate all of the facets of life's functioning as evidence that a divine designer must be responsible. A lot of progress has been made in deciphering life's beginnings and operation, but it's a long journey, and everyone agrees there is a long way still to go. But there is no indication, at this point, that scientists should just throw up their hands and admit defeat.

In the latter part of 2021, Sam Kriegman and other researchers demonstrated an entirely new form of biological reproduction [106]. All known forms of life replicate by growing replicants within themselves or from themselves. Cell fission, for example, occurs when a cell grows large enough to divide. All of the material to form the two new cells is contained within the original, and progeny emerge from the parent. Kriegman and his associates used frog stem cells to form appropriately shaped globules of cells that are self-replicating, self-healing, and replicate themselves in an entirely new way.

In one sense, we could call the conglomerations of cells organisms. Of course, this team of researches did not create the individual cells, but rather harvested them as already existing material. But they did design the end shape of the assembly of cells. And once this arrangement is put into place, these new organisms do something really peculiar: they reproduce not the way conventional lifeforms do, by internal assembly of progeny, but rather by constructing their progeny external to themselves. This is the way molecular chains replicate at the subcellular level, not the way life as we know it replicates.

The point here is that, essentially, a new, simple form of life has been developed. One in which it is clear that simple physical and entropic forces perform the job of self-replication. In other words, the magic is being removed from the process of biological replication.

The research interest in self-replication is sure to intensify in the coming years as it is so important from both the biological and technological points of view. It has huge implications in the field of medicine, but also in how to build machines that replicate themselves on distant planets solely from the raw materials they discover there. The point here is that biologists are not completely ignorant of the facets of biological replication, which is the picture often painted by ID advocates.

7.5 Evolution one more time

As we have seen, a self-replicating organism will change over time due to evolution if the replication process is subject to variation. I do not want to emphasize evolution as the driving process as many authors and textbooks do; instead, I want to clarify that evolution is not a prescriptive process. In other words, the term evolution does not dictate the behavior of replicating entities. Rather, it is an *outcome of the replicating process*. Similarly, natural selection does not command biological processes to behave in a certain way. Instead, replicating entities will experience competition that necessitates that some are successful and others are not—lazily using these words as though some magical process of evolution defines how biological entities must behave leads to much confusion. They are not prescriptive laws; they are descriptions of behavior. Similarly, Newton's laws do not tell the Universe how to behave; rather, they describe the way the Universe behaves. Getting this mixed up leads to a fundamental misunderstanding.

Evolution is not a property of only the biological world. It is much more general. Author Perry Marshall asks in his book *Evolution 2.0* [42] why, if evolution is such a great optimization process, do we not see it being used in engineering in lieu of the processes we currently use? The question reveals a strong assumption about evolution and the natural world, that they have some purpose or goal. We humans are goal-oriented in our thinking. Nature is not. It is far easier, more efficient, and precise to find the maximum of a parabolic function using calculus than to search for the optimum using brute-force search methods. Although we as humans have calculus at our disposal, nature does not. In fact, even saying it this way again misleads one into thinking that there is a problem to be solved by nature; there is not.

The evolutionary concept is so subtle that it is very difficult to grasp. We've already talked about it multiple times previously and are going to do so again here. Seeing it from multiple different views helps in engendering understanding.

There is no "desire" on the part of evolution to produce more efficient organisms. Thinking about it that way anthropomorphizes the process and misses the point altogether. A self-replicating entity that engages in reproducing itself over time represents a dynamic system. If this system has negative feedback, the population within the system will decrease. With positive feedback, the population will increase.

Biological systems (and other systems) grow in size because they are dynamic systems with positive feedback. As long as each generation produces at least the same number of offspring as the last generation, the population's size will remain at least static. If each successive generation produces more offspring than the last, the population will grow exponentially. This is simply a mechanical effect of the operation of dynamic systems. There is no magic here.

Now, here's the key to the entire process. Assume there are two sub-populations, A and B, with feedback gains of α_A and α_B, within the global population of a particular species. Assume that population A has a genetic modification that increases its ability to reproduce, meaning that $\alpha_A > \alpha_B$. In other words, saying that a mutation of organism A led to an increased ability to reproduce is the same as saying that the amount of positive feedback increased for group A. In the long run, the population of group A will overwhelm and replace that of B. The mutation that led to the increased positive feedback for group A will be retained not because nature has any concept of better or worse, but simply because the mutation increased the reproduction rate in group A, leading to faster exponential growth.

Recalling earlier discussions within this book, we as analytical humans do not look for a goal that a river is pursuing when it cuts its riverbed deeper into the Earth and its path becomes more entrenched. We do not say that the river chose that path based on some will of its own. No, the river flows in the channel it cut simply because more water flowed there and cut the path deeper. Again, this is positive feedback. Water at some point started flowing down a small channel, and the more water that flowed, the deeper the channel was dug. The deeper the channel, the more water flowed down it. The more water that flowed down it, the deeper it was cut. And so on. Note that the flow of the river and the cutting of the river bed are in phase with one another, a prerequisite for the positive feedback we discussed earlier. This is simply nature at work and physical processes doing what they do.

The same is true for evolution. Entities that are better at reproducing will reproduce more often than other entities. This is the real secret of the theory of evolution and is no more mysterious in its fundamental operation than that of mindless water constructing its complex river delta. The more effective a particular sub-population is at reproducing, the more of that sub-population is produced. Interestingly, we find the auto-construction of a riverbed perfectly palatable, but the same process applied to biology is a source of constant contention.

Before moving on, I would like to drive another point home in this section: there is no adherence to Darwin's theory simply because it's Darwin's theory. The idea of optimization using random, persistent excitation is by no means unknown today, as we have talked about earlier in this book. Optimization is much better understood today than it was in Darwin's day. Just like the theories of Newton, Einstein, Bohr, etc., no one reveres the theoretical concepts of evolutionary theory as dogmatic requisites. The process is simply the process no matter from where it arrived. We do not believe that massive objects accelerate toward the Earth at 9.81m/s^2 because a book or a person tells us so. We know it because we can test and verify it anytime we wish.

The term evolution is not a trope attempting to define a not-well-understood process. In fact, the word evolution refers, in a contemporary sense, more to a mathematical process that appears in nature. It is a description of positive feedback in dynamic systems and is not just a biological catchphrase. It is a purely mechanistic description of how dynamic systems operate. Moreover, when Mr. Marshall asks why we do not use such a process in engineering. The answer is, *we do*. Identification and adaptive feedback systems make use of random perturbations of system inputs to optimize estimates of state variables and feedback gains. Genetic programming even emulates biological evolution directly. Of course, these optimization processes are not used in engineering when they can be avoided because, compared to direct methods and contemporary gradient methods, they are inefficient. They are utilized in engineering *when intelligent, purposeful methods cannot be employed due to incomplete knowledge of the system at hand*. On the other hand, nature makes use of them because it does not employ design.

Again, it must be clear that making comparisons like this is mixing apples and oranges. Evolution in nature shows up in the form of necessity. In other words, it is simply a derivative of the operation of the natural world. This is fundamentally no different from a riverbed being cut by flowing water. Optimization processes employed by human engineers are different because *they are designed*.

7.6 Dissipative systems

Life seems at first glance to operate in a way that is impossible: it appears to distill order from chaos. Here is how Schrödinger put it in his book *What is Life?* [95]:

> Thus a living organism continually increases its entropy—or, as you may say, produces positive entropy—and thus tends to approach the dangerous state of maximum entropy, which is of death. It can only keep aloof from it, i.e. alive, by continually drawing from its environment negative entropy—which is something very positive as we shall immediately see. What an organism feeds upon is negative entropy. Or, to put it less paradoxically,

the essential thing in metabolism is that the organism succeeds in freeing itself from all the entropy it cannot help producing while alive.

Schrödinger was well aware that life exists on an entropy gradient in saying that life takes in negative entropy and expels positive entropy. So long as the balance is maintained life goes on. (Negative entropy here simply means the incoming entropy is less than the outgoing.)

It is not as some would say that life goes against the second law of thermodynamics; it doesn't. The second law allows for local systems to decrease their entropy by increasing the entropy of their surroundings. Your air conditioner works this way; it cools the interior of your house—lowers the entropy in your house—at the expense of expelling into the atmosphere surrounding your house, not only the heat removed from your house, but also the heat losses in the air conditioner's mechanical and electrical components. The amount of heat that the air conditioner releases into the environment is *more* than the heat removed from your house. That is, the air conditioner's efficiency is less than unity. Saying this another way, the entropy reduction in your house due to its being cooled is more than offset by the entropy increase in the environment.

All real systems lose energy in the form of heat and, thus, increase the entropy of their surroundings. The power grid takes heat energy from coal and converts it to ordered electricity transmission at the expense of dissipating a good percentage of that heat into the atmosphere. Automobiles translate the heat energy of burning gasoline into controllable, ordered vehicular motion at the expense of releasing a large portion of the energy contained in the fuel into the atmosphere.

All systems like this have one thing in common: they all operate at a point far away from thermodynamic equilibrium. What does this mean to operate away from equilibrium? Equilibrium is the relaxed state of a system. If you let a ball roll down a hill into a valley, it will roll past the valley and back up the other side. And then it will roll the opposite way and somewhat back up the original hill. It will oscillate around and finally settle down in the lowest part of the valley. That is the equilibrium point of the system containing the ball. It is the point of minimum energy: it has the lowest kinetic energy as the ball is stopped, and it has the lowest potential energy as the ball is at the bottom of the valley. However, the systems we are talking about here, including life, do not operate at their lowest energy state.

The only way for a dynamic system to operate away from equilibrium is to continue to take in and dissipate energy. We're talking about a dynamic system that remains in motion against any losses. For example, think about a hamster on a wheel running fast enough to remain halfway up the side of the wheel. The hamster keeps climbing and the wheel keeps turning. The hamster is continuously using energy to climb and spin the wheel. The gravitation force pulling the hamster downward is balanced by the friction in the wheel's bearings. The equilibrium point is a stationary wheel with the hamster at the bottom. The hamster running on the wheel is operation away from equilibrium. Energy expenditure is required to operate at that point.

7.6. DISSIPATIVE SYSTEMS

Maybe a better analogy would be a waterwheel. It sits partway up a cliff over which water falls and it absorbs some of the water's energy. This stolen energy allows the wheel to keep spinning and also to do some other work like driving a sawmill or grinding grain. The wheel acts as an intermediary between the energy of the falling water and the energy used to grind grain. There is an energy gradient across it: high energy enters and a lower energy exits. The balance of these two flows goes to offset frictional losses in the wheel's spinning; it is the energy that allows the wheel to do what a wheel does: spin.

Energy searches (yes, again, I'm being a bit anthropomorphic here) for a path to express itself. In a nutshell, that is the essence of the second law. To paraphrase, the second law of thermodynamics states that entropy increases, and entropy is a measure of all the possible combinations that energy can take. Therefore, stating it another way, energy attempts to explore the possibility space of reality. That is simply what it does. This is why there is an intimate connection between energy and entropy.

Going back to the Christian or ID take on the second law, the question comes up as to how life can exist if it is governed by a rule that requires disorder to increase. The answer is to reside in an area where order is increasing at the expense of increasing disorder in another area. An entropy gradient. Recall that we said life is not complete order, like that of a crystal or a rock. And it is not complete disorder like the molecules in a gas. But, if it continually takes in order, how does it not become completely ordered? To see this, assume the inputs to an organism take in entropy flow $\dot{S}_{in}$ and the outputs produce entropy flow $\dot{S}_{out}$. Then the change in entropy of the organism is $\dot{S} = \dot{S}_{in} - \dot{S}_{out}$. If the entropy flow in equals that out, then the entropy of the organism remains constant. Furthermore, if the entropy of the organism remains constant, then its ability to absorb and make use of energy remains constant. This is the trick that life (and many other processes as well) plays to stay between complete order and complete disorder.

Life is not an intelligent design that takes self-serving advantage of the world in which it resides, like a TV plugged into a wall socket. That is an idea of ID. Life is more like the water cycle of our Earth. The water cycle is a very complex phenomenon, but let's look for just a moment at a few of its components. Energy from the Sun evaporates moisture from land, plants, oceans and rivers, and sublimates snow and ice from the poles and mountains, to fill the atmosphere with water vapor. Since water vapor is a lower molecular weight than the dominant constituents of air, it rises in the atmosphere. As it rises, the atmospheric pressure drops, and the vapor cools and condenses into water droplets. These small droplets eventually coalesce and form droplets heavy enough to fall as rain, hail or snow onto the oceans and land. This cycle, which is far more complicated than this simple description, repeats endlessly, powered by the energy from the Sun. There is no static equilibrium here; everything is in complex motion. Energy is continually bombarding Earth from the Sun and then being radiated into space. The energy from the Sun possesses far lower entropy than that which is radiated from the Earth, hence thermodynamic equilibrium is never attained.

The water cycle is an example of complexity generated by the necessity to equalize a temperature gradient. There are many more.

A tornado is an example of a dissipative system that most everyone is familiar with. In the American midwest it happens every year: cold air carried down by the jet stream meets warm air from the south. The cold air above the warm air represents a vast reservoir of potential energy. Sometimes the energy content is too great to be dissipated through conventional means—by the cold air diffusing down through the warm air—and a startling performance ensues: the development of a super cell and a corresponding tornado.

A tornado is more than just a large circular whirlwind. It is the emergence of global organization through the interaction of forces on a local scale. In other words, the tornado is assembled from the bottom up with no intelligent supervision. The molecules in the air of the atmosphere cooperate to form a funnel that maximizes the transfer of cool air from top to bottom and hot air from the bottom to the top. This is accomplished through a small set of simple, local rules. And, yet, just like the cellular automaton, global complexity emerges.

It is very easy to find a video on YouTube of a "tornado in a bottle." [107] The key ingredient in performing this experiment is connecting the spouts of two plastic bottles together, one of them full of water and the other empty, then turning them so the filled bottle is on top, thus allowing the water to flow from top to bottom. If the bottles are simply turned upside-down, the water will drain very slowly into the bottom bottle. However, if the two are shaken slightly to introduce a little random motion into the water, a vortex will spontaneously form that empties the upper bottle very rapidly. This is an example of pushing the system away from equilibrium so that the energy gradient is quickly relaxed [108]. Again, this has nothing to do with design or intelligent organization. The vortex represents the collective behavior of the ensemble of particles in the liquid, driven by natural forces to dissipate the energy gradient.

Another example of emergent order within an energy gradient is detailed in a paper by Schneider and Kay [108] and has to do with creating a heat gradient across a thin film of liquid to produce Bénard cells. As was described in Chapter 2, Bénard cells spontaneously form when the heat gradient across a liquid layer reaches a specific level. At a certain threshold, the liquid no longer relies solely on heat conduction and instead shifts to a new regime where convection in the form of circulating hexagonal cells takes over. This change increases substantially the amount of heat that can be transferred across the liquid layer. The interesting point here is that spontaneous order is generated, increasing the efficiency of the heat transport.

It needs to be understood that the Bénard cells are not a process that emerges on its own. Rather, it is the dispersal of energy through the liquid layer that expresses itself in the form of convective, hexagonal cells. The experimental setup is diagrammed on the left in Figure 7.20, and the temperature profiles for the two operative regimes are shown on the right. At low temperature gradients,

7.6. DISSIPATIVE SYSTEMS

thermal conduction through the liquid is enough to balance the temperature difference. However, as the amount of heat injected at the bottom increases, leading to a larger temperature gradient, the fluid spontaneously forms Bénard convection cells to increase the energy flow and offset the gradient. At that point, the temperature profile becomes discontinuous as shown in the figure.

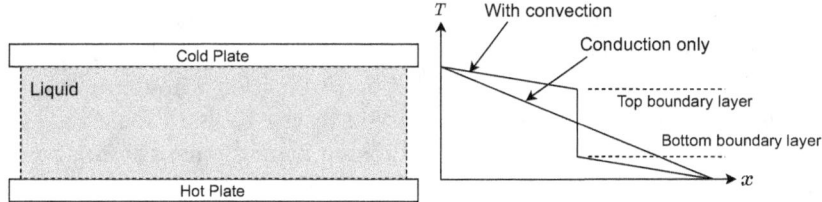

Figure 7.20: A layer of liquid between hot and cold plates. The graph on the right shows the temperature profiles for conduction, and after convective Bénard cells form between two boundary layers.

Again, this is not an intelligent or meaningful process. In fact, it is not even globally controlled. The process relies entirely on local information and represents an emergent phenomenon. It is simply how reality reacts to reduce temperature gradients. Looking at it from another perspective, it is how nature dissipates energy. Thermal conduction alone is not enough to resist the thermal gradient in the liquid. Therefore, convective circulation takes over to dissipate the excess energy. Life behaves similarly. From this perspective, life's sole purpose is to dissipate energy. But even this would be a wrong way of looking at things. It would be like saying that the sole purpose of dominoes are to fall over and dissipate their potential energy. This does not mean that these processes cannot be expressed analytically. They can be.

The point of all this is very subtle. It is quite easy to look at it as though life absorbs energy to perform its required tasks. But this is backwards. Life does not use energy, rather, life is an expression of energy flow. This is really what is meant when we talk about a dissipative system. The flow of energy from a useable form to heat via the second law of thermodynamics produces complexity along the way.

It is important to note that the usual thermodynamic relationships do not apply to these systems as they are not at equilibrium. For example, it is not possible to define a temperature for a system that is not in equilibrium. Temperature is the average kinetic energy of, for example, the particles in a concentration of gas. Unless the kinetic energy is uniform across this gas, it doesn't really make sense to assign it a temperature. In the example above of the liquid between hot and cold plates, there is a temperature gradient from bottom to top and, hence, there is no temperature number that can be assigned to the entire mass of fluid. In fact, it's worse than that. In the presence of a strong gradient, a local temperature is even difficult to define. Of course, one can always the take the global average, for example, if the gradient is linear, then the temperature is the average kinetic energy at the midpoint. It is an average, but it is not what we usually think of as temperature.

Just like the ball in the valley tries to find its way back to the lowest point—the equilibrium point—all dynamic systems tend toward equilibrium. As Schneider and Kay point out, the farther from equilibrium a system is pushed by, for instance, the temperature gradient described above, the more the system will resist the gradient. In the process, the energy consumption will increase as necessary to offset the gradient, and the entropy production will go up in an equal proportion. As has been stated, we don't have the tools to look at this analytically yet primarily because the thermodynamic mathematical models do not hold outside equilibrium. However, it is clear what is happening in some of these situations: life is pulling in ordered information and releasing disorder into the environment. And, as always, there is a price to be paid in under-unity efficiency. That is, the order that is created in the process of life is more than offset by disorder created. As Boltzmann himself put it [109]:

> Between the earth and sun, however, there is a colossal temperature difference; between these two bodies, energy is thus not at all distributed according to the laws of probability. The equalization of temperature, based on the tendency towards greater probability, takes millions of years, because the bodies are so large and are so far apart. The intermediate forms assumed by solar energy, until it falls to terrestrial temperatures, can be fairly improbable, so that we can easily use the transition of heat from sun to earth for the performance of work, like the transition of water from the boiler to the cooling instillation. The general struggle for existence of animate beings is therefore not a struggle for raw materials—these, for organisms, are air, water and soil, all abundantly available—nor for energy which exists in plenty in any body in the form of heat (albeit unfortunately not transformable), but a struggle for entropy, which becomes available through the transition of energy from the hot sun to the cold earth. In order to exploit this transition as much as possible, plants spread their immense surface of leaves and force the sun's energy, before it falls to the earth's temperature, to perform in ways yet unexplored certain chemical syntheses of which no one in our laboratories has so far the least idea. The products of this chemical kitchen constitute the object of struggle of the animal world.

This passage represents an incredibly prescient statement from a genius over a century ago. Life operates on the precipice of change, like a surfer riding a wave.

Creationists argue that life should not be able to exist in a world where the increase of entropy reigns, where increasing disorder is the rule and not the exception. A fallen world. As usual, nature is far more imaginative than this trite scenario. Life performs an incredible act of agility, balancing on the gradient tightrope between stifled order and complete disorder. Leaning to either side results in what we would not call life. A stone is very orderly and predictable, but it is not what we classify as life as it has no variability. A gas has a high degree of disorder as the most probable states are assumed. Somewhere in the middle is what we call life.

It is quite interesting that, as we've demonstrated in this book, randomness is measured in exactly the same way as information: highly improbable events. As Shannon would say, the more "surprising" a sequence is, the more improbable it is, and the more information it contains. On the other hand, the more improbable an event is, the more random it is. The defining characteristic of a random number generator is not that its output is complex in the Kolmorogov sense, but rather that the next state it will assume is completely unpredictable. The only difference between these two is that one has context and the other doesn't. Life gives information or randomness context at the sake of removing the context of the information it produces as a byproduct: this is increasing entropy. Unusable heat energy released from processes that do work is simply energy that no longer can be placed in a particular context.

We're going to take this subject up again in the next section but from a slightly different point of view.

7.7 Life as a dissipative structure

The two defining characteristics of the Universe is that it is action or action in potential. We call the first kinetic energy and the latter potential energy. In a sophomore physics class, one will learn that kinetic energy of a massive object is given by $E_k = (1/2)mv^2$, where m is the mass of that object and v is its speed. Similarly, the potential energy of a massive object above the surface of the Earth is given by $E_p = mgh$, where g is the usual gravitational constant, and h is the height of the object above Earth's surface. Kinetic energy is directed right at quantifying a moving object. Potential energy, on the other hand, is simply saying how an object can move if allowed to do so. That is, the reason an object above Earth's surface is considered to have energy is because it can fall if released.

Energy is a description of the Universe's movement or its ability to move. This motion within reality is truly fundamental. For example, virtually all the laws of physics that deal with dynamics are parameterized by time, that is, by a reference clock. Students working in the sciences just come to naturally assume that there is an "ideal clock" that is always ticking that measures out a constant flow of time. Actually, this is not the case. So how do we tell time? We create clocks, for instance, an atomic clock where a second of time is defined by a certain number of oscillations between two energy states of the caesium-133 atom. But is this really a clock like we usually think of them? Of course not. A measure of time with a clock like this is simply a ratio of the number of cycles of one thing happening to the number of cycles of another thing happening. These things happening—such as the oscillations of an atom—are the action of energy. All happenings in the Universe are driven by energy.

At the point of the Big Bang, the Universe as a whole was infused with a certain amount of energy, both kinetic and potential. (For simplicity, I'm only going to talk about these forms.) This situation could be thought of like a battery that is being discharged. Energy from this initial reservoir diffuses throughout all of the Universe. As it does so, it excites all states it can reach. (Again, the net energy of the Universe is zero. This subject is a little bit complicated, so I'll just talk very briefly about it. The energy from the Big Bang, it appears, will be exactly canceled by the potential energy of the Universe after an infinite time. If the potential energy were greater than this, the Universe would be open. If it were less, the Universe would be closed. It is flat.)

Of course, we could complicate this significantly by looking at it from a quantum-mechanical perspective, but I don't think we need to go there. Even in the presence of quantum mechanics, energy is conserved, meaning that the sum of initial potential and kinetic energy of the early Universe cannot change. It can only diffuse throughout the Universe. In this sense, there is not a particle in existence in the Universe today whose motion (or existence) cannot be traced back to the original energy content of the Universe.

It is similar to kicking off the first domino in the most amazing domino extravaganza that has ever been put in place. This one knocks that one over, it kicks over the next two, they knock over several more chains, etc. This silly domino example will have to be excused, but it isn't that far off. The initial energy of the Universe was kicked into motion and this motion is propagating across all of reality. Not because it has any intent to do so; rather, it is just like the dominoes. One knocking the next over as far as the train reaches.

And, just like the domino stunts that get all of the applause—like climbing over a ramp, the last domino tumbling over the top with a heart-stopping fall before crashing into the waiting string of dominoes below—this energy activates every state of the Universe. But what people remember from the domino show is not the pauses—they remember what happens when the pieces are in motion. And, just like the domino exhibition, the dynamic diffusion of energy pulls off some rather amazing stunts, one of them being life.

Life is different from human-made designs on the macroscopic scale because it operates right down to the single physical or informational bits of the microscopic world. And why would that be surprising? At that level, randomness is a primary component. It no longer shows up as the annoyance of thermal noise in small-signal amplifiers. It is a dominant component. This is the reason Schrödinger suggested that genetic material might be information stored at the molecular level, and not constructed from *intensive* variables.

Intensive quantities in thermodynamics are those that statistically describe a large ensemble of atoms or molecules. Temperature is an example of an intensive variable. Temperature is a scaled measurement of the average kinetic energy of the particles that make up a system, for example, a gas. It has no meaning whatsoever when talking about a single particle.

It must be remembered that at the time Schrödinger wrote about how genetics functioned that DNA had not yet been discovered. By looking at the

7.7. LIFE AS A DISSIPATIVE STRUCTURE

dynamics of genetics, he correctly intuited that the information organisms use for reproduction must be encoded at the molecular level. Consider if the "bits" encoding life were defined by, for example, large ensembles of molecules. If this were the case changing one of the molecules through a single-point mutation would have little effect on the entire group of molecules, thus making evolution impossible. No, it is virtually a requirement that genetic material be encoded at the molecular level to enable evolution through changes in those molecules.

Along the same lines, it is important that biological structures are assembled starting at the molecular level. At the molecular level, as we previously alluded to, objects behave much differently than they do at the macroscopic scale. For example, we use magnetic-field motors for fans, electric cars, etc. Electric-field motors are sometimes used today as well, but they are far less common and only used in special applications. The reason for this is that, at the macroscopic scale, magnetic-field motors are easier to build, more efficient, and capable of higher torque. This changes at the molecular level.

At very small scales, use of the electric field becomes tractable, not to mention that it is difficult for biological tissues to focus magnetic fields. Virtually all functions at the molecular level, including the self-assembly of molecular structures, are based on the electric field.

Before these molecules can assemble they need to be brought into contact with one another. On the macroscopic scale, some means would be needed to bring components together so they could be self-assembled. We saw this with magnetic-sphere examples in the section on self-assembly. If the spheres are simply laid out on a table, separated far enough from one another that the force of the magnetic fields can't pull them together, they will never self-assemble. However, if the table on which the spheres are placed is shaken then they will be jostled into one another, and they will self-assemble. At the microscopic scale, no manual intervention is needed. Let's see how this is accomplished.

First of all, we need to remember what entropy is, but let's start first with Shannon's definition of information. In the simplest terms, he defined the information of a given outcome as the amount of "surprise" associated with it. The more surprising an event is, the more information it carries. In more formal terms, the lower the probability of an event occurring, the more surprising it is, and the more information it contains. Mail showing up today in your mailbox is not that surprising; in fact, it has a relatively high probability of happening. However, a letter with a winning lottery ticket showing up in your mailbox is a low-probability event and it would be very surprising as it would be life changing, and hence the event contains a relatively large amount of information.

The information contained in a specific system configuration or event is defined as the negative of the logarithm of the probability of that configuration existing or event occurring: $I = -log P_i$. The particular logarithm used is arbitrary, the subscript i denotes the event being assessed, and P_i is the probability of that event occurring. The more improbable an event is, the smaller is its probability, and the larger the value of the information calculation.

On the other hand, the entropy a system possesses is the *expected value* of the information the system expresses and is defined as:

$$S = -\sum_{i=1}^{N} P_i \log P_i \qquad (7.5)$$

The expected value is not what most people think about when they talk about the average of a variable. The expected value is the weighted mean—it is the most likely value that will be found when the system is measured. The expected value is calculated in the way it is because certain outcomes may be more probable than others. If the probability of all outcomes are equal, then the expected value does, in fact, just reduce to the standard definition of the average. And so entropy can be thought of the average information of events a system produces. In other words, it is related to how improbable these events are on average.

We discussed probability distributions (PDFs) briefly in Chapter 6. The PDF of a random process gives the likelihood of any specific event occurring. For example, if we have a normal distribution describing a certain process and the mean of this distribution is located at, say, $x = 1$, then the most likely outcome of the process is indeed $x = 1$. If we pick any other value of x, the value of the PDF will tell us the likelihood of that event occurring.

A process has maximum entropy when its corresponding PDF is uniform, that is, constant for every possible outcome. For example, a random number generator that produces numbers between one and ten with a uniform PDF is equally likely to produce any number in this range: the probability of choosing one is the same as choosing any other number. Since there are ten numbers total, the PDF is a horizontal line of value 1/10. The reason the entropy number is maximized for a uniform distribution is precisely because all outcomes are equally probable: you are flying blind and have no idea which might occur. This is the most disordered state a system can achieve.

To illustrate this, consider playing a game with your friend of guessing the number between one and ten that he is thinking of. If he tells you nothing more, the outcome space has a uniform PDF, and all outcomes are equally likely. However, if you know him well and that he chooses a number above five twice as often as a number five or below, the distribution is no longer uniform. You would always choose a number above five to double your odds of guessing successfully. There is less entropy in this case as compared to the uniform distribution because you have some idea as to where the solution lies.

For now, unless stated otherwise, assume that we're talking about processes with uniform PDFs, meaning the entropy measure is the average information of any event. So, when someone says the entropy of a system increases, it means that the number of possible outcomes in the sample space increases—there are more possibilities for the outcome. This has to be the case for a process with a uniform distribution if the entropy is to increase. Think about the number game above and guessing the number your friend is thinking of. If the disorder

7.7. LIFE AS A DISSIPATIVE STRUCTURE

is to go up in that game, the only way is for the number range over which the game is being played to increase. For the ten-number game, the entropy is about 3.32 bits, and for a twenty-number game, it is about 4.32 bits. These numbers make sense, because, in each case, it takes this many bits to represent in binary a number between one and ten and one in twenty, respectively.

Recall that the temperature of a material, say an ideal gas, is the average kinetic energy of the particles that make it up: $\bar{E} = (3/2)kT$. If we place a group of particles in a box and give them some energy, in time, they will all be moving at random speeds and traveling in random directions as dictated by the *equipartition of energy theorem* (we won't go through the mathematics here). The equipartition theorem states that the energy will be shared in all degrees of freedom of the particles. In other words, if we take just the simple case of translational motion (it is actually a little more complicated than this), then the energy is shared between all three axes of motion; that is, there are terms $(1/2)mv_x^2$, $(1/2)mv_y^2$, and $(1/2)mv_z^2$.

If we mix the gas in the original box at temperature, say T_0, with another volume of gas particles at a lower temperature, say T_1, the total energy of the once isolated systems will eventually be shared evenly across the mixture. If the masses of the two volumes of gas were identical, then the final temperature of the mixture will be $T_f = (T_0 + T_1)/2$. So, how does this mixing come about?

Since the temperatures of the original volumes of gas were above zero, the molecules that make them up were moving. And since from the equipartition theorem, the particles in both are moving in all directions, they will inevitably collide. As these collisions take place, momentum (mass times velocity) of the particles is shared between them. Particles with excess energy impart a portion of their energy to the less-energetic particles until equilibrium is once again reached. Energy diffuses throughout the total ensemble of particles, moving from regions of high energy to those of lower energy until the energy content of the gas mixture is uniform.

So, what drove the two temperatures to equilibrium? Why did the energy spread from the higher-temperature gas to the lower-temperature gas? The cause is simply that particles in motion will tend to collide with one another. And as this happens, energy is shared in the collisions. High-energy particles tend to give away their energy and low-energy particles tend to absorb energy until an equilibrium is reached. There is no magic here; there is just arithmetic.

Does the entropy of the system increase when this happens? It seems that it might not because the total energy of the system will be the same before and after the mixing. Let's say a particle has an x-axis velocity component of v_x. If we assume this variable is continuous, that is, the particle can take on any value of speed between zero and $|v_x|$, then the length of the vector is proportional to the number of energy levels this particle can take in the x-axis direction and, hence, determines the entropy associated with the particle. But in a collision between multiple particles, if the energy and momentum are conserved, how does the overall entropy increase?

Let's look at a glancing two-dimensional elastic collision between a particle—call it particle 1—moving with momentum mv_{x1} to the right, and a second particle—call it particle 2—that is stationary. Before the collision, particle 1 has a velocity of v_{1xi} in the x-axis direction, and particle 2 is not moving. After the collision, the first particle's velocity has components of v_{x1f} and v_{y1f} in the x and y directions, respectively, and the second particle has components v_{x2f} and v_{y2f}. This is shown in Figure 7.21. Momentum is always conserved in a collision like this, and since the collision is elastic (no energy lost), energy is conserved as well.

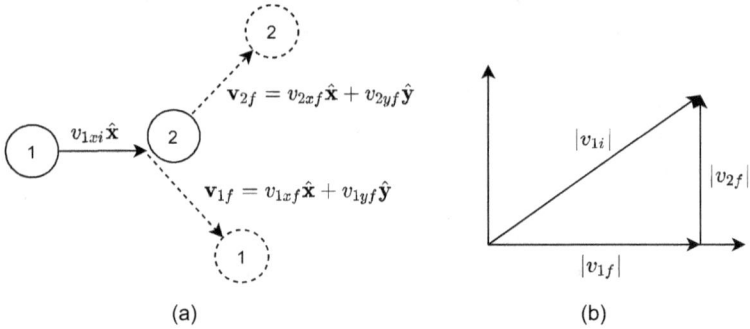

Figure 7.21: In (a), the moving particle, 1, impacts a second stationary particle, 2, elastically. The initial and final speeds of the two particles are shown (b).

The conservation of energy of the system can be written as

$$|v_{1i}|^2 = |v_{1f}|^2 + |v_{2f}|^2, \tag{7.6}$$

and, since this equation is the Pythagorean theorem, it can be graphically drawn as in illustration (b) in the figure. The number of degrees of freedom in the initial system was $|v_{1xi}|$. However, after the collision, the particles can move along both axes. With an application of the well-known triangle inequality, it is simple to show that $|v_{1i}| \leq |v_{1f}| + |v_{2f}|$, which demonstrates that the number of allowable states in the two final velocities is greater than or equal to that for the initial velocity despite the fact that the energy was conserved. Hence, the entropy of the system increased, not because the energy increased, as it did not, but, rather, because the energy diffused into another particle in the system. The combined motion of the two particles provides more possible outcomes than the one particle alone.

This process carries on indefinitely if the system under question is not isolated, which no system in the real world is. Note that the second particle of the system above did not have to be stationary to begin with; the outcome is the same if the two particles had finite, different energies at the start. Energy will make its way from the higher-energy states to the lower and, as it does, expand the allowable number of states of motion. This is increasing entropy. It is also increasing information content simply because it takes more information to describe the state of the final system. So, the entropy of the overall system increased, but what was the entropy change of the first particle?

From the diagram (b) in Figure 7.21 it is clear that $|v_{1f}| \leq |v_{1i}|$, meaning that the entropy of particle 1 either remained the same or decreased. This is an extremely simple example, I know, but it clearly illustrates that the entropy of a local region can decrease at the expense of increasing the entropy of the overall system.

A second thing to note is that the final energy of particle 1 is less than its initial energy, that is, $|v_{1f}|^2 \leq |v_{1i}|^2$. In this case, the loss of energy by particle 1 is what drove its reduction in entropy. Now, don't read too much into this. I'm making two points with this very simple example. The first is that entropy increases because energy tends to explore all possible degrees of freedom of a system. That is its basic function. The second point is that by reducing the energy to zero (this can't be done in practice), the entropy associated with the one-particle system would go to zero as well.

Again, take this example with a grain of salt as the treatment has been very informal. But it is indicative of one thing: zero entropy is not the secret to life. I know all technically-oriented people are aware of this—this is an answer to those who argue that the second law is an impediment to a natural explanation for life.

In order to approach this question of life, it is necessary to start thinking of it in a dynamic sense. Static models will not get us there. In other words, life is not the string of dominoes. It isn't the dominoes at all. *It is the falling dominoes.* In fact, it is just the falling. This is a subtle and, yet, all important distinction.

Entropy is energy seeking to excite all possibilities of a system. When change stops happening in a system, it has reached its entropic limit. This is Schrödinger's entropy death. And this is certainly what happens when two gasses at different temperatures are mixed together for a time. When the temperature equilibrates, nothing changes after that (of course, the particles are still moving, but their mean energy remains constant from that point on). A hot cup of coffee sitting on a table will eventually reach the same temperature of the air around it. When that happens, energy is no longer transferred and things stop changing. This is not what happens with life.

Think about it this way. Let's say we sit this same cup of coffee on a hotplate. Then what happens from an energy and entropy point of view? Does the energy of the cup just keep increasing? No. If it did, the cup would melt. Energy flows from the hotplate, into the cup, and then into the atmosphere. We have an energy gradient between the hot plate and the atmosphere. And what happens in the entropy case? Does the entropy of the cup just keep increasing? No. Since the temperature of the cup does not keep increasing, the entropy remains constant (I'm ignoring things like the water evaporating, etc.). Then where is the increase in entropy that the second law demands? The entropy of the atmosphere continues to increase as long as the hotplate is on, however, the entropy of the cup remains constant.

This may seem to counter what many people in the ID community say about the second law and, in fact, it does. Entropy is not some kind of dark magic that makes everything degrade. It is simply energy finding the maximum number of degrees of freedom of a system. The temperature and physical size of the coffee cup remain constant, therefore there is no entropy increase. And yet, energy is still flowing through it across the temperature gradient. This is what we would call a simple dissipative system.

A *dissipative system* is a thermodynamically open system that is operating out of equilibrium, exchanging energy and mass with its surroundings. Again, examples of dissipative systems include tornadoes, river beds, lightning, snowflakes, even a layer of liquid placed between a temperature gradient. And order emerges within all of these systems to more efficiently remove energy gradients.

For example, one could say that a river bed forms with a complex web of tributaries because this is just how the water flows from higher to lower ground. And this is correct. The complex paths form naturally to maximize the flow of water from a higher to lower energy level. Water will continue to flow as long as there is an energy gradient. But the interesting thing is that there is no intelligence guiding the formation of the river bed, even though the shape of the bed is optimized to maximize the transfer of water from one location to another. Nature finds its own way.

It is not fully known at this time, but it is suspected that metabolic chains in biological life are formed in a similar way. They are the best paths to dissipate energy across a gradient and, as they do this, order emerges.

Now don't get the wrong idea. Life is not optimized by evolution to be simply the most effective dissipator of energy. Dissipative here rather refers to energy moving through a system as a consequence of an energy gradient. This is why, as the definition above indicated, the system is not in equilibrium.

As another illustrative example, non-equilibrium describes trying to run up a hill covered in ice. You keep slipping back down the hill, so you run faster. Children often try this. If one runs fast enough, they might find themselves halfway up the hill, neither moving down it nor continuing to climb. If one can run fast enough, the frictional force of slipping feet matches the downward force of gravity. This runner is able to stay on the side of the hill by continuing to expend energy, resulting in the increase of entropy in their environment through the heat released from the runner's body and the slipping of feet on the ice.

The biological metabolic functioning involves a continual injection of energy. One small break in that energy chain, and the chemical runner will come sliding down the hill. But don't think that the runner in this case has any intent. Energy is simply propagating through these molecular structures, exciting all states it can reach as it does.

The cyclic electron flow during the photosynthesis of adenosine triphosphate (ATP) in plant life is driven by energy traveling downhill. High-energy electrons

7.7. LIFE AS A DISSIPATIVE STRUCTURE

are first generated through photon absorption by chlorophyll pigments within the plant. This is followed by several reactions that claim the energy from the electrons and produce a proton gradient across a membrane. The proton gradient is then used to drive the process that stores energy in the form of ATP. The entire process consists of energy entering in the form of sunlight and then cascading throughout the system, driving the biological processes required by the organism as it goes. Ordered, usable energy enters the system and disordered energy exits. Mass enters the system as water and carbon dioxide and oxygen exits.

Processes like these raise an interesting question: are metabolic chains like this formed in the same way as other natural complex systems? Of course, this question is too simple. The entire biological system is one integral unit and cannot be separated into pieces. The metabolic chains operate in conjunction with the DNA that defines them, along with a myriad of other processes.

It can be looked at like the formation of an early metabolic chain requires a great deal of luck. And that's okay so long as that measure of luck has a sufficiently high probability to be offset by time. In other words, it can't be so improbable as to never occur in an allotted time period. And there's one more piece that is required to make sure that it is not a one-time event.

As Montevil and Mossio have pointed out, one of the outcomes of this particular reaction must be that, in addition to any other products, it produce the constraints necessary for the same reaction to be repeated [98]. This is how the dominoes are set up. This dominoes are not like those in the toy box. This one only has to be built once, and so long as the necessary raw materials are available, it will just keep falling. The fall of the first one prepares the second one for falling, and so on. This is the constraint closure we talked about earlier, and it is the basis of a metabolic chain.

The question is whether this chain forms by random excitation and selection, or as the emergence of complexity via a dissipative system. There is a distinction between these two. The main driver of the diversity of life is random mutation and natural selection, but it isn't the only one. We've seen many other processes that aid in the evolution of life throughout this book. The primary one we're concerned with here is the one that Adrian Bejan dubbed the constructal law. It is a description of how energy, propagating through nature, carves out complexity as a matter of course. The question is, does this tendency of nature hold at the molecular level?

We've already seen that it does. The formation of a snowflake as energy propagates through it is a primary example. A second example is the formation of a phospholipid membrane. In neither case are global rules (or designs) required. Simple entropic behavior—the way energy takes the most probabilistic routes— is enough to explain them. And, although the research needed to confirm it has not been completed, it is not a giant leap to surmise that biological complexity is a result of this same phenomenon.

There is a good chance that these metabolic chains form for the same reason as the formation of a complex river basin. Energy finds the most efficient route for its dispersal. We as humans want to imbue the results of these processes with a special significance because we don't fully understand them at this point. The diffusion of energy as stated by the second law is not a bad thing; it only got that reputation through its being associated with disorder. Disorder is the wrong word. Energy by its very nature propagates through every path it can, like water running down a street looking for a drain.

Just like the fractal river bed that keeps coming up, metabolic chains are formed because energy flows through them. Again, don't get the wrong idea; energy is a descriptive phenomenon. Energy paths that are used more often become dominant. Just like the river's water, where energy flows, more energy flows. The metabolic chains do not form and evolve for the purpose of dissipating energy, rather, they form because energy is dissipated through them. This is why these chains have to be built for continuous energy flow. They do not burn out like light bulbs. And just like the river, their flows do not stop.

Again, this is very subtle and can be confusing because it is turning the picture on its head. The potential for the metabolic path was always there, and it was simple revealed by energy flow. Understanding that a complex river bed can be generated this way is not so surprising, but, apply the same philosophy to a metabolic chain, and people will tend to want to object.

I think a second part of the problem as to why we regard these particular processes as mysterious is because, as we discussed previously, we cannot write down a closed-form solution for them. As with the cellular automata, the only way to propagate a particular initial condition to an outcome is to just do it, either by experiment or simulation. There is no means of predicting the outcome from a mathematical equation. The other reason is that biological systems are complex beyond what we can imagine. Nature has had almost four billion years to develop these systems, and they have become extremely complicated. But it is important to remember that they still have a natural explanation.

It does need pointed out again that entropy increase by the second law is required for life to exist. Increasing entropy is exactly the description of energy seeking out the paths it must in order to bring about complexity. Life is neither order nor randomness; it resides somewhere in between as shown in Figure 7.22.

Life must reside outside of dynamic equilibrium. There are two positions of equilibrium, one on the side of order and the other at maximum entropy. Neither of these is suitable for what we call life. A salt crystal possesses extreme order and is certainly not life. A collection of gas molecules at constant temperature, pressure and volume has reached maximum entropy and is also nonrepresentative of life. Life exists in the continuous transition from order to randomness. It never reaches complete randomness because it takes in energy to prevent it. In other words, it continually runs fast enough to not slip down the hill.

7.8. MAXWELL'S DEMON

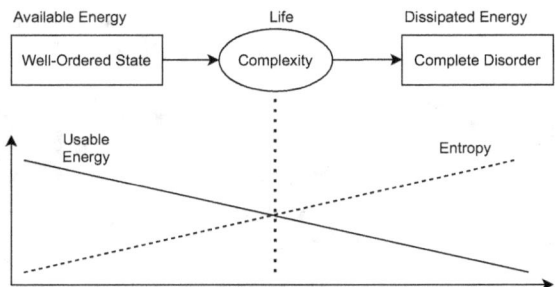

Figure 7.22: A depiction of usable energy being injected on the left and transforming into unusable energy or thermal noise. The entropy of the system increases as this transformation takes place. The processes of life reside on the derivative of the energy and entropy curves; the point between order and disorder.

7.8 Maxwell's demon

Is it possible that there is a designer who played with the second law to enable the emergence of life? This subject has come up many times in the past, but the most famous example is a scenario put forth by James Clerk Maxwell in 1867 called *Maxwell's demon*. Maxwell's demon is a thought experiment that considers a possible means of violating the second law of thermodynamics. Maxwell put it like this [110]:

> ... if we conceive of a being whose faculties are so sharpened that he can follow every molecule in its course, such a being, whose attributes are as essentially finite as our own, would be able to do what is impossible to us. For we have seen that molecules in a vessel full of air at uniform temperature are moving with velocities by no means uniform, though the mean velocity of any great number of them, arbitrarily selected, is almost exactly uniform. Now let us suppose that such a vessel is divided into two portions, A and B, by a division in which there is a small hole, and that a being, who can see the individual molecules, opens and closes this hole, so as to allow only the swifter molecules to pass from A to B, and only the slower molecules to pass from B to A. He will thus, without expenditure of work, raise the temperature of B and lower that of A, in contradiction to the second law of thermodynamics.

Here, we consider a slightly different experiment as shown in Figure 7.23 where a gas at a given temperature is trapped in the left side of a box, and the right side is a vacuum. In between the two is a divider with a massless and frictionless door that is opened and closed by Maxwell's demon to allow particles from the left to escape to the right. The second law says over time, without the divider in place, the box should equilibrate to have the same number of particles on the left as the right, double the original entropy of the system with the gas solely on the left. The intent is for the demon to lift the door at the appropriate times such that all the particles transition to the right side.

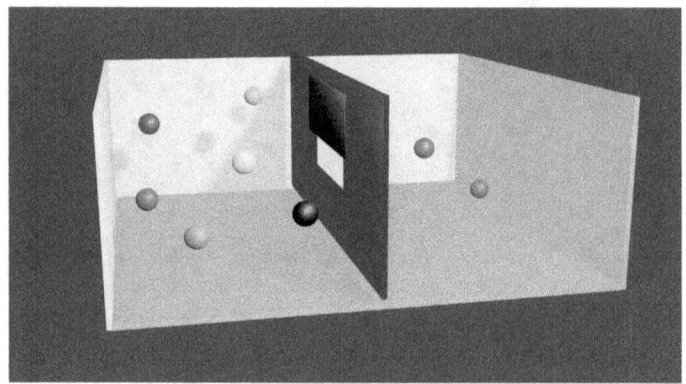

Figure 7.23: A depiction of Maxwell's demon. The door between the two sides of the box is massless and frictionless and opened at just the right times by a so-called demon such that the particles are directed to one side of the box. The intent is to show that the second law of thermodynamics could be violated.

If the demon is fast enough and observant enough, it should be possible for the door to be opened and closed at the appropriate times to corral the particles to the right. The door being massless and frictionless means no work will be exerted in its opening and closing, and that no energy will be injected into the system. If successful, this would be a direct violation of the second law.

Several equivalent experiments have been successfully performed to show it is possible to accomplish this task. They are not exactly like the setup we've shown here, but the effect is the same. It is possible for the demon to open and close the gate such that the particles all end up on the right side of the box. The question is, do these experiments violate the second law? Surprisingly, the answer is no.

In order to work his magic, the demon must develop information about the particles' locations and velocities. Without this information, it would be impossible to appropriately synchronize the door to corral the particles on the right side of the box. And here's the kicker: we've already seen that information is on par with randomness and entropy. Although the particles become more ordered when they are confined to the right side of the box, an equivalent amount of information, or disorder, has been transferred to the mind or memory of the demon. This additional disorder cannot be discounted from the experiment.

One might ask how this information is considered disordered. Consider computer memory for a moment. When the memory is erased, all of the bits are the same value (1 or 0, it depends on the memory type). This is a very ordered structure and takes only one bit to describe, that is, a 1 or 0. However, once the memory has real information placed in it, such as the picture of a sunset, the array of ones and zeros appear to be almost random. The entropy of the memory has increased. In this sense, as Rolf Landauer showed, information is physical [111].

So we end up in a bit of a conundrum: if there is an intelligent designer who is circumventing the second law of thermodynamics, then this designer is suffering the entropy increase, not the life it is aiding. This is a sort of silly implication, but if ID proponents and others are arguing that a god is a solution to the problem of disorder increasing in the biological world, then a subsequent solution must be found for the god.

7.9 Is there design in nature?

Despite the comparisons that are often made, the structures found in nature are not constructed the way engineers design things. At this point in time, nature's systems are far more complex and intricate than that which any electrical engineer can create, but that's not what makes them different. I have very little doubt that with enough time, human engineering will design systems that are comparable in complexity to what nature has put together. No, it doesn't have to do with the complexity of the end products, but rather in the design process itself.

Human engineers require a good understanding of where they are trying to go with a design and how they are going to get there with the functional pieces at their disposal. An intimate knowledge of how the components will work together is crucial. It turns out to be easier than it might seem because many devices that can be bought off the shelf are well characterized in terms of their behavior under a vast number of operating conditions. And, even if the requirements for the design require research to custom design specific devices, the tools of mathematics and physics are always available.

For example, the design of a controller for a particular application requires first a product or device specification that details exactly how the final product will behave and under what conditions. Then, the overall system is broken into modular subsystems and the requirements for each established. This simplifies the design by reducing dependencies across the design so that the engineer is only required to look at one subsystem at a time. It is generally impossible to design an entire system at one time because it is just to complex to keep in one's mind. Modular design also reduces the number of interdependencies across the global system. As these connections increase in number, the complexity tends to increase exponentially, making design and subsequent debugging far more difficult.

Nature does not construct systems this way. The main difference is that there is not a plan, as there is no goal. Human engineers always have a goal; nature does not. This is such a subtle concept and difficult to grasp from the human mind's perspective that, at this juncture, there isn't even really a fully-formed vernacular for speaking about how nature goes about forming the structures it does. Although naturally constructed systems and their processes can be put into a functional framework in an anthropomorphic way, it is incorrect to look at them this way.

As has been mentioned several times in this book, a river does not form by planning. It cuts its own path as it goes with no prior knowledge as to how that path should be constructed to maximize the transport of water. The more water that it is required to carry, the deeper it will be cut, leading to its being able to carry more water. This is a positive feedback condition that is brought about simply by the river's tendency to disperse energy.

Similarly, phospholipid bilayers do not form by design. They are an expression of reality's tendency to disperse energy and to maximize the overall entropy of the water mixture.

Evolution is the hallmark example of nature blindly finding a path, but it is still not satisfying a goal as such. It is very much like the river cutting its own basin. Life is a dissipative system and all of its subsystems are self-constructed because they are effective at dispersing energy. The number of organisms in a species do not increase because of design, but rather due to random mutations that lead to particular individuals possessing genetic material that leads to more individuals having this same genetic material. The more individuals that possess these particular genes the more offspring that will be produced, leading to more of this genetic material being produced. Evolution cuts its own path through the fitness landscape.

Humans do not design systems this way. This is not to say that they wouldn't if they could; many researchers are working intensely in this area. But, even if they succeed, they will have a goal for the end process. Nature does not. If the clock were rewound and life was given a chance to start on Earth again, the resulting lifeforms would look vastly different than what we see today. I don't think anyone would disagree about this when it comes to a riverbed, but I suspect there would be some disagreement in certain circles that life in general would emerge differently.

The point is that even if human designers can harness these aspects of nature to generate new, complex designs, that is obviously not what happened in nature. Nature wandered where it would, demonstrating no impetus from this perspective to conclude an intentioned design.

It is easy to come up with examples of biological functions that were obviously not designed, and they have been the bane of ID advocates for a long time. For example, we have an appendix that was designed for digesting grasses on the savannah many eons ago. It has now become simply a vestigial appendage that only serves as a problem as it often becomes infected and needs removed.

A designer, who constructed the perfect design (and this is what we would expect from a supernatural entity) would have shielded DNA from cosmic rays and provided perfect fidelity in the copying of DNA. These things are certainly done today in contemporary electronics. There are three billion base pairs in human DNA, but this is not such a large data block to copy. We can do it with ease with today's computers. Also, all electronics are shielded from electromagnetic radiation.

It should have occurred to the designer that pain was unnecessary; a heads-up display would have sufficed in most instances. I'm being a bit funny here, but there is seriously no reason for chronic pain. Pain may have been necessary at some point in the past when our ancestors were more animal-like, but it is not needed now. That is unless the usual Christian argument is going to be made that pain is good for us. But, if that's the case, then throw the aspirin out of your house.

I could go on as the list of design deficiencies is very long, but I don't think it's necessary. The point is made.

7.10 Conclusion

The primary component of life is self-replication, aided by, of course, self-assembly. Virtually all aspects of life on Earth rest upon this one property. Without it, evolution, which drives complexity and diversity, is not possible. Life resides at the points of the flow of energy and information gradients. *It is change.*

The beauty of a snowflake and the magnificence of a tornado are both progeny of a gradient; they could not exist in stasis. Life is change. It cannot be understood from a frozen point of view, leading to disbelief that it is even possible without an intervening deity of some type.

I have worked with many engineers who do not understand the link between the time and frequency domains. These two are duals, carrying the same information, and are switched between using a Fourier or Laplace transform. It is certainly possible to be a good engineer while only working in the time domain, but there will come a time when this deficiency becomes a problem. Many dynamic systems are difficult or impossible to analyze in the time domain alone. Looking at life in a static way leads to the same problem.

Our minds attempt to solidify concepts so they are useful to us later, and we have a particularly difficult time in not applying this strategy to all of reality. Dancing is not a memory or photograph of a dance; it is dancing. Life is the same way. It is completely dynamic and any attempt to freeze it in a conceptualization tends to kill it.

This is why making statements like, "It just seems so complicated that I don't see how it came about on its own," make no sense at all. They do not approach the reality of the situation.

I used to be a gymnast when I was younger and doing a full back flip was no problem at all (it is now...). Taking a picture of someone of performing a full back flip while their feet are above their head leads to an apparently impossible depiction. Someone looking at the photograph who had never seen a back flip before would argue that what was displayed was an impossibility: no one can float upside down above the ground. And, this is true, they can't. In a stationary world, it is impossible. But in a dynamic world, such a thing is certainly possible.

This rather mundane analogy is actually very relevant to the ID argument as it is an error in context. And even if they see that some things are possible, like the spontaneous formation of phospholipid membranes, they will say, "Yes, but it isn't possible in a more complex case, such as with actual cell membranes." A gymnast's purpose is to show spectators that the impossible is possible. To do this, they shift the context of the situation from stationary to dynamic.

One has to remember, when you're looking at a photograph of life, you're not looking at life. Life exists in mid-flight not sitting in a metal chair on the sidelines. If you want to see it, you have to stop looking at it in pictures and actually look at what it is doing. It cannot be seen in a frozen frame, even if that frame is a two-thousand-year-old book.

Chapter 8

Intelligence and free will

> "Along the riverbank under the trees, I discover footprints. Even under the fragrant grass, I see his prints. Deep in remote mountains they are found. These traces can no more be hidden than one's nose, looking heavenward.", Zen tradition, *The Ten Bulls of Zen.*

The primary claim of intelligent design is that some supernatural intelligence had to have been involved in the emergence of biological complexity as we observe it today, that this complexity could not have arisen solely by natural means. Intelligent design proponents compare the supposed intelligent designer to a human engineer, who, utilizing this strange power called intelligence, can conjure immensely complex machines into existence via systematic planning, deterministic designing, and maybe some amount of serendipity. But even this serendipitous component appears to fall under the auspices of some idea of intelligence; why else would people get credit for it? Speaking from experience, this is not exactly how I understand the process.

The dictionary defines intelligence as "the ability to acquire and apply knowledge and skills." This definition really doesn't say anything as the entire meaning of the definition rests upon the word "apply." It is clear that it doesn't mean to be trained in the ability to do particular tasks as even obviously unintelligent devices like computers can do this. I can certainly write a program for my computer that will provide it with the ability to balance my checking account. I can even get it to be suspicious of account activity and track down questionable purchases just by employing a particular piece of software. However, I can't get it to spend my money the way I would do so. I can even program it with a desire to buy items, perhaps spend $100 per month in online purchases. What it can't do, however, is make a spontaneous decision concerning what it should buy *without proper reasoning*.

By "proper reasoning," in this context, I mean a set of programmatic rules. Without a set of programmatic statements, or rules, the computer will not be able to arrive at an actionable outcome. On the other hand, this is something that humans apparently do all the time. In fact, even my cat can do this by apparently deciding if he wants to be petted or not. Even a flipped coin can make a spontaneous decision to land on either heads or tails. But a deterministic machine, like a computer, can't accomplish this feat at all. (Of course, the program can act conditionally upon the state of a source of randomness or an unpredictable input, but its programming can't accomplish the task alone.)

What would it mean for a computer to "learn" something new? To come up with a novel idea?

It would be an error. A computer is supposed to be able to follow the code in programs without deviation. This code allows them to perform a specific set of functions. Any operation outside this set of instructions is considered an error and is typically followed by the ominous blue screen of death. The point is that any new behavior for a deterministic system *is a mistake*.

And here is where our story of intelligence starts.

8.1 Intelligence and the brain

The term intelligence is one of the most broadly defined terms in science. In 1994, a consensus report put together by Linda Gottfredson of the University of Delaware was published in the *Wall Street Journal* with the approval of 52 university professors with expertise in the intelligence and associated fields. It defined intelligence as follows [112]:

> Intelligence is a very general mental capability that, among other things, involves the ability to reason, plan, solve problems, think abstractly, comprehend complex ideas, learn quickly and learn from experience. It is not merely book learning, a narrow academic skill, or test-taking smarts. Rather, it reflects a broader and deeper capability for comprehending our surroundings—"catching on," "making sense" of things, or "figuring out" what to do.

This definition is essentially a description of the effects of intelligence, but says nothing about what intelligence is or what brings it about.

Intelligence is typically inferred from the performance of an individual in solving complex problems, or problems that make you think. These can be mathematical in nature, numerical sequences that need finished, spatial reasoning problems, verbal problems, etc. People who do well on one type of test generally do well on all of them and people who do poorly on one of them tend to do so on all of them. However, there are slight differences in a particular individual's performance with respect to each of the categories.

At the turn of the century, psychologist Charles Spearman applied *factor*

8.1. INTELLIGENCE AND THE BRAIN

analysis to extract a general measure simply called g that correlates to all types of tests [113]. Similar to other statistical methods like principle-component analysis and partial-least squares, factor analysis extracts a minimal number of dimensions in the sample space needed to describe the correlations between measurements. A simpler way to think about this is the coordinate system in which the original measurements lie is transformed to a second coordinate system where the same data is expressed within fewer dimensions, or, as in the case of g, only one dimension.

This general factor g is an accurate indicator of an individual's ability to satisfy the definition for intelligence above. It is strongly correlated with a person's level of success in school or in the workforce. But, rather than discuss the effects of intelligence, we want to examine the functioning of intelligence and, in particular, a couple of the requirements that must be met for the emergence of intelligence.

As was mentioned previously, the word intelligence is etymologically borrowed from the Latin *intelligentia* meaning "the act of choosing between."

The term intelligence is used by the ID community as a sort of supernatural property that not only gods, but apparently humans as well, possess that allows them to operate in a manner which the rest of reality cannot. It seems that physical reality and processes do not possess the ability to reason like human beings.

There is no denying that human minds have an ability for abstract thought and virtual simulation of the world around them that is not available to physical objects, or even, maybe, to other animals. I suspect, however, that it is only levels of graduation that separate humans from the rest of the animal kingdom, in particular, animals like chimpanzees and octupi, but this is still an unknown.

The concept of engineering is brought up repeatedly in the ID literature by people who are not engineers, explaining how intelligent entities design electronics and machines of all varieties and how this process is so much different from the random messiness of evolutionary theory. For example, Dembski's description of how intelligent engineers might co-opt components from a microwave to create a TV by reusing common electrical doodads is not entirely accurate. It seems his point is that the precise design process of human engineering does not happen by blind random chance but rather via intelligence and knowledge, and that the engineering of biological systems must happen similarly.

The claim of ID is that evolution is messy or impossible in that it relies on random modifications of DNA and the subsequent evaluation of the performance of these modifications in the real world, with nature keeping the best and leaving the worst behind. The claim is that the engineering human beings use is quite different than that of evolution; it's far more straightforward and guaranteed because intelligence is behind it. And that life must have been designed by a similar intelligence.

In Dembski's words from the preface of his book *No Free Lunch* he describes the design process as follows: "How a designer gets from thought to thing is, at least in broad strokes, straightforward: (1) A designer conceives a purpose. (2) To accomplish that purpose, the designer forms a plan. (3) To execute the plan, the designer specifies building materials and assembly instructions. (4) Finally, the designer or some surrogate applies the assembly instructions to the building materials. What emerges is a designed object, and the designer is successful to the degree that the object fulfills the designer's purpose."

From my experience, engineering does not consist of robbing doohickeys from existing devices and fitting the pieces together like an obvious jigsaw puzzle to create novel, new devices. If truly novel devices are to be created, a predetermined, completely deterministic process without any room for mistakes will not work. In fact, this is impossible for the human mind and represents a misunderstanding of how the design process actually takes place. Here is how it usually goes. Let's say that one wants to create a device that measures the moisture content in materials, such as grain or concrete, to determine how these materials should be used and stored. The first item that needs to be decided is which property of the material will be measured by the device and then correlated with moisture content (maybe it is a direct measurement, who knows). Suppose the lead engineer takes a survey around the office on the technical means that should be used. In that case, ninety percent of the engineers surveyed will not have any idea, two percent will automatically suggest infrared (IR) spectroscopy, seven percent will say dielectric measurement, and there will be a one-percent outlier who suggests using an air-oven direct measurement. These are likely to be narrowed down to IR spectroscopy and dielectric methods, the choice between the two being made by building prototypes and *testing them in practice*. Many iterations and evaluations take place. The winner is the one that performs the best. However, we are nowhere near done at this point.

The dielectric method won out in the real-life story referenced here. Then, based on data taken across a very wide spectrum of the radio-frequency range and the range of moisture levels, the decision was made to rely on measurement near 150 MHz because the imaginary part of the permittivity is minimized at this frequency, which was determined *empirically*, not analytically.

The next step in the process was designing the test cell that must be filled with the material to be measured. The cell could not be any arbitrary shape. The most straightforward geometry was two parallel, conductive plates between which the material resided as the cell was excited with an approximately transverse-electromagnetic wave. Measuring the impedance at the cell input allowed the material's dielectric constant to be calculated, from which the moisture could be inferred. Even with the basic geometry chosen, the precise test cell dimensions could not be calculated analytically. The optimal dimensions were found by repetitive adjustment, taking measurements, and zeroing in on the shape that maximized sensitivity.

The relationship between the real part of the permittivity at 150 MHz, which is the parameter that was chosen as a correlate, and sample moisture was found by measuring thousands of samples and performing a linear least-squares regression. The story contains much more detail that I am going to leave out here, but all of it involved building and testing component prototypes.

The preceding discussion represents a very abridged version of the design of what is now the Federal Standard moisture meter. I'm not saying that the project was all that significant, but the process used to design it is important to the discussion. I bring it up here because ID proponents tend to act as though there is some method by which intelligence bypasses the optimization process, or that the optimization process engineers use is entirely different from the one evolution uses. And in some senses, the two are different. As was discussed previously, evolutionary optimization uses a modified gradient ascent, coupled with additional methods that bear some similarities to genetic programming, whereas more sophisticated and refined techniques are available to human scientists and engineers that explicitly take into account the shape of the function being optimized. Nonetheless, the essence of the methods is the same.

The simple assertion that intelligence provides a means of generating novel designs gets nowhere in explaining how this intelligence accomplishes the task. Worse yet, the term intelligence, primarily how it is used regarding ID, carries a monumental collection of baggage. In particular, the notion that intelligence is somehow magical or divine in its functioning. For example, the belief that the blanket statement of an intelligent designer can be applied as a replacement for evolutionary theory, and that is all that needs to be said or claimed, is analogous to being asked about technological advancement in the world today and replying with the answer "we build stuff". It is certainly true that humans build stuff via technology, but the phrase "we build stuff" tells one nothing about the underlying processes that enable and drive technology's advancement. Asserting an intelligent designer as the cause of biological complexity and diversity is the same as just asserting that we have biological complexity and diversity. The statement is tautological and explains nothing.

The equivocation of the meaning of intelligence is the problem. Intelligent design advocates claim that an intelligence is behind the specified complexity seen within biology, and since the only known progenitor of specified complexity is an intelligence, that biological complexity must have been designed by an intelligence. As will be shown here, the validity of such a statement depends entirely upon the definition of intelligence. If the authors of ID believe that intelligence is a state of already knowing all the answers necessary to achieve a given task, then their definition will not fit biological evolution, neither will it align with human intelligence. This definitional problem arises precisely because the functioning of human intelligence looks very much like the evolutionary process, and that the assumption by the ID community that intelligence arises from divine magic does not correlate with anything in reality.

The new knowledge contained within the so-called specified complexity in biological systems is not pulled from some preexisting pool of information residing in the mind of a supernatural intelligent designer any more than the theory of special relativity was instantly produced by the mind of Einstein. New knowledge is not generated this way.

Intelligence is not a static thing in which answers already exist, like reading a book. Intelligence is a dynamic process of which knowledge is an output. It takes certain inputs and constraints which are embedded in a particular context, that is, knowledge or information, acts upon this knowledge, and then outputs new knowledge or information that is in some way meaningful in the same context as the input information.

The important thing here is that this new information cannot come about in a single instant. Reality doesn't work that way. In the same way the probability is too low for the human genome to shown up in its complete form in an instant, the same is true for other novel information. The creation of new information through purely random chance doesn't happen in either of these cases. There is a process in each instance that creates new information. Evolution uses random changes at its input coupled with a method of choice or selection to select outcomes that matter in some way. In the case of evolution, solutions that increase reproduction and produce more of a particular species are selected. Evolution rests on the fact that a context is already available, and it is completely mechanistic. That is, DNA that leads to more reproduction, which in turn, leads to more of that DNA.

Thinking, on the other hand, appears to be able to choose a generic context. For now, we will go with the assumption that the mind can generate an arbitrary context in which to solve problems.

If, contrary to what ID claims, new designs or solutions do not just spontaneously appear out of the aether, then from where do they originate?

The mind tends to be very chaotic in its stream of consciousness. Unless allowed to quiet, a continual river of seemingly random thoughts flows through it, and a part of the mind watches this flow. The thoughts are not entirely random. They generally center around a mean or average component. For instance, if one is trying to think about types of cars, most of the images that appear in the mind will be about cars. However, many thoughts will also be focused on items associated with cars, including car doors, windows, engines, tires, etc. In addition, there will be those thoughts of items associated with cars but are not directly related to cars, maybe a driver's license, rules of the road, speed limits, and so on. Finally, there will be a few items that do not belong at all, such as the number of calories in a beer, potential presidential candidates in the next election, a TV show that is on later that night, and others. These thought groups form a statistical distribution around the original, central thought.

Thought patterns like this occur in the mind because it is a highly interconnected network. Routes from many areas of the brain crisscross one another, so when one portion is accessed, all are accessed to one degree or another. If you watch closely when you attempt to solve an intellectual problem, the result will be reminiscent of throwing a stone in a pond: ripples will emanate in every direction away from the epicenter, stirring up ideas or thoughts from points far away from the thesis of the original inquiry. Some are relevant to the topic at hand, and some are not. Some have absolutely no relevance, at least not in the

8.1. INTELLIGENCE AND THE BRAIN

way needed. For example, when trying to remember the structure of the covalent bonding of CO_2, it is quite possible that the word "cola" will present itself just because the first letters of the name are "CO.". What was being sought was a connection in the context of chemical bonding and what showed up instead was a candidate solution in the context of spelling.

Calling up a memory or a fact in the mind is not like doing so on a computer. In order to retrieve a value from a computer's random access memory, the exact address to that memory location must be known in advance. Trying to access it based on the type of data stored in it, be it a house address, phone number, etc., will not work. Of course, software can be used that searches through the computer's memory for a particular type of data, and the computer will faithfully respond with all addresses that contain this kind of data. In either case, the action to be performed and the type of data to be requested must both be precisely specified.

Retrieving concepts from the mind is very different. There are no specific locations that we can consciously search. Also, the data or type of data being searched for does not need to be exactly specified. Even a vague notion of what one is looking for will often times get you there. And what will be returned is usually an ensemble of concepts or ideas that link to the search parameter in some way. For example, if you think of your child, you most likely will not see his or her name in your mind. Rather, what will likely be returned is a collection of images of the child smiling, riding a bike, laughing, etc., appearing in the mind's eye.

Memories are distributed all across the mind; memory in humans and other mammals is not centralized. Instead of like a computer's memory, which is located in one or two primary locations, memory in the mind spans multiple regions including the neocortex, prefrontal cortex, basal ganglia, cerebellum, amygdala, hippocampus, and others [60, 114]. Memories with a particular character may be stored primarily in specific areas, but they still rely on other areas. In particular, episodic memories that involve several functions, such as motor functions and sensors, span several modules located in different areas.

It has been observed in some instances, that memories are even stored locally at the sensor apparatus that generates them, such as with the olfactory sense. The very chemical sensor neurons are used in some cases to store the memories of smells. This is another example of signal and function being the same in biological entities.

But how are these memories stored? And how are functions such as motor tasks implemented within the brain?

The human brain is constructed from approximately 100 billion cells called neurons (among other types of cells). These are special cells that specifically handle the processing of information. As shown in Figure 8.1, each neuron has many spikes protruding from its body called *dendrites*. The dendrites act as inputs to the neuron. Although the average number of dendrites per neuron is about ten, this number can be as high as 100,000 [115].

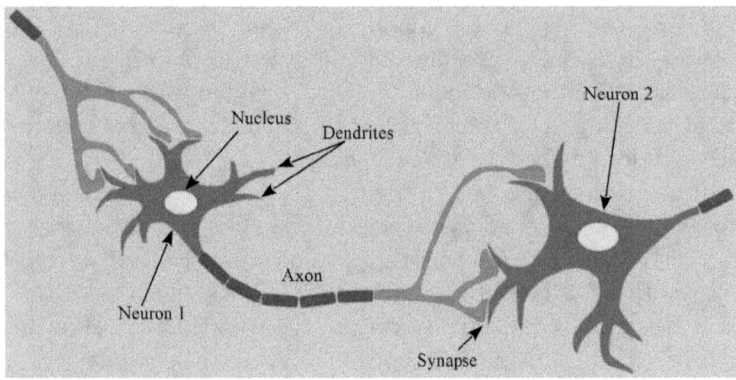

Figure 8.1: The axon tips of a first neuron connect to the dendrites of a second neuron through several synapses.

At the opposite end of most neurons, there is a long transmitting channel called an *axon*. The axon transmits an electrochemical signal from the neuron body to the many branches at its far end, each of which ends in a terminal that rests near a dendrite of another neuron. There is a gap between the terminal and the dendrite called a *synapse*.

When a neuron fires, a pulse of electrochemical energy propagates from the cell body or *soma* of the neuron down the axon to its terminals and interacts with a second neuron via the second neuron's dendrites. If the cumulative signal at the second neuron's dendrites is sufficient, the second neuron will fire, propagating the signal to more neurons further down the line. Note that the dendrites of the second neuron could be receiving signals from hundreds of different neurons. The impact that each of these signals have on the second neuron is weighted by the properties of the associated synapse. Some of them can be strong such that the effect on the second neuron is large, or they can be small such that the signal has little effect.

The signals transmitted down the neuronal axons do not vary in strength. What varies is the weight attributed to them at the input synapses of a subsequent neuron. If that second neuron is to fire, the weighted signals at its dendrites within a given time window must exceed some threshold value. The synchronicity requirement facilitates the use of neuron networks for learning based on correlation. When the input of a neuron is excited at the same time the neuron's output fires, the weight attributed to that input increases; in other words, the effect that input has on the neuron gets larger. In this way, the correlation between the two signals is increased [116]. This is how our brains learn in general.

Of course, this would be of no benefit if the neuron had just one input and one output (except perhaps as a signal buffer). The interesting case occurs when there are inputs from multiple sources as shown in Figure 8.2. In the figure, x_1 and x_2 are synaptic inputs to the neuron with weights of w_1 and w_2, respectively,

8.1. INTELLIGENCE AND THE BRAIN

and y is the axonal output. Let's assume for a moment that the values of the weights are $w_1 = 0.75$ and $w_2 = 0.25$, and the threshold for the neuron firing is 0.5. (This is a very simplified version of what's really happening in the neuron, but it serves our purposes here.) This means that input 1 is capable of causing the neuron to fire, and input 2 is not.

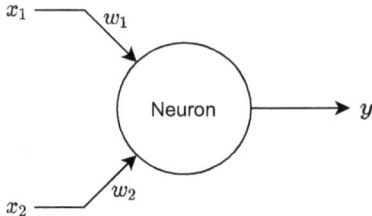

Figure 8.2: A single neuron with two inputs x_1 and x_2, and one output y.

However, if input 2 is excited synchronously with input 1, it will also be synchronous with the firing of the neuron, and the weight w_2 will increase. This is a characteristic feature of neuron behavior and was first explained by Donald O. Hebb in 1949 [117]. Through this coincidence-based enhancement of neuronal input gains an organism can learn about its environment and its relationship to that environment.

In addition to simply changing the efficacy of dendritic inputs, new connections can be established, or old connections pruned, within a biological neural network. Both of these dynamics fall under the umbrella of *neuroplasticity*. Neuroplasticity, in general, is the brain's ability to modify itself in reaction to changes in the behavior or circumstances of the organism. For example, learning a new skill causes changes within the brain that are a consequence of neuroplasticity.

Now we can more readily understand why, for example, the brain will not develop the processing capability necessary to support vision if the animal is not allowed to see as it develops. The nervous-system connections are not programmed into the DNA and will be pruned, or not form at all, if the neuronal pathways are not excited during development. The body solves this complex problem not by following a precise blueprint, but by discovering the neural pathways that are required.

Of course, this explanation is too anthropomorphic. The neurons do not desire anything. Light receptors in the eyes excite neuronal connections and, hence, the connections are strengthened. Again, it is a completely unintelligent process.

Learning in the sense we usually think of when we hear the word is accomplished by association between neurons. For instance, the Pavlovian response in mammals is an example of associative learning [118]. First studied by Ivan Pavlov and published in 1897, the experiment involved Pavlov ringing a bell coincident with feeding his dog every day. At first, the dog would only salivate when food was given to him but, after many times of ringing the bell at feeding time, the dog learned to salivate from only hearing the bell.

The Pavlov response can be demonstrated with the simple neural network of Figure 8.2. Let input x_1 correspond to the dog seeing or smelling food, and let the weight w_1 be sufficiently large to always fire the neuron, and let y drive salivation. Under these conditions, when the dog sees food, causing x_1 to be excited, salivation is the response.

Now, let x_2 correspond to the auditory input of the bell ringing. The weight w_2 would initially be low as the ringing bell has no connection with food. However, if the bell is rung coincidentally with food, this means, using the behavior of the neuron outlined above, that the neuron will fire coincidentally with the excitation of x_2, leading to its associated gain w_2 increasing with repeated trials, until, finally, the neuron will fire with the excitation of input x_2 alone. That is, once conditioned, the dog will salivate when the bell is rung without food.

It is important to note that the conditioned response can be removed by removing the association of the bell with food. Feeding the dog on a repetitive basis without the bell being rung will weaken the conditioned response because the input x_2 is not being excited coincidentally with the neuron being fired.

It is thought this is how a child learns to recognize the world in which he finds himself. Vision, hearing, smell, taste and touch are correlated with one another as they are excited by the environment. For instance, the sight of food becomes correlated with satisfying hunger when a parent feeds him. Seeing objects move as he actuates the muscles in his arms and hands trains his vision to recognize that things in the environment move, and it trains his muscles that they are moving things. This is circular, yes, but it's the best we can do and is the source of the hard solopsism problem. We only have access to our senses to train our senses. The world we are personally familiar with is a world of correlations only.

This is what intelligence looks like from one angle as it relates to animal learning. There is no magic "intelligent force" that human beings possess. The entire world we know is a closed system of senses with varying degrees of correlation between them. In a very real sense we do not truly experience reality.

It is clear that this is a type of memory. For example, in the Pavlovian case, the association that is formed between food and the ringing of a bell can function as a memory to the dog. When we see this coincident behavior, we say that the dog remembers that food is associated with the ringing. In this sense, the associative network connections within the brain implement memories.

The changing of synaptic weights and neural connections provides a candidate for long-term memory storage, but it doesn't address the issue of short-term memory. Short-term memory must be effected by another means since, by definition, brief episodes do not provide the necessary conditioning to retrain the associative couplings of neurons.

Neurons can be used like circuit elements to construct larger networks with varied functionality. As a simple example, the three-element network in Figure

8.3 implements a divergent circuit. A signal input to neuron A is passed on to both B and C. Neurons B and C can then route copies of this signal to different parts of the brain. A convergent function can be created simply by reversing the direction of the elements; that is, two neurons feeding the inputs of a single, subsequent neuron. A simple OR function is built this way.

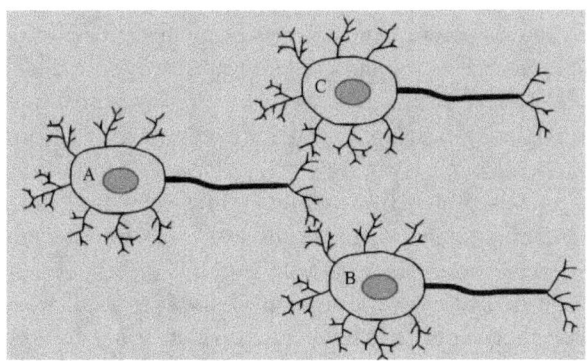

Figure 8.3: A divergent neural network. The circuit duplicates the signal at neuron A to both neurons B and C. Although the signal is duplicated, there is a time-delay for each additional neuron in the chain.

A more complicated *reverberation* network is shown in Figure 8.4. Assume initially that all three neurons are in an idle state with none of them firing. Then, if an input to neuron A causes it to fire, B will then be fired, followed by C firing A again. This cycle will repeat endlessly if the gain of the loop is greater than or equal to one.

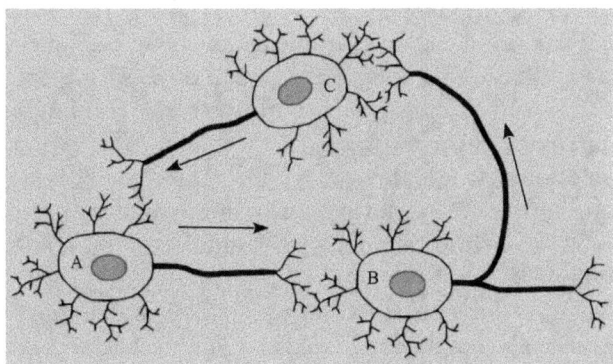

Figure 8.4: A three-neuron reverberating network. Once neuron A is fired due to excitation at one of its inputs, it continues to fire due to positive feedback through neurons B and C.

Much like the digital flip-flop gate that is ubiquitously used in digital electronics, the reverberatory neural network can be used to store information [119]. The circuit is a viable candidate for short-term memory storage in the brain as retention is immediate. The short-term memories then eventually fade on their own as the loop relaxes back to the idle state, or they are copied into long-term memory. The hippocampus and amygdala mediate the storage of long-term memories.

That memory is created in the brain using neural association leads to some interesting phenomena when humans make practical use of their memories. Due to the high degree of interconnectivity within the brain, calling up thoughts in one's mind can, and usually does, lead to a cacophony of responses. When the mind is asked to recall or provide a specific piece of information, it will supply a stream of suggestions pulled from memory locations that are associated with that topic in all imaginable ways. This is directly due to the item being searched for being associatively connected to a vast number of other items in one way or another. It is like a neural network version of six degrees of freedom.

In light of this, how does intelligence come into play to solve a problem or develop a design?

It has been my experience that we can't simply *will new knowledge into existence*, although this might be what one is tempted to think if the concept of intelligence is considered only superficially. When one sits down to solve a difficult mathematical or engineering problem, the mind becomes a boiling pot of random thoughts, with thoughts continually moving in and out of our awareness. As Sam Harris points out in his book *Free Will* [120], we do not choose our thoughts; rather, they simply come and go at random. So how do we generate new, coherent information about any subject?

The design process can be segregated into multiple pieces. First, the problem needs to be understood well enough that it can be placed in a suitable context such that candidate solutions can be evaluated. Next, the problem is divided into smaller modular problems that are easier to solve than attempting to solve the entire problem all at once. Solutions for each of these sub-problems are then found one at a time and the pieces fitted together like building blocks until the complete answer comes into view.

Finding a solution is a hierarchical process; there really is no difference between finding an appropriate solution for a submodule and finding an overall solution. Thoughts come into the mind and candidate designs are selected based on mental evaluation.

When a thought of interest is found, we single it out by giving it our focus or attention, essentially holding it in place. This is not a guaranteed process as thoughts can be very fleeting; they can slip away if our focus is lost, as we have all seen when we are in a conversation and suddenly forget what we were going to say next. The main point is that once we have captured the thought of interest, we are partly on the way to bringing a new idea into existence. I want to reiterate this point to make it clear: when an idea is grasped, *the conscious, thinking mind had no part in bringing it about*. The mind simply watched and

8.1. INTELLIGENCE AND THE BRAIN

waited for an interesting thought to appear. (I highly recommend the reader to Sam Harris's book *Free Will* to find a good description of what I am saying here.)

The single thought that has been acquired is generally not a complete solution and, therefore, must be completed before it is useful in the practical world. The process is continued by holding onto this first thought and looking for an additional thought that complements it. Then using the same process, we watch the myriad of random thoughts passing through the mind's eye until we see one that might fit and move our focus toward it. This usually involves writing things down on paper or typing information into a PC so as to not lose our way.

Integration of the first concept with the second is then attempted in the virtual reality machinery of the mind. Even at this point, it is not always obvious how the pieces fit and work together. They must be manipulated and tried in various ways until they are successfully integrated. How does the mind choose the combinations of adjoining the two concepts in this step? In the same way the original two thoughts were found: by observing random thoughts that describe combining the first two thoughts and then bringing our focus to just one of these combinations. Once an assembly concept is found, the integration of the first two concepts is attempted. If successful (satisfies a specific fitness), the assembled concept is retained, and the process moves forward. If it is unsuccessful, the current thought is discarded, and the process of selection continues.

Each concept found by the mind must be tested for its fitness in meeting the requirement for its particular place in the overall chain. For example, first, it must satisfy the function for which it was being sought. In addition, it must be consistent with reality; that is, physically realizable. Any step of the process can be rejected if it fails either of these criteria. This does not mean that if one component fails, we must throw away the entire chain of concepts and start from the beginning, although that is possible. Instead, it usually means we simply throw away the latest thought concept that would not fit into the solution, keep the rest, and continue searching for additional satisfactory thoughts to test. This process continues until the ultimate objective is met.

It is a very haphazard, random process only directed by the requirement that the individual subcomponents and end assembly are suitable to some level in meeting their objectives. The test for fitness is a simple go or no-go test. In each instance of evaluation, only one or a few candidates are passed; the others are rejected and left behind. It is a survival of the fittest of ideas that are generated randomly by the mind.

Since random concepts drive the process, it is even possible to develop novel solutions to problems different from the one that was being investigated. It is possible that solutions to entirely different problems can be inadvertently found. This is very odd and demonstrates that this process of developing new knowledge is anything but deterministic. It occasionally happens that quantum leaps in technology are made serendipitously because the space of solutions being evaluated is somewhat random.

An example of an engineer looking for one solution and coming up with another is the inventor of the pacemaker. As Wilson Greatbatch was working on the electronics for a device to record the electrical rhythms of the heart in 1956, he accidentally picked up a resistor (an electrical element that impedes the flow of current) of the wrong value and placed it in the circuit. This accidental modification to the circuit allowed it to supply electrical pulses rather than record them, which made him think that it might be possible to drive the heart with such a device. His pacemaker device was successfully tested in 1958 [121].

Serendipitous discoveries occur between a third and half of the time [122]. An engineer or scientist is looking for one solution and ends up finding an entirely different solution, in many cases to a problem they were not even trying to solve.

The associative mapping in the brain, coupled with the inherent uncertainty that we will talk about a little later on, allows coordinate changes in thinking and the generation of entirely new categories. For instance, consider a simple, everyday screwdriver. It is intended to put screws in and remove them. However, if the brain is searching for something to use for a pry bar, it can, via geometrical association, co-opt a screwdriver for this purpose. Similarly, if a tent stake is needed, the mind recognizes that a screwdriver can be employed for this function. Many times I have used a screwdriver as a conductor to short electrical terminals, for instance, to crank the starter motor on a car.

All of these functions are qualitatively different from the screwdriver's intended purpose. Take the example of using the screwdriver for a pry bar. Instead of performing an associative search for the function of inserting or removing screws, the mind implicitly brought forward thoughts about objects that are strong and geometrically shaped like a pry bar. In other words, the search was performed within a context different from the screwdriver's intended purpose, and consequently, the screwdriver was co-opted for a different purpose.

Of course, there were likely many other options that briefly came to mind along with the screwdriver. Even silly ones, like a pasta noodle, which apparently showed up because it has the opposite characteristics of what was being sought. The opposites of conceptual items are associated with those items, and although it is not useful most of the time, those items also show up in the mind's stream of thoughts. The pasta noodle is quickly rejected by the mind's selection process in lieu of a better alternative.

The above discussion was directed toward engineering, but the same process applies to any problem-solving activity. The process of trying to remember where one placed the car keys is identical in function to the above examples. Watch closely the next time this happens and it will be easy to see the random places the mind furnishes for consideration. Keep watching, and the evaluation step will also be visible. It is readily apparent sometimes as the thoughts supplied can be really random. For example, it would not be out of the question if the thought suddenly arose that the keys are on the moon. Oddballs like this one are quickly rejected and will be missed if one is not completely aware, but they do show up. Typically, however, the process goes more like this: pants pocket?

No. In the front door lock? No. In the car's ignition? No. Left on the bathroom sink after handwashing? Yes! The sequence of the mind throwing up virtually random places for the keys can be seen clearly. Furthermore, each thought is of "where the car keys are" is measured to determine its validity and only ceases after a true outcome is found.

In daydreaming, the goal is a little different. One is chasing thoughts for simple pleasure. Thoughts again come and go at random and, when an interesting one is found, focus is placed on it to hold it in the mind's eye. Then other random thoughts are chosen in the same way to join the previously captured thoughts. This stream of consciousness continues to flow as new thoughts are integrated.

The above activities are interesting and educational to try and can be accomplished through simple awareness. Anyone familiar with meditation or Eastern philosophical methods, such as Zen meditation, will be very much acquainted with these concepts.

I am not implying through an analogy that DNA or matter is intelligent, and I am not implying that the mind is not different from simple DNA and matter. There is no evidence for either assertion. I am, however, saying that using the concept of intelligent design is not as straightforward as some in the ID community make it out to be. Engineered designs by intelligent entities (humans) are not done the way ID claims, and one cannot help but notice that real engineering directed by human minds is closer to evolutionary theory than to the apparently all-knowing intelligent designer to which ID refers. I know of no automatic method of design that exists in engineering.

The intricacy and complexity of contemporary engineering designs, such as the cell phone, are amazing, and, in a contemplative moment, one might consider the vast engineering knowledge that went into its construction; the same conclusion can be arrived at when looking at the vastly more complicated structure and function of biological entities. And I believe the general idea of what it takes to engineer either of these complex systems is wrong. It is obvious in both cases that there exists no designer who knows exactly what to do. To know is to be fixed in one place. Nothing new would ever arrive. This has been the point of this entire section. New information does not enter a system, be it electronic or biological, through some mystical, all-knowing process called intelligence. Using the term that way is misleading at best and manipulative at worst. So, where does new information or knowledge originate?

Let's ask another question first. Can a deterministic machine or process gain new capability or knowledge? What does it even mean to do so? I don't want to get too deep in this topic, but a quick glance at it is important to our conversation.

The idea of artificial intelligence is hotly debated today. There is a great deal of disagreement on whether sequential, deterministic computers like the kind we use today will ever be able to show some form of intelligence comparable to that of humans. The question, however, seems to be ill formed. For instance, what made Planck's formulation of a discrete-energy spectrum new and novel in solving the ultraviolet catastrophe? The concept of discrete mathematics had been around a long time and isn't really difficult. In fact, looking back on it from today's perspective, the result seems somewhat obvious. And yet it changed everything we know about physics.

Here's my assertion: what made this discovery (and others like it) so special is that it was a mistake. It was a ridiculous idea that no one would've or should've bought at the time. It fell completely outside the current realm of knowledge and understanding. Don't misunderstand, it wasn't some huge inference made based on the current understanding, and it wasn't a usage of the physical laws known at the time that was so complex no one could comprehend it. The problem was that it strayed entirely outside the realm of the current understanding. We usually subtract ten points from test papers for this kind of thing. We call them errors. But why exactly is it an error?

It is by definition that any concept that falls outside a specified framework is erroneous. It can be no other way. A couple more examples seem in order.

Isaac Newton was one of the brightest people who ever lived. His achievements in optics, mathematics, and physics are unparalleled even today. One of the most difficult problems he encountered was the question of what caused the heavenly bodies, the planets and moons, to move in the way they did. He had a habit of working on a problem night and day by holding it in his mind's eye continuously until a solution presented itself. We cannot imagine today the quantum leap that was required to come to the conclusion that the same force, gravity, that caused massive objects to fall here on earth was the same force that moved the planets in their orbits. Again, at the time, this was regarded as a mistake in reasoning—such an idea had to be wrong as it did not fit within the existing framework of physics.

And, at the sake of treading on sacred ground, consider Jesus Christ. The ideas he espoused (most of them) were so alien to those listening at the time that they were considered by some to be insanity. In other words, they were erroneous statements as the Sanhedrin so forcefully pointed out. I would argue that if such a man showed up today, he would be at least metaphorically crucified again. It is always the case that novel information is considered to be in error and, when coupled with religious dogma, it takes on the guise of blasphemy.

It is for this very reason that computers as we know them today are seemingly incapable of exhibiting intelligence. How could they? It is the job of engineers to ensure signal integrity in such machines. It is expected that a computer user tends to get very angry if the information stored on his hard drive changes randomly. But it is surprising in another way since the information within his own mind is not nearly that dependable; it continues to change randomly to such an extent that it is even observable from his own mind. It seems that to create intelligent machines is to provide them the ability to make mistakes and take advantage of randomized information. How else could new ideas be discovered if only past ideas are considered important? (I think there is a possibility that a sequential machine could exhibit some human-like intelligence once the error

8.1. INTELLIGENCE AND THE BRAIN

rate becomes finite.)

The point is that invoking an intelligent designer does not solve the problem of the diversity of life on planet earth. And I don't mean the hackneyed *reductio ad absurdum* argument that a designer of the designer would have to be explained, although that is certainly a problem which would need a solution. My assertion here is that the designer needs to be imperfect in order to honestly design. Anything less would be stagnation, at best reusing old designs or someone else's designs.

It seems that ID advocates mean to imply that evolutionary processes are imperfect and not up to the task of creating novel, complex systems, and argue that knowledge is the essential ingredient to bring about such systems. In fact, it is just the opposite. Imperfection appears to be the secret of innovation, not only at the biological level but also in the intelligent, human world. Anything that is truly novel will look like noise to one who can't go outside the set of things they "know", but it is only through this "noise" that new information can be found. In my opinion, instead of wasting the time of very bright people on a subject such as ID, their time would be much better spent in investigating where the designer's knowledge comes from in the first place, be it a human engineer, biology or a god.

It is not hard to construct an argument that thoughts which appear at random in one's head are truly random. It is true that the complex neural net of the mind possesses the structure to exhibit chaotic activity, but what mechanism could be the source of truly random thought events? It seems that all random events introduced into this universe originate from randomness innate in quantum mechanical behavior. There is simply no other source that we've pinned down. Chaos doesn't get one there as it only provides diverging results from arbitrarily close initial conditions, but it is not random—it is deterministic. In addition, quantum randomness is not just the appearance of randomness; it is truly random. John Bell demonstrated this clearly as described in Chapter 3.

In many engineering applications a random noise or event source is required. At one time in the past, pseudo-random number generators were relied on for on-the-fly generation of number sequences *close to random*. Pseudo-random number generators are discrete, deterministic systems that compute extremely long sequences of numbers before repeating. They can appear to be aperiodic for very long runs of output data, but they do eventually repeat, meaning that they are not sources of random data.

One of the best means of creating an electronic generator of randomness is a simple diode. A diode is a device that only lets current flow in one direction, and are often used to create constant-voltage sources from the 60 Hz alternating voltage provided by electrical utilities.

The important thing here is that the diode, when reverse biased, produces a barrier to the flow of electrons or holes. In order to cross this barrier, an electron would need an energy great enough to cross the electric field at the junction but, by definition, an electron within the diode only has the energy that the applied voltage provided to it, and thus cannot cross the junction barrier. This is similar to a batter in a baseball game trying to hit a ball over the outfield wall. If the batter cannot provide the ball with sufficient energy to go over the wall, then it doesn't happen. Enter quantum mechanics again.

If one solves the Schrodinger equation for an electron at the energy barrier of the diode, the solution will extend partially past the wall of this barrier. That is, there is a finite probability that an electron, which does not have the necessary energy to pass through the barrier, can find itself suddenly on the other side.

I'm not going to go through the math here, but let's assume that $\Psi(x)$ is the solution to the Schrodinger equation for a barrier at $x = x_0$. That is, approximate the barrier as a wall at this location. Then the probability of finding the electron past the wall, $x > x_0$, is the integral of $\Psi(x)$ from x_0 to $x = \infty$ squared. Since quantum mechanics only reveals probabilities of physical variables, we cannot solve for the exact location of the electron, but can only find its probability of being within a region. What we just calculated is the probability of finding the electron on the other side of the barrier.

Now, if we reverse bias a PN junction and build an electronic amplifier to measure the current flowing through the diode, what we will find is that a small, *random* current will be present. This current comprises charge carriers that *quantum tunnel* through the energy barrier, and this happens in a completely random way. Noise sources designed in this way are not like pseudo-random generators, which only approximate true randomness. These sources are truly unpredictable.

So what does all this have to do with intelligence? Synaptic junctions within the brain do not behave like diodes, but their size is small enough that quantum-mechanical effects come into play in partially dictating their functioning.

Casey Blood from Rutgers University provides a very clear and easy to follow argument for the behavior of neuronal synapses in the human brain being dictated by quantum mechanics [123]. The transmission of impulses from neuron to neuron is mediated by the behavior of the synaptic junctions between, which, in turn, function by passing calcium ions (there may be additional ion types as well) across the synaptic gap. The question that Blood answers is whether the wave function associated with the calcium ions expands such that it engulfs the synapse. If it does, then it is quite conceivable that the calcium transition event is a linear combination of crossing and not crossing the junction, the outcome being dictated randomly.

Blood starts with Heisenberg's uncertainty relation $\Delta x \Delta p \approx \hbar = 1.0546 \times 10^{-34}$ J·s. Next, consider a calcium atom initially at the origin and let $\Delta x = x$ and $\Delta p = mv = m\, dx/dt$. Substituting these two equations back into the uncertainty relation, we get $\hbar = mx\, dx/dt$. Assuming the initial spread of the wave associated with the calcium atom is $x_0 = 10$ nanometers, which is large enough to encompass the atom, then upon integrating we have $x(t) = x_0^2 + (2\hbar/m)t$.

If we assume a lower limit on the time between synaptic events of 0.1 mil-

8.1. INTELLIGENCE AND THE BRAIN

liseconds (this is very conservative), and the mass of a calcium ion as about 66×10^{-27} kilograms, the expected spread of the ion's wave function is about 500 nanometers. This is well large enough to encompass the synapse. Therefore, we can conclude that it is certainly plausible that synaptic events exhibit quantum-mechanical behavior, that is, randomness in their firing.

With this last piece in place, we can now finish our story of human intelligence. Intelligence is the outcome of a combination of simple processes: the random generation of concepts (thoughts) in the mind, and a means of selecting between these thoughts based on their functional optimality. The overall process, at least in its basic functioning, is virtually identical to that of biological evolution.

When the solution to a problem is being sought by the mind, thoughts appear in the mind. Some of them are intimately connected to the problem, others are more remote or orthogonal, and some are just plain random. The thinker does not consciously choose these thoughts, but they are consciously evaluated, and this is the part that allows useful solutions to be found. Similar to the case of evolution, creativity is injected via what appear to be random thoughts, and context is given by conscious selection between these ideas.

Isaac Newton once related that his success in formulating radical new ways of looking at the Universe stemmed from his ability to place the problem in his mind's eye and incessantly focus on it for days or weeks on end until a solution was found. I believe most of us are familiar with the notion of trying to solve a problem for a long time and then all of a sudden, it just clicks, and the solution presents itself. It might be easy to just say that the subconscious solved it in the background, but the question is, how? I contend that it is by the method outlined here.

There are some authors, Sam Harris being one of them, who state that since you don't author your thoughts, that when your mind solves a problem, you shouldn't really take credit for it. (I'm not talking about taking personal credit here; I'm referring to the cause of finding the solution to the problem.) We need to be clear on what we are talking about when we say "we did it" or "we did not do it." It seems that the general consensus is that if we consciously solve a problem or come to a conclusion then it has more significance than if the solution presents itself to us out of the subconscious.

This is a strange way of looking at the human mind and would mean that we are only responsible for what we consciously decide or act out. David Eagleman in his book *The Brain, The Story of You* relates a story about a man who sleepwalked and killed his in-laws without being consciously aware of it. After a great deal of expert witness testimony and evidence of sleepwalking running in his family, the man was acquitted of murder. Apparently we are not responsible for our actions that are driven by our subconscious.

Sam Harris goes a step further and argues that we can't even be logically held accountable for our conscious actions as how our minds are wired is beyond our control, and we do not consciously determine the thoughts that arise in our minds.

I would argue that we are not responsible in either case. We're going to talk about free will in the next section, and so I'll leave out that facet for now. However, the inconsistency should already be apparent: we are either responsible for our actions, or we are not. From a Christian point of view, accountability is a requirement, otherwise the entire religious foundation falls apart. They argue that there is something special, otherworldly, about the human mind and its intelligence. The argument goes that our minds are somehow capable of making decisions that are the sole responsibility of our own person. This does not seem plausible.

This notion comes from the assumption that we are completely autonomous entities. We are not. No one thinks this about plants. That would be silly because plants obviously do not have this soveriegn nature; they are affected by the direction of the Sun, the nutrients in the soil, the ferocity of the wind, etc. A plant does not choose to be healthy or unhealthy. But I just have to ask: how much choice do you think you have in determining your actions or physical state? If these are determined by your thinking—your thoughts—then you have very little control. Positing a magical soul that no one can prove to exist is immoral and, in certain instances, cruel. That's how the blameless are made guilty. A man who commits murder because of a brain tumor is not morally culpable. He needs help. Christianity may recognize this today, but in the past this man was an evil monster. I can understand it took science a while to catch up but this is something that a god should have known in advance.

The information that we have on hand in our minds to make decisions is not under our control. Even if an individual has written down all the pros and cons concerning an important decision, there is no way all of these concepts can be held in the mind and a meaningful decision made. But that's not the main problem. We cannot even guarantee that we have considered all the facts relevant to making decisions as we do not determine the thoughts that arise in our minds. The thoughts are pulled from a complex, chaotic and indeterministic neural network that does not possess the hardware to guarantee complete coverage of a given topic.

Even the process of making an *intelligent* decision is not well-defined. It is a somewhat hodgepodge method of evaluating the random premises that arise in our minds about making the decision one way or the other. As it turns out, in the end, we usually make a snap decision that we cannot fully justify anyway.

Nowhere have I found a definition for intelligence (beyond the symptomatic list given at the beginning of this section) that describes what it actually is and how it functions. The usual implication in the ID literature is simply that it is something that does what dumb, natural processes can't. And yet, when we look at examples of this type of intelligence, such as in the creation of biological complexity or the functioning of the human mind, dumb, natural processes effecting the outcomes is what we see. Intelligence is an emergent behavior, not a magical capability that has no explanation.

I'm not saying that DNA is intelligent and that it forms itself. Quite the contrary. If we define intelligence as its effects, such as the ability to form complex structures, we first must admit that this is different from the definition most people and the ID community are thinking of when they speak of intelligence. But, if this is how we define intelligence, then, yes, biological complex structures are built by an intelligent process, but that process is composed of only dumb, regular, natural laws. If the ID community wants to define it another way then they will need to provide the definition, but I presume it will have a supernatural, magical character to it.

8.2 Free will

Most of us believe that when we make a choice, for example, to buy a new car, that it was our decision alone. We could have decided either way, but in the end, it was solely our decision. But how true is this? Many religious people believe in a dualist picture where the human brain does exist and does control the body in the sense that science claims, but there is also a component such as a soul or the like that is primary. It implements free will and provides the justification for holding persons accountable for their choices. (Ironically, if you are Christian and ask for forgiveness, there is a way out of that accountability.)

Without this accountability, concepts such as sin and divine punishment would be non sequiturs. It would be the same case for our legal system, which is just now taking the nascent steps of coming to grips with this question. Unfortunately, the legal system is saturated with individuals that still cling to Christian ideas of absolute autonomy. The need to swear on a Bible is evidence of this.

It is not clear where the brain ends and the soul begins, that is, which part is responsible for which functions. In this section, we try to run down the issue of free will and determine if it exists and, if so, in what form.

When visiting a car dealership, the first thing that may interest you in a car is that it possesses 325 horsepower. The salesman will hasten you to take the vehicle for a test drive so you can actually experience the feeling of that power. And the car's list price of $24,995 doesn't seem terrible. After you finally decide to buy the car, the sale, as usual, takes hours, with the salesman offering you additional warranties and interior stain-resistant treatments, requiring you to pay taxes and title, and a few other miscellaneous items. Finally, as you are leaving the dealership, the salesman tells you that you made a good decision.

Despite that he appeared to be your friend throughout the arduous ordeal of buying the car, sales tactics that you probably didn't even pick up on were being utilized the entire time to drive your psychological state toward the decision to buy the car. The salesman's first persuasive action was to get you to drive the car, allowing you to experience it for yourself. The human mind has a tendency to value more highly those items that we presently have access to over those to which we could have in the future [124]. Once you are in the car and feeling it pull away from a red light, your thinking tends to become more enthralled with the car's performance and the way it looks than your limited budget.

Even the odd price is designed to draw in your mind. It is well known to those in sales that potential buyers round odd-ending prices down, meaning, for example, that a price for a car of $24,995 is far more enticing than $25,000 despite there being only a five-dollar difference between the two [125, 126]. Finally, the sales experience is intentionally drawn out by the salesman to put pressure on you to pay for items that you might ordinarily dispute or decline to purchase. The sale has already taken a long time, and there is a psychological aversion to going through it again, possibly at a different dealership.

So, did you want to buy the car or not? Some might argue that despite the salesman's tactics skewing your level of desire to purchase the vehicle, in the end, it was still your conscious choice. Consider the probability distributions plotted in Figure 8.5. The horizontal axis shows how many cars an individual buys (on average) and the vertical axis is the number of people who purchased cars at a particular dealership. The solid line is the distribution when the salesman uses no tactics to sell cars. The result is that, on average, half of the customers buy a car and half do not. On the other hand, the dashed line represents the distribution when sales tactics are used. Here, customers are fifty-percent more likely to buy a car.

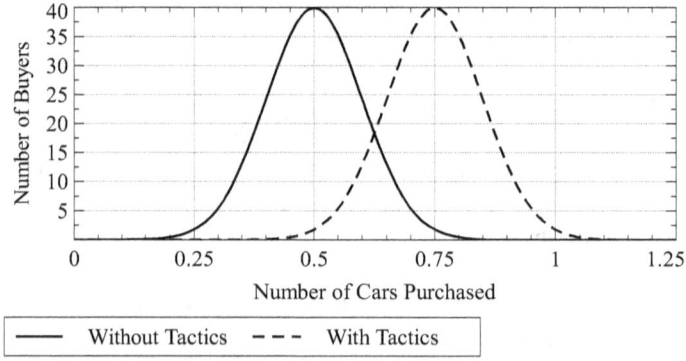

Figure 8.5: The probability distributions for buying a particular car at a dealership. The solid line is the distribution when the salesmen do not use sales tactics, and the dashed line is the case when they do.

In light of the shift in the statistical distribution with sales tactics used, does it seem that the customers are still making free-will decisions about buying a car? Although it can't be conclusively tested for any specific instance, with the salesman's coaxing a person's free-will choice can be modified from a no to yes. Does this mean that the person could not freely decide or that with updated information—the salesman's actions—the person simply changed their mind and made a completely free decision to buy the car? It's starting to get difficult to pin down exactly what a free-will decision is.

What happened as the salesman's words and actions registered in the mind of

8.2. FREE WILL

the potential buyer? As we discussed previously, these were stored in the short-term memory of the brain, and they also elicited additional memories. All the while, the buyer's collective, cognitive picture of the car and their relationship to it is changing. Perhaps while on the test drive with the salesman, they passed a billboard advertising a particular college, and the customer remembered the importance of an education and decided not to buy the car. Was that a free-will choice or did the salesman inadvertently destroy his commission by preventing the buyer's actual choice?

Would this picture change if the salesman had a neurosurgeon friend who owed him a favor, and they tweaked a few neurons in the buyer's brain leading to an extreme desire for the car? Would this still be a free-will choice since the customer would still be physically free to not buy the car? Really, there is not much difference between the salesman altering neurons in the customer's brain with words and him doing the same with surgical tools; the latter just sounds more inhumane and like it stunts the customer's freedom of choice.

In a video posted on YouTube, the Indian-American neuroscientist Vilayanur Subramanian Ramachandran describes an interaction he once had with a patient that had their corpus callosum surgically severed to mitigate epileptic seizures [127]. The corpus callosum is the large bundle of nerve fibers that connects the two brain hemispheres and allows them to exchange information. Once this neural bundle is cut, the vast majority of the communication between the two sides of the brain is cut off.

The brain is connected up to the body in a contralateral fashion such that the left hemisphere handles the motor functions on the right side of the body and vice-versa. Also, the left field of vision (not the left eye) is processed by the right side of the brain, and the right field of vision by the left side. Lab experiments with so-called split-brain patients have demonstrated that if you show their right field of vision an object like a ball, they can verbally tell you what it is, as usual. However, if you show the same ball to their left field of vision, they will not be able to tell you, but can write the word "ball" with their left hand. This is a direct consequence of the language processing center (in Broca's area) residing in the left hemisphere. In the first instance, the left side of the brain has "seen" the ball and can talk about it. In the latter case, the right hemisphere has "seen" the ball and it has no ability to talk about it, but it can write about it with the left hand.

This is a clear demonstration that the two halves of the brain are functioning for the most part independently of one another. Both halves have the processing power to generate a personality and to behave pretty much as normal [128]. In the case of Ramachandran's experiment, things got very interesting when he asked the patient if they believed in God. The patient's right hemisphere said yes, and the left hemisphere said no. It seems that the ability of this patient to freely choose to believe in God was impaired. Which hemisphere would a Christian claim is the one that moral or other responsibility is to be attributed?

The point of this digression is that, once again, the evidence points toward our choices and preferences being generated within our brains alone and that there is no supernatural soul or power that ultimately decides.

Let's say that our previous car-buying customer has decided to buy the car and the salesman has printed out all the paperwork and pushed it across the desk for signing, and the customer suddenly says they can't go through with it and decides not to buy the car after all. Which one of these was free choice, or were they both? Did the customer make a free choice to buy the car and then make a second free choice not to buy the car? Had the customer signed the papers and then decided not to buy the car, would the original free choice have still been a free choice? How do we measure when a free choice has been made?

As another example, consider someone who is deathly afraid of heights. They would not get into a plane if their life depended on it. This is obviously their choice. Now, assume that there is another person who flies all the time and is not afraid of heights. If the neurosurgeon friend from the example above surgically modifies a few neurons in this person's brain such that the fear of heights is installed, is this an instance of precluding free will? This person could still choose to get on a plane, but the likelihood they will is vanishingly small.

How about if the salesman lies to the customer? Then is the customer's ability to make a free choice negated? Many would claim that it is not as the customer could still make a free choice but based on erroneous information. But this means that the customer made a choice opposite of what he would have wanted to make had he had the true story. Going down this road implies the free will concept can be, and potentially is, completely disconnected from reality. If we take this to a limit, there is no way we can absolutely know everything that influences a particular situation and, therefore, cannot make a completely informed decision.

It is apparent that the concept of free will or making a free choice continuously contracts under scrutiny. One can't make the decision too soon beforehand as one might change one's mind. Also, one can change their mind after a free choice and the original free choice still holds in its own right—unless one decides later that they did not indeed make a free choice previously.

This is getting confusing, so let's hear it straight from the horse's mouth. How many of us, I wonder, have actually stopped and watched a conscious choice being made in our own minds? How do you differentiate a conscious choice from one that is not conscious?

Watching the mind work is a fundamental exercise in Eastern philosophies such as Zen. Anyone who has spent time in meditation, or zazen, is very aware there's no time that can be pinned down as to when a conscious decision occurs. Let's say that someone is trying to make a decision about participating in a particular activity. Thoughts continue to run through the mind concerning the good and bad things about the activity, images appear in the mind about the activity, such as one's self or other people participating in it in the past, and all kinds of other anecdotal information show up that are of varying degrees of relevancy. Since the mind is reviewing material that would be necessary in

8.2. FREE WILL

making the decision, it seems that it is busy getting ready to make that decision. However, when the time comes, there is no conscious event that you will be able to pin down that can be labeled as "having made a decision." Instead, in an instant, the mind goes from a state of indecision to resolution. The mind is now ready to participate (or not) in the activity. However, indecisiveness and determination lie on two sides of an empty space. There was never a recordable event of making the decision.

What is a free-will choice then? The most common definition of free will is called *libertarian free will*, which means that human beings are autonomous, and can make decisions that are not contingent upon anything else. This is a very strong statement. It claims that the above customer, despite the salesman's efforts, still had the ability to make a free choice, and that the choice made was completely independent from the salesman's tactics. Libertarian free will stands in opposition to determinism, or the belief that every action has a cause and could be predicted if that cause (or causes) was known.

Neuroscience, starting with the work of Benjamin Libet who, in the 1980s, did research using an electroencephalogram to investigate the existence of so-called *readiness potentials* in the human brain that demonstrated detectable neurological precursors to conscious decision-making, has provided evidence that what we normally consider to be conscious decisions are actually made ahead of time in the subconscious, and that we are completely unaware of this process [129]. Since then, these types of experiments have repeatedly shown that we do not necessarily make decisions consciously the way we think we do [130]. Functional magnetic resonance imaging of the brain during experiments in which a subject is asked to make free decisions can predict the outcome with some degree of accuracy full seconds ahead of time. Research like this has seemingly thrown the ball back into the court of determinism.

The biggest problem with the entire situation is the false dichotomy between determinism and free choice. These are not opposites of one another. The entire discussion hinges on what exactly is meant by a free choice. Certainly, libertarian free choice rules out determinism. In fact, reality itself rules out determinism. We have already seen that the behavior of matter and energy in the Universe is not deterministic. Therefore, unless one is going to claim that the brain and mind somehow behave differently than the rest of reality, then their operation certainly can't be one-hundred percent deterministic.

However, just because determinism is ruled out does not mean that we have free will like most people think: the will to make a choice that is not contingent upon anything else. It is even difficult to fathom precisely what this statement means. Without any preference, which would be a contingency, how would a decision come about? There is no impetus to choose A or not A. From the position of no preference, both choices are the same.

If one says that a free choice means that one could have, under exactly the same circumstances, including the state of the brain, have chosen differently, then I call this randomness. In fact, this is the definition of randomness: varying outcomes under exactly the same conditions. This is an extremely subtle point and easy to miss. The magic that is presumed to be the freedom to act any way one wants under a set of circumstances is essentially the description of random behavior. If it weren't random, it would be contingent.

A free choice, or we can just label it as a signal, say the result of a binary choice with outcomes denoted true and false, is either contingent or deterministic, or it is random or unpredictable. It matters not whether the source of this choice is an immaterial soul or a physical brain.

Many would argue that the human mind obviously doesn't just make random choices. The mind is perfectly capable of working out complex mathematical problems and performing quite logically. On the other hand, it certainly isn't error free either. The fact we all don't agree with one another is evidence enough for that. Given the same problem to solve, two different people can come to entirely different conclusions. In fact, the same person can come to different answers at different times. Although the mind does appear to consist of a neural network that can exhibit chaotic behavior, and it also has neural machinery built in to overcome null decisions, what we see in human behavior is more than these two processes can explain. Undoubtedly, the human brain operates at some level in a random manner. For example, if the brain makes $999/1,000$ decisions completely deterministically, and $1/1,000$ in a random manner, then the brain does make use of randomness to arrive at conclusions.

Michael Egnor, a professor of neurosurgery and pediatrics from the University of New York, who has written for the Discovery Institute's *Evolution News*, provided what he considered the best definition of free will from the philosopher and theologian Thomas Aquinas [131]:

> My free will is inclination based on abstract reasoning that arises wholly from me. Nothing other than me determines my will. I determine my will and my will is an immaterial power of my soul. In this specific sense, I have free will.

At this point, it has become clear once again that the entire crux of the discussion rides on what the term "free will" means. Is the free will of the person who gets up in the middle of the night and raids the ice cream from the fridge when they are on a diet just not strong enough? Is it malfunctioning? How about the agoraphobic that we mentioned earlier who is so afraid of heights that they would never get on a plane; does the fact that they would like to fly but can't mean their free will is on the fritz?

This line of thought seems a bit silly because it is. There is a better explanation for all of this, and it doesn't involve a supernatural power that can't be measured in this reality and for which there is absolutely no evidence. Perhaps we get our inclinations and make our decisions using a large neural net that weighs competing objectives and sometimes provides certain answers and sometimes not so certain answers. With a diet on one hand and a carton of ice cream on the other, the mind's neural machinery does its best to compute the most

8.2. FREE WILL

beneficial path. But this is a computation that is plagued by a myriad of exogenous disturbances and seemingly random thoughts, leading to the decision's optimality being questionable.

Taking this point of view, all the available data can be collected under a central thesis. On the other hand, the only reason that ID and religious proponents support the existence of a magical power that really doesn't make logical sense in its own right, is that a god must be able to punish people justifiably. Despite this, all the evidence points toward a biologically complex neural network, shaped by evolution, that still attempts to achieve its perceived goals. On top of that, however, it can process logic and abstract thinking. Of course, it can also process illogical, motivated thinking.

As we discussed above, the common concept of free will entails the ability to decide in various different ways under the same conditions, and as was pointed out, this is the definition of random behavior. It certainly appears that randomness, originating biologically, enables the unpredictability of free will.

There are two characteristics that determine how a choice will be made. The first is a deterministic component and the second is a random component. It can't be any other way unless we're going to admit into the argument a supernatural power like the Christian soul, which we are not going to do. If a large neural net, like the ones used in today's advanced computer systems, is used to come to a marketing decision let's say, it matters not at all how sophisticated this machine is, it is still deterministic, and so its behavior is determined beforehand, and it has no choice in its selection. However, adding a bit of random perturbation within the system allows indeterministic outcomes to occur. Some might argue that this still doesn't allow true free choices of the type most people think of to be made, and I agree with these dissenters. One more time, this discussion hinges totally on what "free choice" actually means.

A complex neural network, despite the fact that its dynamics can exhibit chaotic behavior, is still deterministic. A system like this would not allow the generation of new information; it would not be capable of intelligence in the sense we generally think about. A unique thought in such a mind would, just like with a conventional computer, represent an error. For example, if you are working with a word-processing program and during the spellcheck, instead of accessing the memory location for the word "electromagnetic," it instead accesses the word "electrostatic," this would be considered an error. It would drive the writer crazy trying to find out why the program keeps flagging a correctly spelled word. However, when a random event such as this happens in the mind, as described previously, it can be a fortuitous situation. The mind's neural net is malleable enough to integrate this mistake into a new way of looking at things.

These types of random events are also evident in our decision-making, especially when making a decision with competing objectives that are very close in value. Although as mentioned above, the brain does have tie-breaking circuitry, as the differences between competing objectives become narrower, the ultimate decision maker is chance.

I want to be clear that I am using the term random in a very general sense that includes thermal noise, dynamic neuronal wiring errors, potential synaptic quantum effects, exogenous sensory stimulus, etc. In other words, all the sources of unpredictability within the brain.

In order to retain the concept of free will, the definition is going to have to change. We are certainly capable of abstract thought and logic; we do it whenever we work out any mathematical problem. This activity requires extreme consistency. However, we are also capable of making choices that make absolutely no sense at all. And we are certainly capable of disagreeing with others who also claim they have found a rational solution. Logic and analytical thinking are not guaranteed. Our neural net was not designed, it evolved, leading to a system that is rapidly adaptable and well suited for surviving on Earth, but not optimized necessarily for rational thinking. The brain has to be trained for this endeavor.

In addition, we are capable of making knee-jerk decisions that just come out of the blue. In fact, this is how we make a lot of our decisions. After much deliberation, a snap decision is made, and we have no idea why we chose the way we did. The subconscious does a lot of processing in the background looking for optimal solutions to problems involving competing objectives.

Finally, we can also make decisions like the ones that put ourselves at risk for the sake of others. These are generally unthinking situations where the conscious mind is not involved at all.

Decision-making does not involve an immaterial soul. There is absolutely no evidence for this as it falls into the same untestable category as the ID hypothesis. Just like ID claims that evolution can't account for biological complexity, those who claim the existence of a soul provide as evidence an assertion that the human brain can't account for the characteristics of consciousness. Neither of these claims constitute evidence. They are simply unproven assertions to support presuppositional ideas.

Choices are made using an incredibly sophisticated neural net in the brain that draws on immense amounts of data. This system is not perfect—it is hard to imagine what perfect would even mean here. Despite the fact that computers can be perfect because they are deterministic, it is good that human brains are not deterministic. This neural net is subject to chaos and inherent noise. When choices can be made through clear analysis, the brain is capable of doing just that, even with mathematical precision. However, when the problem to be solved is not so clear-cut, the solution is likely influenced by noise, internal and external, being amplified through the large sensitivities of chaotic dynamics, allowing one to tumble to unorthodox solutions. A brain that is locked up in a state of indecisiveness is not evolutionarily favorable.

The operation of the brain between order and randomness is what allows the development of novel information and the rapid adaptation to new situations. Random events within the brain are also what brings about free will, just not in the way most people would like it to be. *People do not possess autonomous free will.* They cannot. By definition, determinism cannot lead to a free-will decision.

Determinism would require that under the same circumstances that the decision always be the same. However, this is contradictory to the usual definition of free will. On the other hand, indeterminism goes contrary to the usual definition as well: how can anyone be held responsible for their actions if they are not the author of their actions. This problem stems from an assumed definition of free will. Let me clarify this.

We are all accustomed to the familiar legend that love has the power to move us to do what we normally would not do. So, what would we normally do? What we normally do is adhere to the deterministic logic built into the neural nets of our minds. The result is self-centered logic in our actions. Logic will not step in front of a bullet meant for someone else. It will argue that there is no assurance that the bullet will hit the target, but, if I try to put myself in front of the bullet, the odds become 100% that I will be hit. Hence, the best decision is to not get involved. This is extremely logical and the best approach from a deterministic, analytical standpoint, but it is not what we think of when we consider love. Love can overwhelm this logic and take a bullet despite the logical arguments not to do so. This is a decision not made by the analytical mind. So where does it come from?

Take another example that Jesus gave: the parable of the lost sheep. The sheppard will potentially forsake his entire flock to save the one sheep that is lost. This goes entirely against logic, and I think it is meant to do so. Where does the ability to do this come from if the mind is really deterministic?

Yes, free will exists, and completely noncontingent decisions can be made; they just don't come from what you think is you. Free-will decisions are not just noncontingent on the rest of reality; *they are also not contingent upon you.* And I'm not attributing right or wrong to any of this—those are anthropomorphic concepts that do not belong in this discussion. It appears that you cannot be congratulated any more than you can be condemned for your decisions. However, I can't confirm this as the subject gets very complex in the middle. What I can say at this point is if your response is deterministic, you are not responsible. And, if your response is indeterministic, you are also not responsible. And this is exactly where we need to go in terms of research. I can't say what happens in the middle. I'm taking a risk here because I know simplistic minds will read into this and say, "That's just what we meant." That is not what you meant. We're on the path to finding the truth now, and it has nothing to do with the nonsense that has been going on for 2,000 years.

8.3 Artificial intelligence

In the late 1980s, a group of graduate students at Carnegie Mellon University constructed a computer nicknamed Deep Thought intended to compete against chess masters [132]. Although the machine lost its first two competitions against Garry Kasparov, the world chess champion at the time, the capability of Deep Thought so impressed IBM that they hired these graduate students to work on a successor computer named Deep Blue. This next-generation chess-playing computer also lost to Kasparov at a 1996 tournament in Philadelphia, but it did win one out of the six games. Spurred on by that humble one-game victory, the team returned and triumphed over Kasparov in 1997.

Then, in 2016, Google's AI system AlphaGo beat the GO master Lee Se-dol in a tournament in Seoul four games to one. AlphaGo is a system that was actually built by the same group that created Deep Blue—DeepMind Technologies— which had been acquired by Google in 2014 [133]. This victory by the machine was astonishing for two reasons. First, the game of GO is more complex than that of chess by many orders of magnitude. Secondly, whereas Deep Blue had relied primarily on processing speed to explore possible outcomes of the game ahead of time to achieve its success, AlphaGo did not.

Anyone who has played amateur chess understands what "looking ahead" means. Beginners look at the state of the board and play sequences of moves in their heads, trying to decide which starting move they should go with. Unfortunately, the number combinations of move patterns is astronomically large, around 10^{123} [133]. However, the number of combinations for GO is on the order of 10^{360}, which is vastly larger than that for chess. The playing space for either one of these games is so large that no computer could analyze them all in a reasonable amount of time, but, even so, the playing space for chess doesn't even approach that of GO. Masters of either of these games rely heavily on intuition and not analytics for this very reason.

Whereas Deep Blue relied primarily on the advantage of computational speed to look ahead and choose the most advantageous moves, AlphaGo, instead, relied heavily on what is termed today as *artificial intelligence* (AI), a term coined by John McCarthy to describe machines that can supposedly think autonomously [134]. Let's take a look at what AI really is and if it has any connection to the natural intelligence exhibited by humans.

Artificial intelligence is usually based on *artificial neural networks* (ANNs). These are computational systems composed of a series node layers such as those shown in Figure 8.6 [135]. The leftmost row of vertical nodes (the circles) represents the inputs to the network. The rightmost node is the output node. Either of these layers can have different numbers of nodes depending on the application at hand. The center two columns of nodes are called hidden layers and they perform the primary computation function of the network.

The network takes one or more inputs, performs a mathematical manipulation within the hidden layers, and provides an output that is functionally related to the input. As a simple example, let's model a vehicle traction controller with the network in the figure. Assume that the three input nodes receive the vehicle's left front wheel speed, the vehicle's right front wheel speed, and the actual vehicle speed taken from a rear wheel sensor. The output node is a signal that reduces the engine torque (scales the value coming from the accelerator pedal). The idea is the output torque should be reduced when either of front wheel speeds differ from that of the rear speed sensor as a difference in the speeds would

8.3. ARTIFICIAL INTELLIGENCE

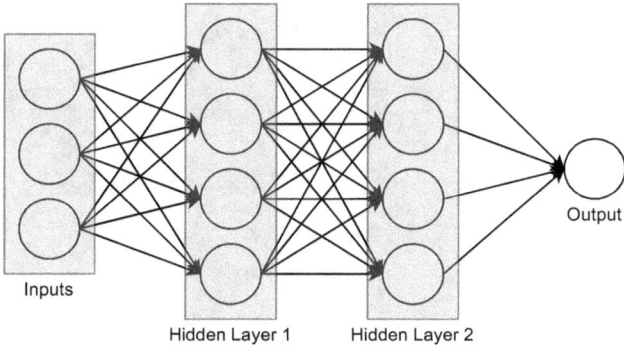

Figure 8.6: A simple neural network with three input nodes, two hidden layers, and one output node.

indicate that a wheel is slipping. And so we have the model $T_w = f_h(v_{lf}, v_{rf}, v_{rs})$ where T_w is the torque output of the network, and v_{lf}, v_{rf} and v_{rs} are the left-front, right-front, and rear wheel speeds, respectively. The function $f_h()$ models the computation of the network and is what must be solved to implement the controller in a vehicle. (This is a mathematically complex process and we're not going to cover it here.)

The arrows in the figure represent connections between the nodes. Each of the nodes or circles in a particular column output information to every node in the subsequent column. In addition, each of the arrows that link the nodes has a certain scaling, called a *weight* in the literature, associated with it. The weights scale the values that are passed from one node to another. Each of the nodes, also called *neurons*, has an activation function. The complete functionality associated with a hidden-layer network node is shown in Figure 8.7 and has two main parts. The first part in the rectangle sums scaled versions of the inputs from the previous nodes. Once these scaled inputs are added up, a constant b is added to them, and the result is provided to the activation function. The weighted inputs from previous nodes are intended to simulate biological axons from previous neurons in the brain.

The second part of the hidden-layer node is shown in the circle in the figure and is what is called an *activation function*. This name is taken from the activation potential that is present in biological neurons. In the biological setting, when the potential of a neuron is brought beyond a particular threshold via the inputs from axons from surrounding neurons, the neuron will "fire"; that is, send a signal down its own axon to a following neuron. When the activation potential is passed, the neuron switches from the inactive to the active state. Neurons in ANNs mimic biological neurons by adding an activation function after the weighted sum of inputs.

An example of an activation function used in ANNs is given in Figure 8.8.

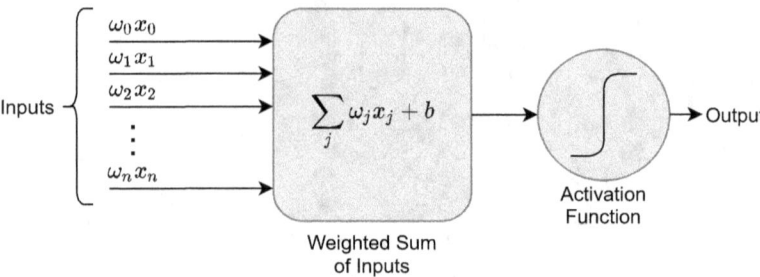

Figure 8.7: The structure of a neural network node. The weighted inputs are summed and passed through an activation function to the output.

It is based on what is called a *sigmoid* function defined by $f(x) = 1/(1+e^{-x})$. The input is on the horizontal axis. As the input increases from -10 the output doesn't change much at all but, as it crosses zero, the output makes an abrupt transition from zero to one. As with a biological neuron, the weighted sum of inputs has to pass a certain threshold before the output changes state. This emulates a digital process (we'll see in a minute why it can't be exactly digital, but can only approximate it).

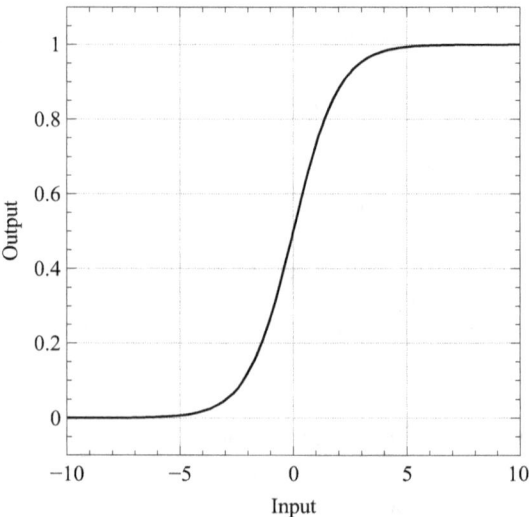

Figure 8.8: The sigmoid activation function $f(x) = 1/(1+e^{-x})$.

It turns out an ANN with at least one hidden layer that can grow infinitely wide can simulate any analytic function. This is very important and allows the neural network to perform arbitrarily complex calculations. These networks are used where functions describing a system cannot be solved in a convenient manner. There are many instances in engineering where systems can be characterized analytically using physical laws. For example, many times an electronic circuit can be reduced to a set of mathematical expressions that can be used to calculate outputs as functions of the inputs. However, it happens just as often that systems are analytically intractable. (This doesn't necessarily mean an analytical solution doesn't exist; it may just mean that it is too difficult to find.) In these instances, methods like neural-network modeling come in very handy.

An example of where a neural network can be useful and effective is in spectroscopy measurements for material analysis. For example, in inferring the moisture of materials using dielectric spectroscopy, the sample being interrogated is subjected to an electric field at many different frequencies. The amount of moisture the sample contains directly determines how the electric field interacts with the material. How then do we come up with a function that relates the set of measurements to the moisture of the sample? If this can be done, we can create a moisture meter that measures a sample at these frequencies and then calculates the moisture. The problem is that it is not clear what type of function should be used. This is where a computational method like a neural network comes in. The measurements at the different frequencies are supplied to the inputs of the network of Figure 8.6 and the moisture is calculated at the output node.

This all sounds good except for one problem: we need to tune the network for the specific function of spectroscopic moisture measurement. The only parameters we have to tweak in the network in the figure are the scalings ω_j that feed values from nodes to subsequent nodes. Here's how this is done. Many samples are subjected to the spectroscopy process allowing a large number of input and output values to be collected. Then all that is needed is to calculate the appropriate weights such that when a particular set of inputs from a specific sample are input to the network, the appropriate moisture is provided at the output.

The process of calculating these gains is called *back propagation* and involves taking derivatives to minimize the error of the network as it is applied to the data. It's a little more complicated than taking a single derivative to minimize a function like $y = x^2$, but the idea is the same. This is why the activation function used in the network nodes needs to be continuous and cannot be digital in nature. If that were the case, it would not be possible to take the derivative.

I'm not going to go through how these gains are calculated, but once they have been, the neural net can be used like any other estimator: independent variables are provided to the inputs, the hidden layers are calculated, and the estimate is provided at the network output.

Note that the network is completely deterministic in the sense that its behavior is entirely determined by the inputs and the weights linking the hidden layers. This deterministic functionality will not change with time. The same output will always be produced given the same inputs unless the weights are re-tuned.

Network functions like this (there are many different types) are at the heart of artificial intelligence. This includes *deep learning* [136], a term that is used a lot today for marketing emphasis, but really just means that there are multiple, sometimes many, hidden layers embedded within the network. Coupling deep networks with vast amounts of high-speed processing and memory allows computers to perform tasks that, without close analysis, look like the kind of cognitive functionality humans possess. Despite this, they still would not pass a Turing test.

The so-called Turing test was described in a seminal paper by Alan Turing in the quarterly review of *Mind* in October 1950 [137], although he referred to it as the "imitation game". The game involves testing the ability of a machine to emulate a thinking person. It comprises a conversation between two people and a machine. One of the persons is an interrogator who, without being able to view the other person and machine, must, by simply asking questions, determine which is the person and which is the machine. If the interrogator cannot determine which is which, then the machine is said to possess the ability of human thinking.

The Turing test suffers from not being able to sift out what exactly the properties are that it is trying to test [138]. For example, most people can't solve incredibly complex mathematical problems, but one might suspect that an advanced machine intelligence could do so. Therefore, in order for the machine to successfully emulate human intelligence, it would need to curtail its computational abilities, otherwise it would be found out. Also, the error rate for a digital computer is very, very low compared to that of a human. It does not make mathematical mistakes, does not forget, and can store vast amounts of precise data. This is in stark contrast with human beings who tend to be very forgetful, distort their memories, and make mistakes all the time. And, as was discussed in the previous section, the machine would also have to make decisions between near-identical competing objectives. In fact, it would need to do so even if these objectives possessed identical fitness.

For the machine to pull off its intended subterfuge, it would have to make mistakes. But how does one design a machine that makes mistakes? The network must admit random errors in order to emulate a human being.

The neuronal network in the human mind exhibits chaos and is noisy [139–141]. The combination of these two allows the human brain to display intelligence and decision-making abilities that a conventional, deterministic computer cannot. Of course, we do not have a definitive description of how the mind is laid out or how it precisely operates, but it is certain that it behaves in a chaotic manner that allows large jumps in the cognitive phase space with only minute changes in initial condition, and that these jumps could be driven partially by noise-induced changes in state. In other words, an incidental thought can lead to a leap of epiphany.

There isn't really a clean way to implement a "noisy" digital computer, but *quantum computing* is set to disrupt conventional deterministic computing in the coming years. It is much faster at computing some problems, such as prime factorization of large numbers, than a digital computer. The ability to do so stems directly from a few properties of quantum-mechanical systems, notably *superposition*. Superposition allows a quantum system to exist in multiple states simultaneously.

The dual-slit experiment that was described in Chapter 2 is a perfect example of superposition. If the firing rate of the electron emitter is slowed so that only one electron is traveling to the screen at any given time, the interference pattern on the backdrop behind the two slits will still be displayed (on average). The single particle's wave function is interfering with itself as it passes through both slits simultaneously. The electron is in a superposition of two states: one where it passes through the left slit, and one where it travels through the right slit.

Using this property, the bits in a quantum computer, *qubits*, effectively compute multiple values simultaneously. The outcome of these computations, when measured, have a degree of randomness to them as the wave function associated with each collapses. Many such runs have to be performed, and their states sampled to get an average of the probabilistic states of the qubits. These average values are what actually provide the desired answer. However, the operation of the quantum computer is indeterministic from the point of view of the instantaneous state values. These computers are built to extract only the average values of the outcomes, which are dictated by the Schrödinger equation, and are deterministic. It'll be interesting to see how long it takes before someone starts using them in a way that allows some indeterminism to creep into the computations.

Perhaps in the future, it will be possible to construct a quantum computer that is intelligent on the level of the human brain. If and when this happens, what would ID advocates have to say about such a development? Would such an intelligence deserve the same rights as every human on the planet, or would it be considered as inferior because it doesn't have a soul? This seems a legitimate question when considering how we humans have treated other intelligent species on this planet like primates, whales, and octopi, not to mention each other.

This is an example of how such nonsensical ideas can lead to harm. An inherent prejudice against artificial intelligence would be the immediate consequence. And yet there would be no evidence or rationale to treat these intelligences any different from how humans are treated. At this point, would we admit that we should get rid of hypotheses that are not justified, like ID and the idea of an immaterial soul? I highly doubt it. My guess is that the story would simply mutate to integrate any new developments.

One of the most important concepts from this section is that AI will require some degree of uncertainty to mimic human intelligence. Deep learning so-called AI machines used today are excellent pattern-recognition devices, but they do not create novel information.

8.4 Intelligent evolution

Biological evolution that occurs naturally has two components: a random modification of the genome, and natural selection of the beneficial mutations through enhanced reproduction. Without both of these, evolution will halt.

In today's society, at least in most developed countries, healthcare is available to virtually everyone, and the standard of living is high enough that most people can live in comfort. Death from disease or hunger is no longer the dire situation it has been in the past. Almost anyone can start a family and take care of that family. In this sense, there is no longer any selective pressure on humans.

Humans still play a part in driving evolution, however. In many instances, they are unintentionally applying extreme evolutionary pressure to other species. Evolutionary pressure applied by human beings to some effect, like creating larger carrots through selective breeding, is termed *artificial selection*. On the contrary, when humans unintentionally affect the evolution of other species it is typically labeled *unnatural selection*.

Humans, through their negligence and need to acquire more and more, have unintentionally altered the evolution of many species of animals. The harvesting of wild animals for food and sport has substantially changed the morphologies of some species, for example, fish and big horn sheep [142]. Commercial fisheries typically extract the largest animals from a population, leaving behind the smaller fish to reproduce. An immediate consequence is the genetics that produce smaller fish are more easily propagated through the population. This leads to the average size of fish shrinking. Since fisheries still want the largest fish, they continue to gather the largest, and the average size perpetually contracts.

Another example has to do with big horn sheep. Hunters pay extravagant fees to hunt these sheep for trophies. Of course, the males with the largest horns are what the hunters seek, leading to the horn size of these sheep getting smaller over time. Professor David Coltman, at the University of Alberta, has studied the sheep of Ram Mountain in Canada for 40 years, and has observed the average horn size shrink by 25% over that time [142]. And, whereas natural evolution favors animals with the genetics for larger horns, due to human intervention, smaller horns are now favored. Larger horns are a liability as they tempt hunters to kill these animals.

Finally, due to decades of poaching them for their tusks, the elephants of Mozambique are gradually losing their tusks [143]. Whereas tusks used to be an advantage for defense and competition for mating, they now lead to a less advantageous position within the evolutionary landscape as they are a beacon for poachers looking for ivory. Those elephants with genetics that are predisposed for not developing tusks more easily transmit their genetic material through the population as the elephants with this trait are less likely to be killed and, therefore, have a better chance at reproduction.

Of course, humans also intentionally alter the evolution of animals for specific purposes. For example, cows, pigs and chickens are selectively bred to produce better food. The same is true for the plants we eat or need for other purposes, for example, corn and beans. In fact, we routinely go a step further with plants and genetically modify them using gene splicing—manually altering the genome. Genetic manipulation is only going to become more prevalent in the coming decades as the world's human population continues to increase and resources become more scarce.

8.4. INTELLIGENT EVOLUTION

As mentioned above, the selective pressure for humans is lessening all the time. Anyone has a chance to procreate and live a long life, particularly in developed countries. At the present time, biological evolution has changed for humans in that it has no definite direction set by natural selection. One of the primary jobs of medical science is to keep people alive and allow them to live normal lives in the presence of genetic variation. The evolutionary landscape has become flat—medical science is smoothing the peaks and valleys, allowing random genetic modification to wander as it will. We are no longer at the mercy of random variation and natural selection, and this is a good thing.

Modern medicine has given us tools to fight infections, mitigate illness and has answered questions about how to live healthier lives, all of which has improved our lifespans and quality of life. However, it is the tools and technological innovations that are coming in the future that are really going to change things.

One of the most promising techniques is in *gene editing* using a process first published in 2012 by Jennifer Doudna and Emmanuelle Charpentier called CRISPR-Cas9 [144]. (CRISPR is an acronym for clustered regularly interspaced short palindromic repeats.) The discovery earned them the Nobel Prize in chemistry in 2020. It is based on a process that some bacteria and archaea use to generate immunity against bacteriophages (viruses that attack bacteria and archaea).

When attacked by a new bacteriophage, a bacterium can incorporate fragments of the virus' DNA into its own genome as templates for their immune system. If the bacteria are attacked by that same virus again, this genetic template is used to lock onto and cleave the DNA of the virus. And, since the fragments are incorporated into their own genome, it is passed on during reproduction, leading to a defense system that propagates through the population. Doudna and Charpentier realized that this same process could be employed by scientists to cut DNA strands in very precise locations, allowing for sections of the strand to be removed or new sections to be inserted, effectively modifying the genome of an organism [56, 145].

The selectivity of CRISPR is such that a single nucleotide can be altered in a complete DNA strand. Research in animals indicates that the method may be highly effective in treating diseases including cancer, sickle-cell disease, Huntington's disease, cystic fibrosis and heart disease.

So long as these types of medical procedures can be accomplished without endangering anyone's life, they should be used. There is no need for anyone to live with a defect in their genetic code if they don't have to do so. Methods like CRISPR are the first steps in humans taking control of and guiding their own evolution.

It is not as if we could just come up with a design for the genome appropriate for a being that could live on Mars and yet retain the remaining characteristics that make us distinctly human such as intelligence, emotions, curiosity, etc. Recall that the genome is not like a set of switches, each of the switches labeled like those in a home's breaker box. The wiring in a home is direct, with a one-to-one correspondence between switch and function. The genome's relationship to phenotype is chaotic and complex: change one thing in the genome and it results in potentially ten changes in the phenotype. Everything is connected. As we learned from looking at cellular automata, there are some systems whose outcome we cannot predict—the only way to know the outcome is to put the pieces on the board and play the game. This is the case with genomic manipulation. Experimentation is required, and so, in some regards, we end up right back where we started, varying the genetic makeup and analyzing the results, looking for correlations.

Despite this difficulty, it can be assumed that with time, modern science will figure out a way to make predictions about genomic changes in general. At that point, we will truly be able to guide our evolution. This will be a much different evolution than what has so far taken place naturally. Natural evolution simply works to increase reproduction. Our evolution will no longer be governed by this positive-feedback mechanism. With the latest genetic tools, it will not be necessary to bring about changes in the genome during mitosis; we will be able to rewrite the genome of any human being.

There are capabilities that we haven't even imagined yet that may be possible by reprogramming ourselves. We have already done this the hard way via bionics and artificial implants. Hearing aids and heart pacemakers are two areas where artificial electronic implants have helped many people live normal lives again. We have come to accept these as standard pieces of equipment such that we pay no attention when a human is augmented with either of these technologies. However, as we move further into the future, and as the functionality of these devices is extended, it begins to conflict with what some people believe should be done.

The bionic eye has left science fiction and has become a part of reality [146–149]. A *charge-coupled device* (CCD) array is used as a camera at the front of the eye. Signals from this CCD array are then translated to appropriate levels and injected into the neurons at the back of the eye. The arrays used today are about 10x10, or have about 100 pixels. Therefore, they certainly do not provide the resolution of a biological eye, and provide only the ability to differentiate between light and dark, sense flickering light conditions, or provide a very rough, pixilated picture, but they are of definite value to the person who cannot see. From an evolutionary perspective, it is far more advantageous for a person to be able to simply detect the presence of light than not (just like in the natural evolution of the biological eye). In the future, we will see the technological advancement of these basic artificial eyes into something much closer to our biological eyes. In fact, the ability is there to surpass the sensitivity of our biological eyesight, to see into the infrared and ultraviolet regions of the spectrum.

We have the potential for creating electronics that far surpasses our present physical capabilities. For example, exoskeletons are being developed presently to increase our physical strength far beyond that given to us by nature. An example of this is the exoskeleton marketed by Sarcos for industrial use. A suit

8.4. INTELLIGENT EVOLUTION

like this has obvious benefits such as assisting the wearer to lift up to 200 lB without fear of injury. Along the same line, we can't forget that the military is actively working on exoskeletons to enhance the physical capability of soldiers in the field.

Elon Musk's *Neuralink* device is intended to connect the human brain to a digital interface, allowing direct access to the internet with an incredibly high bandwidth [150]. As it stands, humans have a very low communication bandwidth. Most of us can't type any faster than about 300 characters per minute on a PC keyboard, while an ethernet connection can communicate faster than one billion characters per second. The Neuralink would involve a series of very fine connections to individual neurons through a surgical implant in the brain. The neuroplasticity of the brain will hopefully sort out the signals passing through this interface, and the human subject would be permanently connected to the internet or other information store and processing center. The messages would appear as simply thoughts in the mind. Instead of having to type in a search on the web, one could simply try to remember the subject being searched for and the retrieval of the information over the communication channel would be like remembering something you already knew. One vision of a device like this is to be connected to an artificial intelligence to allow humans to solve problems much faster than they can alone.

As we move down the list of electronic devices intended to be augmented to the human body, many people tend to become nervous. And I do agree that these devices should be proven safe long before they are integrated into humans. However, safety is not one of the reasons cited by many people. These reasons have more to do with worries about their religious views. For instance, one does not have to dig very deep on the web to find opinions that any electronic device implanted in the human body is a candidate for the so-called mark of the beast—the branding of Satan. I'll repeat, I think these devices should be vetted to the fullest extent for safety, and I don't think anyone should be forced to accept them, but there is no place at all for superstitious nonsense in the discussion. These kinds of beliefs, secular or religious, are what drove a large percentage of the population to refuse Covid-19 vaccinations in 2020. These ideas are dangerous and serve no useful function whatsoever.

Where we go as a species from this point is a continuation of the evolutionary process that got us to where we are now. I understand that new things elicit fear in people—the same is true with me—but I don't see the logic in drawing a line in the sand and saying what came before was natural, and if we meddle with it, that is unnatural. It seems customary to separate out what humans do in constructing things and being creative as not natural. It is perfectly natural, as we are part of nature.

The religious belief that some areas are sacred or untouchable, and ID's contention that some things simply can't be understood, are two methods of bringing research and innovation to a halt. I'll say it for a third time: I don't think any of this should be done in a hasty manner. Human beings have repetitively shown that we can't always be trusted to do what should be done. We have done potentially irreparable harm to Earth's climate, driven other species to extinction, and been at continuous war with one another for thousands of years. We need to show more care in the future. And this is not because of some silly notion like we are fallen creatures. It's because we haven't been at this for long—about two-thousand years—and we are still learning. It would be more tragic if we stopped learning now and just sat around waiting for armageddon to arrive. We may have a shot at redemption if we keep going, but it won't come through religious dogma.

8.5 Simulated reality

I would be remiss if I didn't mention one area of science that is currently considering an intelligent designer, albeit one of a form different from what we have been considering in this book. This hypothesis stems from the conjecture of a *simulated reality*. One of the allusions to this idea is from the *Zhuangzi*, an ancient Chinese text containing stories describing Taoist philosophy [151]. The most widely known story from the *Zhuangzi* is "Zhuang Zhou Dreams of Being a Butterfly":

> Once, Zhuang Zhou dreamed he was a butterfly, a butterfly flitting and fluttering about, happy with himself and doing as he pleased. He didn't know that he was Zhuang Zhou.
> Suddenly he woke up and there he was, solid and unmistakable Zhuang Zhou. But he didn't know if he was Zhuang Zhou who had dreamt he was a butterfly, or a butterfly dreaming that he was Zhuang Zhou. Between Zhuang Zhou and the butterfly there must be some distinction! This is called the Transformation of Things.

The point of this story, of course, is how does one tell the difference between the waking and dreaming states? To one who is immersed in either, the task would seem impossible. The same question arises in the topic of a simulated reality: is it possible that we are entities living in a simulation? How would we know?

The concept has come up in serious scientific circles in the past two decades, elicited in 2001 by the philosopher Nick Bostrom publishing a paper [152] titled *Are you living in a computer simulation?* Even Elon Musk has stated that the odds that we are not living in a simulation is one in billions.

Our technology has been accelerating at an exponential rate over the past two-thousand years after taking tens of thousands of years to implement civilization. Logic came on the scene around 350 BC, but it was not until about 1100 AD that formal science arrives. In the 17th century, Isaac Newton and his contemporaries developed what would become physics and calculus. Then, in the early 20th century, physics was remade with quantum mechanics and relativity theory. We could barely get radio working at the start of the 20th century, but a mere one-

8.5. SIMULATED REALITY

hundred years later, we have cell phones and internet connections that would've appeared as magic to people living a century in the past. The number of new discoveries and technological advancements now made every day are impossible to keep up with. And, it seems likely that the developments in the next one-hundred years will look like magic to us today.

The speed at which technology is accelerating leads to the question of whether we will have the capability to generate video games in the future that are so lifelike that we can't tell the difference between them and reality. Will such an environment allow us to be completely immersed in it? There are two possibilities here. The first is that we would be completely immersed in the game, feeling the digitally-simulated wind in our hair, being able to move just like we move in actual reality, and possibly have our memories purged so we come to believe we really are a character in the game. The second possibility is that the computer processing power reaches a level where it is possible to create conscious, virtual players in the game. Instead of being mechanistic entities in the games sold today, these programs would be sentient.

Although conjecture, it seems that these are real possibilities given enough time for our technology to advance sufficiently. And, if these are real possibilities, what's the odds that such a computing environment has already been created and *we are the simulated characters in the game*? This is the crux of the simulated reality hypothesis. If it is possible to achieve one or both of these outcomes, and given that a great amount of time was available for the advancement of technology, could not another race of intelligent beings have already achieved this long ago, and we are the contents of the simulation they fashioned? It is even conceivable that, if we are living in a simulation, the designers of this simulation could also be living in a simulation that was designed much further back than them, and so on [153]. It is easy to envision an endless regress of simulations backward in time.

The next question is, why would an advanced race of beings want to create such a simulation? It seems logical that the answer might be for the same reasons we create simulations. We watch television because we like to immerse ourselves into a reality that is fun to be in. We play video games for the same reason. We are capable of losing ourselves in either one of these virtual worlds. It seems to stem from trying to find that most important thing in life: happiness. This is a reasonable goal.

A second reason for wanting to perform these types of simulations is research. We already discussed how, like cellular automata, evolutionary outcomes cannot be predicted in general. Perhaps they could be predicted by simply just simulating them, and we are the simulation. On the other hand, perhaps we are not the primary purpose of the simulation, but just a functional artefact. Perhaps we are like the subroutine in a microwave oven that calculates the amount of time left on the timer, but in a sufficiently advanced system that such monotonous functions are actually sentient.

A simulated reality would explain several things. Recall that quantum entanglement allows multiple particles to exchange information instantaneously across arbitrarily large distances. Whereas this is difficult to accept in our physical reality, it is easy for a simulation to explain it: the locations and properties of the particles are just information stored in memory and they can be accessed at any time; a large distance in the simulated world does not correspond to a large distance in the simulator.

Random events could also be explained as they would not be random from the perspective of the simulator, only to the simulation.

Also, a simulated reality would explain the discrete nature of the Universe at the quantum level. Simulating a continuous universe would require an infinite resolution. A discrete simulation would be far less computationally intensive. When we run simulations of continuous systems on today's computers, the variables making up the simulation space must be discretized. All of our simulation programs have a minimum resolution which is the smallest number that the computer can represent. For example, the resolution of a 8-bit microcontroller is $1/2^8 = 1/256$. The Planck length (1.616255×10^{-35} m) is the smallest distance that can be resolved in our reality. Quantum mechanics forbids a definite measurement of any shorter length.

One could even surmise that like a contemporary video game, in order to reduce computational requirements, only the portions of the game that are being observed by the player need be rendered. There is no need to expend the resources to compute a three-dimensional landscape that isn't even visible to the player, and today's video games do not. For the same reason, perhaps the quantum wave function collapses to a definite outcome only when it is necessary to do so; that is, when an observation is made (or when it interacts with a element of reality that is being observed). Along the lines of the old adage, it appears that a tree falling in the forest makes no sound if it is not observed.

Our reality at the microscopic level appears to be only information. That is, there are no such things as little spheres called protons, neutrons and electrons. They have no substance. A wall is not solid but rather is composed almost entirely of empty space. The reason you hurt your head when you run into it is the photon mediators of the electric fields in you and the wall inform both of you that you cannot pass through one another. It is the same in a video game. The reason the ball in the game breakout breaks the bricks is not because they actually impact one another, but because the positions of both are in memory locations in the game, and there is a function that causes the brick to explode when the proximity of the ball to the brick falls below a preset level. A video game is simply played out in a field of information. There's nothing real about it. Our reality is very similar.

Of course, all of this is just conjecture, but it is being looked at by several researchers, including those attempting to propose a means of testing if we are living in a simulation [154]. One cannot miss that the simulated reality conjecture sounds very much like an ID hypothesis, and it is. Nevertheless, it is far different from the ID hypothesis we have been critiquing in this book. For one thing, if there is no means for testing the simulated reality hypothesis, it will be rejected completely, just like ID. No scientist will argue for it if it is unfalsifiable. They know better.

The second difference is even if there is a designer of simulated reality, that designer will still reside in the "natural". Not one person who has contemplated a simulated reality has the notion that it indicates something supernatural. On the other hand, the ID hypothesis is built upon the idea of a supernatural designer.

Finally, the simulated reality hypothesis makes no outlandish claims about observable reality. Any composition would have to fall within the science that exists today. In contrast, ID makes unsubstantiated claims about irreducibly complex systems and impossible probabilities of specified complexity. For example, no one would say that the improbability associated with biological complexity means that we must be living in a simulation. From this perspective, these two hypotheses are vastly different.

It is my suspicion that the ID community definitely would not endorse a simulated reality even though it fits as a conclusion to their so-called evidence better than any other alternative. The reason is it would go against the Christian version of reality.

Creating a simulated reality for ourselves is certainly one avenue for us to continue to evolve, but I think it would bother a good many people and likely for irrational reasons. Reasons like a simulated reality not being real. But just how real is this "real" reality in which we find ourselves?

In the simulated reality field, the "real" reality is called *base reality* to distinguish it from all the possible recursive hierarchical simulations thought possible. If we are living in the base reality, as demonstrated in the list of points above, it is a rather strange one. It is discrete, comprised of information alone, has nonlocal connections, and does not appear to be really there unless it is observed in some way. In this reality, the water you drink is not a mathematically-continuous liquid; it is closer to drinking a glass of microscopic marbles. The space you reside in is also discontinuous.

It seems that most people would balk at the opportunity to happily reside within a simulated reality because "it wouldn't be real." This attitude is easy to understand and it stems from a notion that there is something more. We don't want to live in a "dream world." Undoubtedly, the idea of a "real" world grows in some way from teleological belief. It appears, oddly enough, that being happy is not the most important thing. The ironic item concerning the whole idea of a simulated reality is that we essentially live in one now. Our perceptions and the subsequent neural processing of those perceptions clearly generate an artificial reality that is only partially related to the real thing. We have made up stories that we try to convince ourselves are real to act as a balm to the fears we have of not knowing where we came from, what we are, or where we are going. It is very telling that scientists can't prove that we're not living in a simulated reality. Going hand-in-hand with this notion, Christians believe that this world doesn't matter nearly as much as the "real" one that is to come later.

8.6 Conclusion

Intelligence is not some magical property that allows novel solutions to simply emerge upon demand. When examined closely, we see that our so-called intelligent thinking minds are something different. Intelligence is built from non-intelligent processes within the mind.

We have no ability to control the thoughts that come into our minds. This is just a fact that anyone can demonstrate to themselves quite easily. The way we design things is to choose concepts from a stream of essentially random thoughts. The thoughts that are relevant and satisfy certain criteria are selected from the ensemble for further consideration.

It is strange that we can't decide the thoughts we think and, therefore, do not really have control over what path our "intelligence" proceeds down. But, at least we have the power to select which of these random thoughts we believe, right? Actually, no, we don't even have a choice there. Can a person really decide to believe something? I'm not talking about saying you believe it. I mean, *can you actually choose to believe something that you don't*? When analyzed closely, the answer is obvious: I can only believe what I believe.

Combining these two observations, it seems that we are not in control of ourselves; we do not have free will. If that is the case, then what is the intelligence we've been talking about?

We are natural processes. We are products of all the things we've been talking about in this book: random excitation, chaotic dynamics, the working of energy to explore all possibilities, etc. We are apparently unique in the animal kingdom in that we are self-conscious. I'm not saying self-consciousness is a magical thing; I'm saying it is just an emergent property of the natural systems that make us up. Is it any surprise that when we try to define intelligent action that we arrive at a conclusion very similar to that which created life itself?

If life resides on the entropic gradient of energy dispersal—energy, or action, searching out all possibilities to be found in reality—then so do we as humans, and so do our minds. We are dynamic, in every sense. As is well known in eastern philosophy, and even in western psychology, what we call our identity continually changes: there is no perpetual "I" that one can pin down.

When searching for a design solution, an engineer does not just conjure it into existence. If she tries, all she will get a is stream of thoughts, some of them strongly connected to the design and some not. Then she will have to select which thoughts are worth keeping. But, actually, the choice of selection is made by default, and she only executes the result of that selection. Intelligence is simply a natural process unfolding in time. No magic is required.

Chapter 9

The universal method

> "Few people realize the number of things that are possible.", Richard P. Feynman.

We have already seen how complexity is built in the biological world by natural selection applied to random genetic mutations. Furthermore, it has been shown that optimization methods such as gradient ascent or genetic programming use similar methods to great effect. The purpose of this chapter is to demonstrate that systems making use of the information contained within random processes are ubiquitous in society's functioning.

Random processes coupled with a selection mechanism show up in the areas of technology, engineering, finances and the economy, religion, and the mind, among others. The operation of each of these systems involves random changes that are subsequently evaluated in terms of some performance metric. The changes that cause an increase in performance are generally kept, and those that produce a negative effect are weeded out of the system. Advances in technology include performance increases, the development of disruptive functions, cost reduction, and more. Designs or products that do not meet the set of desired objectives fail in the marketplace. Similarly, financial instruments that are less efficient than their competitors are discarded.

As we dissect this ensemble of subjects in the following sections, it will become clear that new information arising from random processes, contrary to being an impossibility, is actually *the universal* means of generating useable, novel information. It is not some novel, unbelievable idea exclusive to the development of biological complexity that has been co-opted to describe the dynamics of these other areas. Rather, it is an important means which nature applies in many different situations.

This chapter serves only as an introduction to this material. Each of these subjects could be a book of its own. I briefly examine them to demonstrate that evolution is a ubiquitous process.

9.1 The evolution of technology and science

Technology is the umbrella we use to describe the process of manipulating physical reality to meet our wants and needs as individuals and as a society. Designing new types of cellphones and more fuel-efficient automobiles are examples of advances in technology. Science is the means by which we advance our understanding of physical reality. For example, the discovery of the quantum-mechanical model of reality was a radical step in the evolution of science and gave us a much more accurate picture of how matter and energy behave, which allowed understanding, for example, the physical processes behind the functioning of the transistor. Science and technology are two subjects inextricably linked through the field of engineering. Engineering uses the statements from science to enhance its ability to meet particular design objectives, and science uses the skills of engineering to facilitate experimentation necessary to validate its hypotheses of the Universe's functioning. Together, they heighten our ability to understand and manipulate the surrounding reality.

Science and technology have a symbiotic relationship. Science enables technology by providing the understanding of the physical world necessary to engineer machines and devices that manipulate that world. The physics of, for example, the transistor was developed through the use of science, and the ways in which the transistor is used to facilitate the construction of devices like cellphones and computers are the domain of engineering. Conversely, it is engineering that enables the precision design of fast electromagnets and imaging devices necessary to implement, for example, a particle accelerator that leads to further scientific discovery. Physics is primarily concerned with learning about the physical world, and engineering is directed at using these discoveries to develop new technologies. One works hand-in-hand with the other, and they are difficult to separate in many instances.

Technology and science are developed from the ground up. Although designs and experiments may be planned and developed from the top down, the areas of technology and science evolve globally from the bottom up with smaller subunits building into conglomerate structures. As W. Brian Arthur points out in the book *The Nature of Technology* [155], devices are engineered by making use of previous devices. An example of this would be the use of the MOSFET. Invented at Bell Labs in 1959, the MOSFET has been employed to develop a good deal of the digital technologies we use today including the random-access memory and microcontrollers in our personal computers, cellphones, and most other digital devices. Memory and microcontrollers are created through the use of millions of individual MOSFETs laid down on a silicon die using a lithographic process. Each year the MOSFETs are made smaller and, hence, faster, and employed to develop the latest and fastest digital devices.

An example of scientific development is the discovery of electric and magnetic fields leading to Ohm's law, Faraday's law of induction, Ampere's law, Maxwell's equations and finally, with quantum-mechanical theory, to the unification of the

9.1. THE EVOLUTION OF TECHNOLOGY AND SCIENCE   367

electric and magnetic fields. These hierarchical concepts are also typically built from the ground up as well.

The driver of both technological and scientific development is human wants and needs. Everyone wants faster computers, cellphones with more features, automobiles that are more fuel-efficient, larger television screens, better healthcare, etc. Advancements in technology are what make these things possible. Advancements in science, on the other hand, allow us to understand better the physics required by engineers. Without it technology would be severely limited, and we would know much less about our place in the Universe.

New ideas, designs and hypotheses in these fields are created by the minds of human beings. Not all of these ideas will be successful; in fact, the majority of them will be proven wrong or fail in time. Scientific theories are proposed all the time, but in any specific area of study, only the theory that matches reality survives. Examples of such failed theories include cold fusion, the luminiferous aether (which we discussed in Chapter 3), Einstein's hypothesis that the Universe was static, and the spontaneous generation of life (commonplace before Louis Pasteur's famous experiment). All of these theories were proven wrong in time. On the other hand, many technologies also fail. The Tucker automobile in 1948 is a candidate for the most infamous technological failure of the past 100 years. It was a futuristic car at the time and only 51 were ever built. The problem was that the feature set developed for it was not only impossible to manufacture, but also unwanted by customers. Another failure was the Bubble Memory of the 1970s. Bubble memory was going to be the next great thing in solid-state memory until it was promptly outdone by competing technologies. The Google Nexus Q is a final example of a technology that failed spectacularly.

If these hypotheses and technologies are failures, why were they ever introduced in the first place? It's because our intelligence, as we discussed in the previous chapter, is not perfect, and this is reflected in the many failed attempts at scientific theories and new technologies, with a few successes emerging every year. It is difficult to grasp complicated technical issues rapidly and understand all their implications, therefore proof of concepts are used in engineering and experiments in physics to validate concepts. These are not just guesses. They are hypotheses generated from an intimate knowledge of the relevant subject matter. However, they are not proven concepts and thus must be tested with the distinct possibility of failing the tests.

Scientific hypotheses are tested for validity against reality. For example, as we discussed in Chapter 3, the first test of Einstein's theory of general relativity involved Eddington measuring the bending of starlight as it passed by the Sun during a solar eclipse. If the starlight had not demonstrated the curvature of space due to the Sun's mass, the theory of general relativity would have needed revising or outright rejection.

Once theories are integrated into the scientific framework and come to support other theories, their consistency is implicitly tested on a regular basis. For example, the correctness of theories like Newton's laws or special relativity is required to perform experiments in other areas. Science builds upon itself and it uses what it has learned to perform the experiments, be they physical or theoretical, required for it to learn more. A single experiment can have a vast number of interconnected pieces that stem from previously learned physics. If this chain breaks at any point, it is enough to require going back and correcting where the theory broke. Similarly, there are long chains of physical principles that go into every engineering design and, if there is a problem discovered in the physics, this means going back and investigating. In this way, both scientific research and engineering continually validate what we have learned about reality.

Technologies and technological devices are tested by their usefulness in serving the purposes for which they were intended. The economy is the primary testing ground for new technologies. If a device designed for sale in the economy does not sell then it fails and goes by the wayside—at least in that role. Most new technologies are in direct competition with existing technologies, and the winner is the one with the best performance. The performance metric could be any of a number of things including fuel economy, price, speed, and many others. For example, a newly released cellphone must be able to compete at some level with existing cellphone products or it will not sell. There are also disruptive technologies that enter a niche with little competition, but even these must be accepted by buyers as fulfilling their intended purpose.

The evolution of both science and technology is driven by the ideas generated within human minds. These ideas, once formed, are translated into physical experiments or devices for validation. After validation, the results are codified into physical laws on the scientific side of things, and core knowledge sets on the technological side. These concepts, no matter how they are recorded, are analogous to a biological genome. From the previous discussion on the operation of the mind and intelligence, the ideas that drive this "technological genome" are apparently random. After random generation, they are either sorted out by selection in the mind or during device testing. It is similar for scientific theories.

The technological or scientific genome then drives the physical construction of experiments or the development of technological devices which then compete for success. One could think of these experiments and devices as making up the scientific and technological phenotypes. The phenotypes compete either for their robustness in predicting the behavior of reality as with scientific theories, or in their ability to satisfy the demands of the market as with technological advancements.

An example of technological evolution is shown in Figure 9.1 [156]. Here, a partial evolution of the bipolar junction transistor (BJT), starting with its invention at Bell Labs in 1947, is shown. The entire tree starts with quantum-mechanical developments in physics that led to a much better understanding of semiconductor behavior. Then, with one experiment in 1947 that verified an idea, the entire electronics industry was changed. This modification propagated through the entire technological world because the transistor completely overwhelmed its competition which included electronic tubes and relays at the

9.1. THE EVOLUTION OF TECHNOLOGY AND SCIENCE

time.

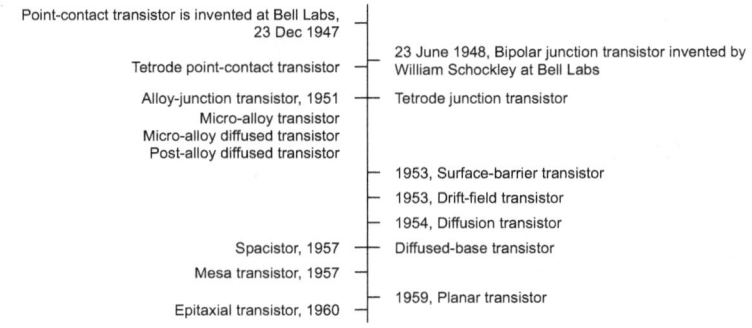

Figure 9.1: Evolutionary timeline of the BJT transistor.

It is worth noting that the progression of transistor advancement was erratic and unpredictable as it led up to the modern version. Many small changes were made over time that represented movement in the fitness landscape. Some of these changes proved advantageous and were retained, and some were abandoned for better alternatives. However, without the knowledge beforehand as to the optimal form of the transistor, this sequence of hypotheses, testing, and course corrections, led inexorably toward the optimum. This process is still going on today in the semiconductor industry and what it means to be optimal is changing continuously.

A second point to consider is the staccato character of the evolution of technology. There are long periods of time through the history of industry where only small incremental changes are made. Nevertheless, once in a while a change takes place that leads to sweeping change through the entire industry, after which, again, slow progression proceeds. This is very similar to long periods of gradual change interlaced with short periods of rapid change in biological evolution.

Figure 9.2 displays a similar example of evolutionary change in science. It delineates the understanding of the gravitational force, starting with Aristotle claiming heavier objects fall faster than lighter objects (which they do if the light objects can be affected by air resistance), and proceeding to today's ongoing research on quantum-gravity. Again, this is not a systematic, predictable process. It progresses by people coming up with ideas, which are a combination of essentially random thoughts that are selectively filtered to produce viable hypotheses. These hypotheses are then tested to see if they match reality. If successful, they are retained, and, if not, they are rejected.

Advancement in both science and technology progresses by testing (selection) of a continuous stream of virtually random ideas. These ideas, on the first level are tested in the mind, and either accepted as viable or rejected, and then, if they are accepted, they proceed to being physically implemented and tested. At this stage, if they pass, they either become a part of the scientific body at large,

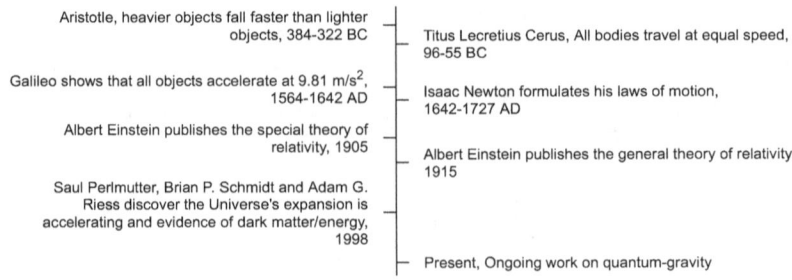

Figure 9.2: Evolutionary timeline of the theories of gravity.

or they become viable products. It is interesting that science and technology evolve using the same rules apparent in biological evolution.

9.2 The financial system and the economy

One of the very first forms of money used was wheat in ancient Egypt. Although it wouldn't have been thought about the same way money is today, it was intended to serve the same purpose as what most of us think money is today. Originally, the main purposes of money were to allow efficient trade and to mitigate the inherent uncertainty in life. In ancient Egypt, wheat was a staple food and was needed by everyone at all times. It could be traded for other items at any point in time as there was always a demand for it. As it could be successfully stored for long periods of time in granaries, it represented wealth that could be maintained over time. These are two of the most important characteristics of a satisfactory money (and will be the only two considered here). Money that is an actual good is typically called a *commodity money*.

Wheat from farms all around was stored mostly in community granaries with receipts being given to the owners as their crop was deposited. It could be retrieved at any time and used to "purchase" other necessities.

Over thousands of years, many different forms of money have been tried, including livestock, grain, cowry shells, cigarettes, silver, gold, copper, bank receipts, government notes, and digital entries. Virtually anything that was in demand at a particular time has been used as money.

Money was intended to be an accounting method to measure the energy expended by people to produce certain goods or perform specific services. Allowing the trade of energy, or generic value, meant that exact trades weren't required. For example, in a world of barter, a shoemaker who is needing to acquire bread must find a bread maker who is in need of shoes. And these two must find one another simultaneously. This is a very inefficient way of trading goods due simply to the low probability of the coincidence of wants. Money gets around this problem because everyone wants it all the time, and they are willing to exchange

9.2. THE FINANCIAL SYSTEM AND THE ECONOMY

items with intrinsic value for it.

Money has continued to evolve since ancient times and has taken many forms as it works its way toward the optimal form. As was mentioned, the first forms of money were tangible goods. This implicitly prevented cheating. For example, it was impossible for the ancient Egyptians to counterfeit wheat—you either possessed it or you didn't. However, even something like wheat had disadvantages in acting as money. First, although it could be stored, it didn't last for very long periods of time. Second, it was difficult to transport, especially in large sums. Its value to volume ratio was not high enough to maximize convenience.

Due to these deficiencies, people eventually concluded that precious metals would be a better choice for money. They are long lasting, have a high value to volume ratio due to their rarity, are readily separated into arbitrary sums, and are easily transportable. It must be understood that there was no universal committee hundreds or thousands of years ago that got together and decided what should be money. (Even when that was done later with governments mandating fiat money, it would not be the end of money's evolution. We'll get to this point a little later.) Monies like those made of precious metals came into use because people started using them precisely *because they possessed the above qualities*.

In other words, people would try using a new commodity money for transactions because it was a matter of convenience. Then, over time, this candidate money, if it performed better than the others, would replace all other monies. It was a sort of survival of the fittest money, where selection was based on how fit a particular commodity was in satisfying the qualities above. People whose neighbors and trading partners were using the new money would've been far more likely to use it as well. It's kind of like today where it is expected that paper money will soon get left behind, being replaced exclusively by digital currency. This is a positive feedback effect.

So, we see new money coming into existence through happenstance followed by selection based on convenience, with the process being hastened toward equilibrium (one or only few types) by positive feedback.

Once again, although precious metals acting as monies were far superior to the commodity monies that had came before, with their extended use their innate weaknesses began to shine through as well. Metal-based money being easy to transport or transfer also made it easy to steal. This led to a need to protect one's money. There were already gold smiths around that had vaults for protecting valuables from theft, and these were co-opted to function as the predecessors of what we know today as banks. Bringing about this change required only a small shift in society's thinking: gold smith vaults were already there to protect valuables from theft, and money was valuable. Therefore, although unplanned, it was a simple matter to adopt the protective function of the gold smith vault to that of protecting money.

The gold smiths would provide written receipts to depositors so they could prove their ownership of their money the smith was holding. It wasn't long before people—unplanned again—began simply trading the receipts instead of actual gold. The receipts were after all a representation of the actual gold. At this point, we see the emergence of something much closer to contemporary money. But it was not consciously planned.

Soon the receipts came to be regarded *as the metal-based money they represented*, leading to the receipts coming to be actual money. This was advantageous to the smiths and to society, especially once the smiths realized that they could create receipts out of thin air and people would treat them as if they represented actual value. At that point, we see money transition into something very different: it assumed its original intent as an accounting mechanism, but no longer required actual value to perform this function. The ability to create money, as it was needed to drive the corresponding creation of real wealth, led to an economy that grew much faster. For example, money could be created to sponsor the building of a house, and this money would be paid back through the real value created by the owner over time. This idea was certainly not planned, but it made possible a vast expansion of the economy.

Later this concept would be co-opted by governments to mandate fiat money; that is, money which by definition has no value backing it other than the government's say so. With this step, money truly assumed its originally intended function, existing only as an accounting function. Whereas previously the integrity of the system had been guaranteed by the inherent value of a commodity money, it was now guaranteed by a government. The real value of the commodities previously used to back money was now freed up for constructive use.

Today, we see further evolution of this idea in the move to a purely digital money world. No one knew they wanted a debit card until they had one. Credit cards were originally designed to allow banks to make money through lending. Later, this idea mutated into a debit card, which gave people immediate access to a number that represented the value they possessed.

Money is simply intended to keep track of how much work a person does and provide a means for that work to be traded for the work of others. As long as the integrity of this accounting method can be maintained, there is no need to trade wheat or gold coins; all that is required is a number that represents an amount of work.

This evolution of the money system was not planned. It was driven by essentially random ideas of how to do things more conveniently. These ideas were tried and, if successful, they were retained, driving the system toward the optimal form of money: a simple number in a computer.

Of course, the use of money adapted in other ways and is still doing so today. Right now, we see cryptocurrencies trying to be born in opposition to the unfair practices inherent in conventional banking. This is an example today of a new process, that again was unplanned by society as a whole, being tested on its fitness to overcome inefficiencies and obvious unfair practices in commercial banking. We'll see how it goes in hindsight.

In addition, we see the development of every kind of financial instrument imaginable, bonds and derivatives being just a couple of examples. Things are so complicated that one requires an investment advisor to simply live today. It is just another example of the emergence of complexity from a few simple rules. And anyone who thinks I'm being overly dramatic here, remember that money is supposed to be simply a mechanism to facilitate trade. It has evolved well beyond that today, and it did so from the ground up. The financial system we have today was in no way planned from the top down. It is exemplary in demonstrating system evolution and the emergence of complexity.

9.3 Religion

There are few things that are immutable in this world, but if any exist, it seems a common-sense notion that religion would be one of them. I am going to focus primarily on Christianity here, but the results apply equally well to all religions. Here, I'm using as the definition of religion *the institutionalized belief in a supernatural power that has divinely supplied a written set of governing rules or principles for human conduct.* Since these rules are provided in a written language, they are susceptible to interpretation. Also, many of the religions around the world, Christianity in particular, are driven to spread their beliefs to others.

As I mentioned before, religions are essentially top-down tautologies, proclamations of perfect truth. And yet I've included religion within this chapter describing processes that are driven by random inputs and an optimizing selection method. How does that make sense?

The answer is straightforward: the contents of any holy book such as the Bible are supposedly static (although transcription errors and insertions are known to have happened), but the interpretation of these books is quite malleable and is completely contingent upon the individual's predilections and personal beliefs, and these beliefs can change over time as the individual has new experiences, thinks new thoughts and is exposed to new ideas. Concepts that do not derive from the original text can be interwoven with it to precipitate hybrid beliefs; that is, interpretations.

Although the proclamations in the Bible are there for everyone to read and analyze, they are inherently not as precise as mathematical or logical statements. In fact, they are not even in the same ballpark.

From the individual's point of view, these exogenous sources of change are virtually random, and the individual selects the effects that he or she deems acceptable based on their own belief system they have built up over time. In other words, a religious codex provides random (or nonspecific) excitation and the individual's preferences sorts it into parts that are accepted or rejected.

A (very) simplified flowchart of western religion from Mesopotamia to present day is shown in Figure 9.3. It's interesting that it has a tree structure which, when inverted, looks very much like the biological evolutionary tree. There is a reason for this: both are driven by a similar underlying process. An individual is subjected to various ideas and then selects for acceptance those particular concepts that align with their own psychological programming.

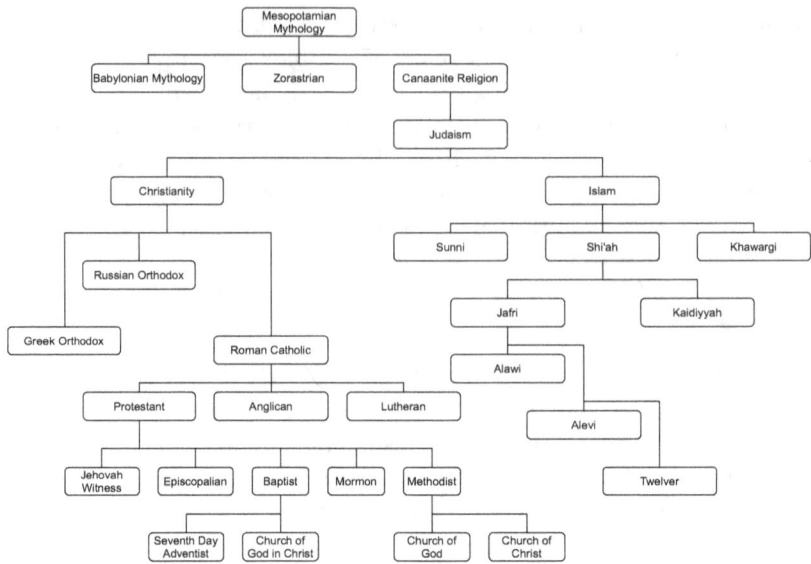

Figure 9.3: A simplified block diagram of western religion from Mesopotamia to present day. (Note that it has been simplified to fit on the page. The diagrammatic tree is actually much larger than is shown.)

It is estimated today that there are as many as 30,000 different Christian denominations throughout the world [157]. Differences in denomination are dictated by parameters such as name and organization, but primarily by doctrine. Given that all stem from the same holy book, this is an incredible amount of diversity.

Figure 9.4 shows a second evolution, this time of Christianity alone. Progression, along with bifurcations, are shown in the solid lines, and unions are denoted by dashed lines.

So where do the changes in interpretation come from? We are all human and subject to errors in our perception and understanding. Worse, we're really not talking about genuine errors here. No written book can be absolutely specific, certainly not to the extent that its meaning is always transferred perfectly. We all have views of the world that a virtual infinitude of experiences has shaped; it would be astounding if interpretations didn't vary among people. Our history of experience acts as a programmed filter within our minds that colors perceptions as they are acquired. Some of these preferences can be attributed to physical differences between individuals, but I suspect the bulk of them are derived from the unique events each of us experience throughout our lifetimes. For example, many people enjoy chocolate and some would prefer to avoid it. Some like classical music and others prefer country music. Some are democrats and others are republicans. All of these small or large differences in people drive them to

9.3. RELIGION

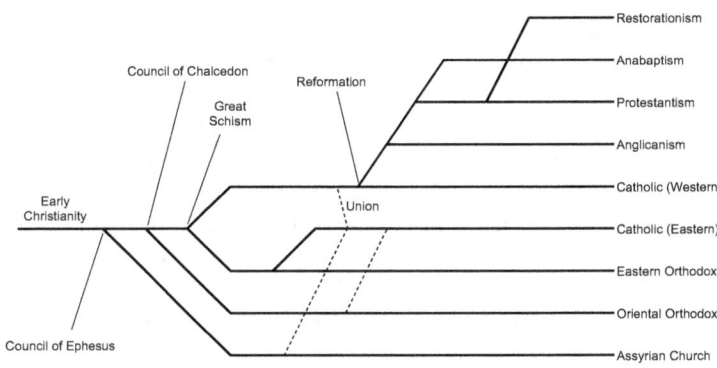

Figure 9.4: The evolution of Christianity.

interpret information differently. It is these random differences in the individuals within society that lead to us believing different things. This is sociological random mutation.

As with the chaotic systems discussed previously, very small disagreements on doctrine can lead to large schisms within the church due to the propensity of religious rules and laws to be all or nothing—that is, breaking the smallest of laws in the sight of God is equivalent to breaking the largest. In more mathematical terms, even a small deviation from the law has the same effect as a large deviation. If one believes that misinterpreting even minimal aspects of the holy book can lead to an eternity of misery, there is no room at all for mistakes, hence the sensitivity to even a minor issue tends to infinity. There is no room for compromise. This is the extreme sensitivity we see in chaotic systems and I contend it is the very reason we see so many varieties of religion in the world.

Of course, the sensitivity isn't infinity. Our limited analytical capabilities coupled with our desire to associate with other people creates a region of convergence for each of these groups that varies in size; so long as the individual's thinking falls somewhere within this region, they can be incorporated into the body of the particular faction. A larger region of convergence allows acceptance of more followers but a region too large will not bind the group tightly enough together to ensure survival and members will be siphoned off to surrounding groups, or spawn entirely new factions.

As we saw in the previous chapter, complexity is an innate component of the mind and its associated thinking process. It seems when the imagination gets involved, complexity follows. As a simple example, consider the concept of original sin, conceived to account for death and evil in the world (and possibly a myriad of other reasons) with no real basis in the Bible. However, once this concept was in place, the question of why innocent children get sent to Hell arose, which required the invention by the Catholic Church of Limbo (the concept of which has been subsequently withdrawn) that appears to be an analog of the

Bosom of Abraham (although there is no specific connection mentioned in the Bible), a place for those who died before Christ came on the scene. The countermand of Limbo resulted as a recommendation by the International Theological Commission when they "[...] took a fresh look at the question" [158].

Conjecture or extrapolation at one point creates boundary conditions for the next or adjacent conjectures. This is akin to Kauffman's vision of biological coevolution; complexity is built up one piece at a time, the system providing its own boundary conditions. A similar thing happens in contemporary fiction. One story sets the stage for the next, the next being obligated to respect these prior boundary conditions, each free to add additional imaginative features so long as they fit coherently in the whole. In the physical world, these constraints are intrinsically satisfied; in the fiction world the writer is compelled to enforce them: if you have original sin and God cannot punish the innocent (the first block), then a Limbo or some other explanation is required (the second block). These blocks can be continually stacked upon one another to form more complicated structures.

The result is that the complexity within religions continues to build, one brick at a time, reaching amazing depths of dogma and liturgy, and, as it builds, its rate of change increases. Along with this comes a multiplication of factious denominations and variations, each competing with one another and claiming to have reasoned out the correct doctrine. Just as with biological and other complex systems, religion lives at the border between order and chaos.

As we've learned throughout this book, once we have a set of different and competing entities, all trying to satisfy the same objective, fitness selection can take place. In this case, the objective is generally to maximize the number of adherents in a particular religious group (note the similarity in maximizing reproduction in biological evolution). It can also serve to satisfy secondary objectives such as maximizing the group's monetary donations. As an example, the Catholic Church has been very successful in satisfying both of these objectives. To do so it has relaxed the rule set, thus broadening the region of convergence for acceptable adherents. For instance, the Catholic Church even accepts the theory of evolution today. This is in stark contrast to other variations of Christianity. On the other hand, the number of restrictions imposed by the Catholic Church are greater than those of, say, Protestantism. Dying as a Catholic without a priest at hand is not a good thing.

Finally, the vast majority of religions are self-assembling organisms, particularly the evangelical varieties. The individual members of many Christian groups feel a built-in obligation to proselytize people outside their particular faith. Once converted, these new members will do the same, a sort of ecclesiastical autocatalysis. Each new block in the chain is built by the previous. Along the way, just as in the over-referenced telephone game, or the biological evolutionary game, each new individual brings new variations to the group. As these variations add up over time a peculiar thing can happen: a new species of the church can come into existence.

An event such as this is very similar to biological speciation; there has never been a case where a dog birthed a cat, and there has never been a case where a Catholic birthed a Mormon. As we've already talked about, biological speciation doesn't happen in one step even though it is conceivable that the final discontinuity that prevents interbreeding might happen in one generation. Similarly, Luther's 95 theses did not immediately create Protestantism but rather was the initial impetus which eventually led to its appearance. It was only long after his defiant act of 1517 that the Protestant faith became incompatible with Catholicism, the evolution of each eventually widening the breach.

Take, for example, the following quote from Perry Marshall's book *Evolution 2.0: Breaking the Deadlock Between Darwin and Design* [42]:

> If God made whales and other mammals with identical parts, that was one thing. But if living things had the innate ability to change their own genomes and generate new species, if the whale's ancestors had the capacity to transform into another species, that was fantastically more impressive. If evolution were true, *God could teach us more principles of engineering through nature* than if it weren't. If evolution makes you uncomfortable, trust me, I understand. But stick with me, because this will make more sense as the story unfolds. Yes, I am suggesting that you can understand God better by studying evolution!

This is a very clear example of religion mutating, preparing to expand its core beliefs.

Religion evolves as new ideas are randomly tested within the interpretations of its followers. If these new ideas are accepted by enough of the followers, they are retained; if not, they are rejected. In addition, if the changes are successful in bringing about more conversions, they will spread. The spreading will lead to yet more conversions, and the changes will spread further. And again, we see the same positive feedback loop that is modulated by the effectiveness of changes in the "doctrine genome" leading to more conversions.

Additionally, the evolution of religion follows the same template of slow progression over long periods of time interspersed with short periods of rapid change. For example, the split of Christianity from Judaism and the split of Protestantism from the Catholic church represent two events which lead to rapid change. Between such events change is still occurring, but it is much slower.

It is interesting that the biological evolutionary tree looks so much like the evolutionary tree of western religion. The processes that drive changes in these two are not so different I think.

9.4 Conclusion

The discussions of topics in this chapter were not intended to be exhaustive as that would be well out of the scope of this book. The reason I included them was to further dispel the notion that evolutionary, or evolutionary-like, processes are "ludicrous and too improbable." The evolutionary process is not the exception; it is the rule.

Optimization processes like evolution are ubiquitous in nature. They also emerge in the functioning of society. This is not surprising as society is a part of nature. It is not surprising that since the mind functions in this way that the derivative functions of society would work that way as well. Capitalism *is* identical to biological evolution. It is based explicitly on competition and surviving. Contrary to being a novel concept describing the generation of biological complexity, evolutionary-like optimization processes show up everywhere.

Chapter 10

Last thoughts

The overarching thesis of this book has been a rejection of ID in favor of a scientific perspective. A lot of material has been presented to that end. Along the way several fundamental concepts in relation to not only evolution, but also intelligence have been addressed. I'm not going to go back through all of them in this summary, but I would like to polish some of these ideas a bit more to make them a little more transparent. And although I'm not going to reiterate all the deficiencies with ID, I would like to touch on what I consider the root causes of the disagreement between ID and the scientific community.

One of the biggest problems with ID is the subterfuge that it's attempting. Endorsing the idea that everyone has an innate design-sensing intuition appears to encourage people to continue to believe what they have already been told from the religious community despite evidence to the contrary. Our "common sense" just does not work to decipher the physics of reality because that is not what our brains evolved to do. And after examining the so-called "wedge document", it seems all the more convincing that the intent is to convince people that the Christian view is correct rather than the objective truth.

This is pointed out in a recent article in *Yahoo! news* [159]. The article is concerned with a study between NASA and religious leaders about how religious people would take the news that alien life exists. It seems to me that this should not even be a question that needs deliberation. What could possibly ever be wrong with knowing the truth? (I'm not saying that alien life exists, but if it does, shouldn't we know?) And what's the alternative? Should we pretend we don't know and go ahead convincing ourselves that a particular religion is accurate? We should never sacrifice the truth for any real or imagined greater good.

In addition, from the technical side of things, the ID hypothesis fails completely. It is basically an argument from incredulity. Finding it difficult to believe the one theory that does have supporting evidence does not mean that your hypothesis is viable. Evolution and ID do not represent a true dichotomy; proving evolution false will not prove ID correct. But this is the picture that the ID community attempts to paint to the general public. The reason the scientific community views evolutionary theory as successful is because there exists evidence for it. On the other hand, there is no evidence for ID. Worse yet, there can't be any evidence for it because it relies on a supernatural explanation.

We live in the natural world. The term supernatural means not existing in this world. I'm not sure why this is so difficult to grasp. If there really was a designer, that designer would be brought under the umbrella of the natural and, when that happened, science would attempt to analyze it. Again, I'm not saying that there aren't personal spiritual experiences. There are. What I'm saying is they, by definition, cannot be part of the realm of science. Science only deals with things that can be observed objectively in reality. Not understanding this leads to a great deal of confusion. Science only deals with the observable. If ID has no observable evidence to link it to reality, then it is not science. This is why even if a deity is behind the functioning of the natural laws, it won't matter. Science will continue to view these phenomena as natural laws.

Designers who leave no traces in reality do not exist, just like clockwork deities don't. They have no impact on reality. Their actions will get lumped right in with natural law. But we have to be careful here. If we are going to allow theories with no evidence, the door is open for anything. That is the problem.

Opinions are personal beliefs that are not founded upon any proof or conclusive evidence. We all have opinions and they're generally different for everyone. One of the tasks of science is to objectively sort out what is true about reality, regardless of opinion. But it does not do this the way religions do, and I think this is a second source of confusion for a lot of people.

No matter how desperately a person who has had a personal experience wants it to be true for everyone else, it isn't. And I don't think there's a problem with people having personal experiences; I've been there myself. But I think it is a problem when someone wants to use their own personal experiences as a template for how others should live. And it's not enough to just ask persuasively; Christians will make it quite clear that you're going to Hell if you do not heed their warnings. And yet, they don't know.

I give everyone the right to play the game of Pascal's wager if they so like. What I don't want is any infringement upon anyone else because of this fear. And that's what we're talking about, right? Referencing the article above, why should a government hide knowledge of extraterrestrial life to protect the safety blanket of a religious majority at the sake of the truth? Again, I'm not saying alien life exists or not. In this case, it doesn't matter.

Covid-19 arrived in late 2019. At first, it didn't seem to be that dangerous. And then, it began spreading exponentially until shutdowns became a regular occurrence. The country would end up being irrevocably changed. Finally, a vaccine was developed and there was some hope that the pandemic would be eradicated and things would go back to normal. Unfortunately, this didn't happen because many people would not, and still won't, take the vaccine. Christians represent the largest percentage of this group [160]. Although there are valid reasons not to take the vaccine, there are also ignorant reasons not to do so. Not taking it because of an immunological deficiency is a good reason. Not taking it because you're afraid it is the mark of the beast is not a good reason.

I'm not saying that religious beliefs were the only reason the Covid-19 vaccine was not embraced by the entire populace; there were many different reasons, including political. In addition, the wealthier countries of the world did not ensure the poorer countries received enough vaccine. Having a world partially vaccinated applied evolutionary pressure to the Covid-19 virus to mutate, which it did. As it stands in December, 2020, we are right back where we were the year before due to mutated variants. Evolution does work.

It is unfortunate that a large percentage of the general public has come to believe that science is just another opinion, and that religious beliefs should stand on the same footing. But this is mixing proverbial apples and oranges. Religious belief is personal, and the right to have these beliefs is guaranteed to you by virtue of your personal sovereignty. Science derives its validity from the objective, observational consensus of reality's behavior. These are two vastly different things that we will discuss right now.

There are two different worldviews that are being considered when we talk about religion and science. Religion, by its very nature, requires what is called faith. Faith can be defined as belief without evidence. The criteria for evaluating the validity of faith is the individual's own mind. This is the only place it could be confirmed as there is no external evidence. If there is real evidence then faith is unnecessary. Therefore, the belief that faith inspires or, equivalently, the faith that belief inspires, is determined by our thoughts. Outside evidence is not the governor of this process.

If you talk to a religious person and ask where their faith originated, undoubtedly, the most prolific answer of all is they were raised that way. This is an interesting thing: people who do not know the answer teach their children as if they really do know the answer. I'm very familiar with this process. We, as Americans, have become so acclimated to things being this way that it doesn't even surprise us. In fact, it goes against social etiquette to even criticize it. Certainly one can talk *for* it, but not against it.

These people wouldn't even think of making other decisions in their life based on faith. But, when it comes to the supposed most important question in life, that is exactly the approach they take. And this is okay. The problem is that they attempt to enact policy based on beliefs for which they have no evidence. This is how authoritarianism starts. Freedom cannot, is not, provided by religious dogma. The religion itself does not allow it. The only way freedom is admitted to a particular denomination is to generate a new species of that denomination, and this has been happening for thousands of years as was discussed in the previous chapter. This is exactly what we would expect from a non-evidential belief system. This results in a continual expansion of modified, or mutated, beliefs. How could it go any other way?

It seems the hardest thing for most people to accept is that they may be wrong, especially if they have no evidence to indicate that they are right. Just as much as a Christian believes they are right, so to does a Muslim. How can this be explained and, more importantly, how can it be reconciled?

The first thing we need to do is define what truth means. Many, or most, people might assume that there is some definitive truth that exists out there, some absolute fact. But we have already come to the conclusion that we can't defeat hard solipsism, and we can't prove the logical absolutes, among other items. This is a strong hint as to just how non-concrete truth is. We'll get back to this notion in a moment, but first we need to find this elusive definition of truth.

Contrary to popular belief, there is no book from which we can find absolute truth. The first problem is that there is no way *to demonstrate that the writings in such a book are true.* We might say that the probability is high that some things happened, but certainty is an entirely different matter. Let me give a simple, almost comical example to show the reader where I'm headed.

When an engineer needs to calculate the circumference of a circle based on its diameter, she will, of course, use the constant pi. For example, assume that the diameter of a circular pool is d and she needs to know its circumference to buy enough material to construct a safety fence around it. The circumference is easily calculated as $c = \pi d$. So, she grabs her calculator and calculates c and rounds the answer to the nearest quarter of an inch. This is quite reasonable.

As far as we know, the value of pi is irrational, which means it cannot be represented as a ratio of integers. In 2019, a team at Google calculated pi to 31.4 trillion digits. That's an impressive number, but since the actual value, if truly irrational (this has not been proven conclusively), is a never ending series of non-repeating digits, 31.4 trillion digits doesn't even come close. What does this say about what we can know about reality?

All calculations made in engineering and physics are done so with a particular precision, this precision being chosen specifically to meet the requirements of the problem or design at hand. It is done this way because these people know that they cannot pin down any value absolutely. Nothing in physical reality can be measured or calculated with absolute certainty. It seems we are banned from doing so. Even if there were some magical tape measure that could measure a length of pool deck 2x4 lumber exactly, we would not be able to find the edges of that piece due to the quantum-mechanical uncertainty relation. *The result is we can never know anything with absolute certainty.*

To further complicate this matter, even if the subject you're considering—perhaps a mathematical expression like $c = \pi d$—is exact, there is no means by which you can absolutely prove it. In other words, even if you could measure the physical world exactly (which you can't), you would still be left with uncertainty.

For example, let's consider for just a moment a 2,000 year old book called the Bible. Can anyone out there say without a doubt that they know its contents are absolutely correct? If your answer is yes, then what's the need of faith? The real answer is that you can't know with certainty. Don't feel bad; no one else can either. Faith is the finger in the dam that keeps this illusion from being washed away. But it is an illusion. Having faith in something does not make that something true. But, in the proper doses it will shield you from the unknown.

So, if we can't say anything is true with absolute certainty, then what do we mean by truth? How would you know truth if you saw it? The only way anything is ever known is that it has a measurable effect in this reality. Again, anything that does not manifest in reality is not real, by definition. Now we're getting someplace. *Truth is what manifests in reality.* Well, if that's the case, how do we know what manifests in reality? We've already said that we can't trust our senses. So how do we calibrate or check the accuracy of our senses or perceptions? We compare them to the perceptions of other people. This is our best shot at truth and it is what is called science.

This makes the truth we can know a subset of reality or the Universe. Since science is based in observation, and the events which are observed are derivatives of reality, it is impossible to make absolute statements about reality. It is effectively the difference between a what and a why. Science cannot tell one why the natural laws exist, it can only describe those natural laws.

We never know anything for sure. Although there may well be an absolute truth, we cannot know it. The best we can do is approach it. In this sense, science builds models of reality that are consistent and mathematically correct. These models are extremely effective in predicting the behavior of this Universe, but they are not statements of absolute truth. Science makes no proclamations of truth. It simply provides the best explanations of reality, which are contingent upon the observed behavior of reality.

Many people have the notion that science is some kind of highfalutin ideal. It isn't. And it doesn't belong to some aristocracy or upper-class cognoscente. It is simple observation coupled with reason. But here is where it becomes important: the laws it develops through observation are real, independent of any one person's preferences, or a god's preferences for that matter. For example, as far as we know the Earth *does* revolve around the Sun. The Catholic Church imprisoning people for stating this fact was a tragedy. But the biggest tragedy is that today the same thing is still happening. The lesson the Church should have learned from that encounter has still not been learned. The subjects are different, and the words are different, but the effects of unjustified beliefs are still the same. I hope this book has went a little way in making that clear.

Bibliography

[1] ACLU. The trial of kitzmiller v. dover. https://www.aclu.org/other/trial-kitzmiller-v-dover, 2021.

[2] Roger G. Gallop. *Evolution, The Greatest Deception in Modern History*. Red Butte Press, Inc., Ponte Vedre Beach, FL, 2014.

[3] Scott M. Huse. *The Collapse of Evolution*. Baker Books, Grand Rapids, MI, 2011.

[4] Douglas Axe. Author douglas axe presents his book undeniable. https://www.youtube.com/watch?v=SC9Hx3WpsCk, Jul 2016.

[5] Douglas Axe. *Undeniable, How Biology Confirms Our Intuition that Life is Designed*. HarperCollins, New York, NY, 2016.

[6] Atanu Chatterjee. Frontiers of aging science. https://www.researchgate.net/publication/312464687_Frontiers_in_Aging_Science/download, Aug 2016.

[7] Wikipedia. Ultraviolet catastrophe. https://en.wikipedia.org/wiki/Ultraviolet_catastrophe, Jul 2021.

[8] John D. Cutnell and Kenneth W. Johnson. *Physics*. John Wiley & Sons, New York, NY, 1998.

[9] Anand Awasthi. Derivation of planck's law. https://www.physicsvidyapith.com/2021/04/derivation-of-plancks-radiation-law.html, Apr 2021.

[10] Wikipedia. Louis de broglie. https://en.wikipedia.org/wiki/Louis_de_Broglie, Jul 2021.

[11] U.S. energy Information Administration. New york electricity profile 2019. https://www.eia.gov/electricity/state/newyork/, Nov 2020.

[12] Wikipedia. Cantor's diagonal argument. https://en.wikipedia.org/wiki/Cantor's_diagonal_argument, Feb 2021.

[13] Sylvain Saurel. The story of the wise man sissa will help you to understand the magic of compound interest. https://sylvainsaurel.substack.com/p/the-story-of-the-wise-man-sissa-will, 2021.

[14] Karl R. Popper. *The Logic of Scientific Discovery*. Martino Publishing, Mansfield Centre, CT, 2014.

[15] Christian Research Institute. What is the biblical definition of faith? https://www.equip.org/bible_answers/what-is-the-biblical-definition-of-faith/, Aug 2010.

[16] Mary Fairchild. How does the bible define faith? https://www.learnreligions.com/what-is-the-meaning-of-faith-700722, Jan 2021.

[17] William Lane Craig. Do the laws of logic provide evidence for god? https://www.reasonablefaith.org/writings/question-answer/do-the-laws-of-logic-provide-evidence-for-god, Dec 2016.

[18] The Late Show with Stephen Colbert. Ricky gervais and stephen go head-to-head on religion. https://www.youtube.com/watch?v=P5ZOwNK6n9U, Feb 2017.

[19] Wikipedia. Michael faraday. https://en.wikipedia.org/wiki/Michael_Faraday#Electricity_and_magnetism, Aug 2021.

[20] David J. Griffiths. *Introduction to Electrodynamics*. Pearson Education, Singapore, 1999.

[21] Claudio G. Carvalhaes and Patrick Suppes. Approximations for the period of the simple pendulum based on the arithmetic-geometric mean. *Am. J. Phys.*, Dec 2008.

[22] Jeremy S. Heyl. The double pendulum fractal. *Am. J. Phys.*, Aug 2008.

[23] Wikipedia. Mitchell feigenbaum. https://en.wikipedia.org/wiki/Mitchell_Feigenbaum, Mar 2021.

[24] Benoit Mandelbrot. How long is the coast of britain? statistical self-similarity and fractional dimension. *Science*, May 1967.

[25] Stephen Wolfram. *A New Kind of Science*. Wolfram Media, Inc., Champaign, IL, 2002.

[26] Wikipedia. Conway's game of life. https://en.wikipedia.org/wiki/Conway's_Game_of_Life, Sep 2021.

[27] Wikipedia. Tower of hanoi. https://en.wikipedia.org/wiki/Tower_of_Hanoi, Aug 2021.

[28] B. Podolsky A. Einstein and N. Rosen. Can quantum-mechanical description of physical reality be considered complete? *Phys. Rev.*, May 1935.

[29] Amit Goswami. *Quantum Mehcanics*. Wm. C. Brown Publishers, Dubuque, IA, 1992.

[30] J. S. Bell. On the einstein podolsky rosen paradox. *Physics*, Nov 1964.

[31] Brian Greene. Your daily equation #21: Bell's theorem and the non-locality of the universe. https://www.youtube.com/watch?v=UZiwtfrisTQ, May 2020.

[32] DrPhysicsA. Bell's inequality. https://www.youtube.com/watch?v=qd-tKr0LJTM, Apr 2012.

[33] Alain Aspect, Philippe Grangier, and Gérard Roger. Experimental tests of realistic local theories via bell's theorem. *Phys. Rev. Lett.*, 47:460–463, Aug 1981.

[34] Alain Aspect, Jean Dalibard, and Gérard Roger. Experimental test of bell's inequalities using time-varying analyzers. *Phys. Rev. Lett.*, 49:1804–1807, Dec 1982.

[35] Wikipedia. Bell's theorem. https://en.wikipedia.org/wiki/Bell%27s_theorem, Aug 2021.

[36] Wikipedia. Reductionism. https://en.wikipedia.org/wiki/Reductionism, Jun 2021.

[37] Michael J. Behe. *Darwin's Black Box*. Free Press, New York, NY, 2006.

[38] Stuart A. Kauffman. *A World Beyond Physics; The Emergence and Evolution of Life*. Oxford Univ. Press, New York, NY, 2019.

[39] Wikipedia. Kalam cosmological argument. https://en.wikipedia.org/wiki/Kalam_cosmological_argument, Aug 2021.

[40] et. al. W. W. Zhu. Testing theories of gravitation using 21-year timing of pulsar binary j1713+0747. *The Astrophyscial J.*, Aug 2015.

[41] Edwin Hubble. A relation between distance and radial velocity among extra-galactic nebulae. *Proc. Nat. Academy of Sci.*, Mar 1929.

[42] Perry Marshall. *Evolution 2.0: Breaking the Deadlock Between Darwin and Design*. Benbella Books, Inc., Dallas, TX, 2015.

[43] Robert Stanley Reppa. A maximum power point tracker for photovoltaic applications. *Science*, May 1996.

[44] Akinwale Adio Taofiki Adewole Philip and Otunbanowo Kehinde. A genetic algorithm for solving traveling salesman problem. *Int. J. Adv. Comp. Sci. and App.*, Jan 2011.

[45] et.al. Gregory S. Hornby. Automated antenna design with evolutionary algorithms. *Am. Inst. of Aeronautics and Astronautics*, Jun 2012.

[46] Derek S. Linden. Antenna design using genetic algorithms. *Proc. 4th Anual Conf. Genetic and Evolutionary Comp.*, Jul 2002.

[47] Richard Dawkins. *The Blind Watchmaker*. Penguin Books, Harlow, England, 2006.

[48] Wikipedia. Evolution. https://en.wikipedia.org/wiki/Evolution, Aug 2021.

[49] Charles Darwin. *On the Origin of Species by Means of Natural Selection*. John Murray, London, 1859.

[50] Mike Cassidy. *Biological Evolution; An Introduction*. Cambridge Univ. Press, New York, NY, 2021.

[51] Douglas Axe. Evolution and the experts - douglas axe at dallas science faith conference 2020. https://www.youtube.com/watch?v=1pVulmHTd74, Apr 2020.

[52] Wikipedia. Convergent evolution. https://en.wikipedia.org/wiki/Convergent_evolution, Aug 2021.

[53] Andreas Wagner. The role of robustness in phenotypic adaptation and innovation. *Proc. Bio. Sci.*, 279(1732):1249–1258, Apr 2012.

[54] Philip Ball. Phyletic gradualism. https://nautil.us/issue/41/selection/the-strange-inevitability-of-evolution-rp, Oct 2016.

[55] William A. Dembski. *The Design Revolution*. Intervarsity Press, Downers Grove, IL, 2004.

[56] et.al. H. Lodish. *Molecular Cell Biology*. W. H. Freeman and Co., New York, NY, 2016.

[57] Adrian Bejan and J. Peder Zane. *Design in Nature; How the Constructal Law Governs Evolution in Biology, Physics, Technology, and Social Organization*. Doubleday, New York, NY, 2012.

[58] Dan-E. Nilsson and Susanne Pelger. A pessimistic estimate of the time required for an eye to evolve. *Proc. Biol. Sci.*, Apr 1994.

[59] Wikipedia. Evolution of the eye. https://en.wikipedia.org/wiki/Evolution_of_the_eye, Aug 2021.

[60] James H. Schwartz Eric R. Kandel and Thomas M. Jessel. *Principles of Neural Science*. Appleton & Lange, East Norwalk, CT, 1991.

[61] J. F. Willot. Physiological plasticity in the auditory system and its possible relevance to hearing aid use, deprivation effects, and acclimatization. *Ear Hear*, Jun 1996.

[62] et.al. Philip M. Mwachaka. Effect of monocular deprivation on rabbit neural retinal cell densities. *J. Ophthalmic Vis. Res.*, Feb 2020.

[63] et.al. Ke Chen. Monocular visual deprivation and ocular dominance plasticity measurement in the mouse primary visual cortex. *J. Vis. Exp.*, Feb 2020.

[64] Alok Jha. People at darker, higher latitudes evolved bigger eyes and brains. https://www.theguardian.com/science/2011/jul/27/higher-latitudes-bigger-eyes-brains, Jul 2011.

[65] Wikipedia. Phyletic gradualism. https://en.wikipedia.org/wiki/Phyletic_gradualism, Jun 2018.

[66] Niles Eldredge and Stephen Jay Gould. Punctuated equlibria: An alternative to phyletic gradualism. *Models in Paleobiology, Thomas J. M. Schopf (ed.)*, pages 82–115, 1972.

[67] Understanding Evolution. What has the head of a crocodile and the gills of a fish? http://evolution.berkeley.edu/evolibrary/news/060101_batsars, Aug 2008.

[68] Neil H. Shubin Edward B. Daeschler and Farish A. Jenkins Jr. A devonian tetrapod-like fish and the evolution of the tetrapod body plan. *Nature*, 440:757–763, Apr 2006.

[69] Kate Wong. Newfound fossil is transitional between fish and landlubbers. *Scientific American*, Apr 2006.

[70] Wikipedia. Archaeopteryx. https://en.wikipedia.org/wiki/Archaeopteryx, Sep 2021.

[71] Wikipedia. Chromosome 2. https://en.wikipedia.org/wiki/Chromosome_2#:~:text=Human%20chromosome%202%20is%20a%20result%20of%20an,but%20they%20are%20found%20in%20two%20separate%20chromosomes., Nov 2021.

[72] William Paley. Natural theology: or, evidences of the existence and attributes of the deity. http://darwin-online.org.uk/content/frameset?itemID=A142&viewtype=text&pageseq=1, 1809.

[73] Wikipedia. William paley. https://en.wikipedia.org/wiki/William_Paley, Aug 2021.

[74] Discovery Institute. A brief history of discovery institute. https://www.discovery.org/f/4441/, Jan 2008.

[75] Discovery Institute. The "wedge document":"so what?". https://www.discovery.org/m/2019/04/Wedge-Document-So-What.pdf, Jul 2005.

[76] Mark J. Pallen and Nicholas J. Matzke. From the origin of species to the origin of bacterial flagella. *Perspectives, Science and Society*, pages 784–790, Oct 2006.

[77] Luitfried von Salvini-Plawen. Photoreception and the polyphyletic evolution of photoreceptors (with special reference to mollusca). *Evol. Biol.*, 1977.

[78] Patrick Callaerts Georg Halder and Walter J. Gehring. New perspectives on eye evolution. *Current Opinion in Genetics & Devel.*, 1995.

[79] William A. Dembski. *The Design Inference, Eliminating Chance Through Small Probabilities*. Cambridge University Press, New York, NY, 1998.

[80] William A. Dembski. *No Free Lunch, Why Specified Complexity Cannot Be Purchased without Intelligence*. Rowman & Littlefield, Lanham, MD, 2002.

[81] Wikipedia. Human genetic variation. https://en.wikipedia.org/wiki/Human_genetic_variation#:~:text=According%20to%20the%201000%20Genomes%20Project%2C%20a%20typical,51%20SVA%20insertions%2C%20204%20NUMTs%2C%20and%2010%20inversions., May 2021.

[82] RationalWiki. Borel's law. https://rationalwiki.org/wiki/Borel%27s_Law, Jun 2021.

[83] Wikipedia. Emile borel. https://en.wikipedia.org/wiki/%C3%89mile_Borel, Oct 2021.

[84] Lawrence M. Krauss. *A Universe from Nothing*. Blackstone Audio, Inc., Audible, 2012.

[85] Matt Slick. Discussion on logical absolutes as a proof for god's existence. https://carm.org/dialogues/discussion-on-logical-absolutes-as-a-proof-for-gods-existence/, Nov 2008.

[86] Matt Dillahunty. Atheist debates - argument from design, part 2: What are the odds? http://tinyurl.com/prnfbx5, Oct 2014.

[87] Wikipedia. Dark matter. https://en.wikipedia.org/w/index.php?title=Dark_matter&action=history, July 2021.

[88] Ethan Siegel. How the anthropic principle became the most abused idea in science. *Forbes*, Jan 2017.

[89] Natalie Wolchover. The hoyle state: a primordial nucleus behind the elements of life. *Sci. Am.*, Dec 2012.

[90] Robin Collins. The fine-tuning design argument. https://www.discovery.org/a/91/, Sep 1998.

[91] William Lane Craig. The existence of god (fine tuning). https://www.reasonablefaith.org/media/reasonable-faith-podcast/the-existence-of-god-fine-tuning, Jun 2007.

[92] Wikipedia. Regenerative circuit. https://en.wikipedia.org/wiki/Regenerative_circuit, May 2021.

[93] Stuart Wachowicz. Entropy and creation. https://www.youtube.com/watch?v=brKQSVoYU_w, Mar 2019.

[94] Conversations with s. logue, 2021.

[95] Erwin Schrodinger. *What is Life?* 1944.

[96] S. Egri and Gábor Bihari. Self-assembly of magnetic spheres: a new experimental method and related theory. *J. Physics Comm.*, 2:179 – 191, Sep 2018.

[97] Universitaet Mainz. Scientists confirm original tetrahedral model of the molecular structure of water. *Phys.org*, Feb 2013.

[98] Maël Montévil and Matteo Mossio. Biological organisation as closure of constraints. *Theor. Biol.*, 372:179 – 191, 2015.

[99] Wikipedia. Von neumann universal constructor. https://en.wikipedia.org/wiki/Von_Neumann_universal_constructor, Nov 2021.

[100] Wikipedia. Artificial life. https://en.wikipedia.org/wiki/Artificial_life, Oct 2021.

[101] V. Norris and D. J. Raine. A fission-fusion origin for life. *Origins of Life and Evoution of the Biosphere*, 28:523–537, 1998.

[102] Michael P. Robertson and Gerald F. Joyce. The origins of the rna world. *Cold Spring Harb Perspect. Biol.*, 4, May 2012.

[103] Wikipedia. Coacervate. https://en.wikipedia.org/wiki/Coacervate#Coacervates_hypothesis_for_the_origin_of_life, Sep 2021.

[104] et.al. Alan Ianeselli. Non-equilibrium conditions inside rock pores drive fission, maintenance and selection of coacervate protocells. *bioRxiv*, Jul 2021.

[105] Romain Attal and Larent Schwartz. Thermally driven fission of protocells. *Biophysical J.*, 120:3937–3959, Sep 2021.

[106] et.al. Sam Kriegman. Kinematic self-replication in reconfigurable organisms. *PNAS*, 118, Jul 2021.

[107] LittelScientistSVideo. Tornado in a bottle. https://www.youtube.com/watch?v=8Ft5I8mO11Q, Aug 2013.

[108] E. D. Schneider and J. J. Kay. Life as a manifestation of the second law of thermodynamics. *Mathl. Comput. Modelling*, 1994.

[109] D. Reidel. *Ludwig Boltzmann, Theoretical Physics and Philosophical Problems*. 1974.

[110] Wikipedia. Maxwell's demon. https://en.wikipedia.org/wiki/Maxwell%27s_demon, Nov 2021.

[111] Rolf Landauer. Information is physical. *Physics Today*, May 1991.

[112] Linda S. Gottfredson. Mainstream science on intelligence: An editorial with 52 signatories, history, and bibliography. *Intelligence*, 24:13–23, 1997.

[113] Linda S. Gottfredson. The general intelligence factor. http://www1.udel.edu/educ/gottfredson/reprints/1998generalintelligencefactor.pdf, 1998.

[114] Queensland Brain Institute. Where are memories stored in the brain? https://qbi.uq.edu.au/brain-basics/memory/where-are-memories-stored, Jul 2018.

[115] Wikipedia. Dendrites. https://en.wikipedia.org/wiki/Dendrite, Sep 2021.

[116] Wulfram Gerstner and Werner Kistler. *Spiking Neuron Models; Single Neurons, Populations, Plasticity*. Cambridge Univ. Press, New York, NY, 2002.

[117] Richard E. Brown. Donald o. hebb and the organization of behavior: 17 years in the writing. *Molecular Brain*, 13, Apr 2020.

[118] et.al. Zheng-Hua Tan. Pavlovian conditioning demonstrated with neuromorphic memristive devices. *Scientific Reports*, 7, Apr 2017.

[119] Arash Hadipour Niktarash. Discussion on the reverberatory model of short-term memory: A computational approach. *Brain and Cognition*, 53:1–8, Oct 2003.

[120] Sam Harris. *Free Will*. Free Press, New York, NY, 2012.

[121] Wikipedia. Wilson greatbatch. https://en.wikipedia.org/wiki/Wilson_Greatbatch, Jun 2021.

[122] Wikipedia. Role of chance in scientific discoveries. https://en.wikipedia.org/wiki/Role_of_chance_in_scientific_discoveries, Mar 2021.

[123] Casey Blood. Quantum physics, the brain and free will. http://implications-of-quantum-physics.com/qp40_quantum-physics-brain-free-will.html.

[124] David Eagleman. *The Brain; The Story of You*. Vintage Books, New York, NY, 2017.

[125] Eric T. Anderson and Duncan I. Simester. Effects of $9 price endings on retail sales: Evidence from field experiments. *Quantitative Marketing and Economics*, 1:93–110, 2003.

[126] Michael F. Fox Philip Gendall and Priscilla Wilson. Estimating the effect of odd pricing. *J. Product & Brand Management*, 7:421–432, Oct 1998.

[127] V. S. Ramachandran. Split brain with one half atheist and one half theist. https://www.bing.com/videos/search?q=vs+ramachandran+split+brain+paper&docid=608005457593463072&mid=10B9200AD60A90D2F2ED10B9200AD60A90D2F2ED&view=detail&FORM=VIRE, Jun 2010.

[128] Mike Corayer. Split brains: what happens when you sever the corpus callosum? Psych Exam Review, Jan 2017.

[129] E. W. Wright B. Libet and C. A. Gleason. Readiness-potentials preceding unrestricted 'spontaneous' vs. pre-planned voluntary acts. *Electroencephalogr Clin Neurophysiol*, 54:322–335, Sep 1982.

[130] et.al. Chun Siong Soon. Predicting free choices for abstract intentions. *PNAS*, Apr 2013.

[131] Michael Egnor. Physicist sabine hossenfelder and biologist jerry coyne, who deny free will, don't seem to understand the neuroscience. Mind Matters News, Oct 2020.

[132] Larry Greenemeier. 20 years after deep blue: How ai has advanced since conquering chess, Jun 2017.

[133] Christof Koch. How the computer beat the go master, Mar 2016.

[134] Darrell M. West. What is artificial intelligence? https://www.brookings.edu/research/what-is-artificial-intelligence/, 2021.

[135] Shengyang Sun. Csc411 tutorial #5 neural networks. https://www.cs.toronto.edu/~jlucas/teaching/csc411/lectures/tut5_handout.pdf, Oct 2017.

[136] Wikipedia. Deep learning. https://en.wikipedia.org/wiki/Deep_learning, Aug 2021.

[137] A. M. Turing. Computing machinery and intelligence, Oct 1950.

[138] Wikipedia. Turing test. https://en.wikipedia.org/wiki/Turing_test, Aug 2021.

[139] et. al. Kristen A. Richardson. Encoding chaos in neural spike trains. *Phys. Rev. Letters*, 80:2485–2488, Mar 1998.

[140] Kelly Clancy. Your brain is on the brink of chaos. https://nautil.us/issue/15/turbulence/your-brain-is-on-the-brink-of-chaos, Jul 2014.

[141] E. Roderich Gossen Richard B. Stein and Kelvin E. Jones. Neuronal variability: noise or part of the signal? *Nature Reviews*, 6:389–397, May 2005.

[142] Adam Hart. Are humans driving evolution in animals?, Feb 2016.

[143] Caitlin O'Kane. African elephants are evolving without tusks because of poaching, Mar 2019.

[144] Wikipedia. Crispr. https://en.wikipedia.org/wiki/CRISPR, Oct 2021.

[145] Wikipedia. Crispr gene editing. https://en.wikipedia.org/wiki/CRISPR_gene_editing#CRISPR_screening, Oct 2021.

[146] Abhinn D. Suthar and Tejas R. Suthar. The bionic eye...a new vision of the future. *IJSR*, 9:691–697, Jun 2020.

[147] S. L. Ananthu Suresh. A study on bionic eye technology. *IJRESM*, 3:386–388, Jan 2020.

[148] K. Vasantha K. Pradeep and C. Sunitha. An overview – artificial eye (bionic eye). *NCNHIT*, pages 23–26, 2013.

[149] Deeksha H N and Sandeep S. Bionic eye – an artificial vision & comparative study based on different implant techniques. *IJEEER*, 6:87–94, Aug 2016.

[150] Elon Musk. An integrated brain-machine interface platform with thousands of channels. *J. Med. Internet Res.*, 21, Oct 2019.

[151] Wikipedia. Zhuangzi. https://en.wikipedia.org/wiki/Zhuangzi_(book)#%22The_Butterfly_Dream%22, Oct 2021.

[152] Nick Bostrom. Are you living in a computer simulation? *Philosophical Quarterly*, 53:243–255, 2001.

[153] et.al. Tom Campbell. On testing the simulation theory. *Int. J. Quantum Foundations*, 3, Jun 2017.

[154] Anil Ananthaswamy. Do we live in a simulation? chances are about 50–50. *Scientific American*, Oct 2020.

[155] W. Brian Arthur. *The Nature of Technology; What It Is and How It Evolves*. Free Press, New York, NY, 2009.

[156] Wikipedia. Transistor. https://en.wikipedia.org/wiki/Transistor, Nov 2021.

[157] Fruit of Spirit. How many are denominations of christianity? https://fruitsofspirit.com/how-many-are-denominations-of-christianity/, 2021.

[158] Joe Heschmeyer. Is limbo closed? (where do unbaptized babies go?). http://shamelesspopery.com/is-limbo-closed-where-do-unbaptized-babies-go/, May 2015.

[159] Erin Snodgrass. How would humans respond to the discovery of aliens? nasa enlisted dozens of religious scholars to find out. https://www.yahoo.com/news/humans-respond-discovery-aliens-nasa-024208831.html, Dec 2021.

[160] Emily Brown. Why won't christians get vaccinated? https://www.relevantmagazine.com/magazine/why-wont-christians-get-vaccinated/, Sep 2021.

Index

Amperé, 52
Antenna design, 129
Anthropic principle, 250
    Fred Hoyle, 251
    strong, 251
    weak, 251
Appearance of design, 9
Archaeopteryx, 178
Arthur C. Clarke, 259
Artificial intelligence, 350
    AlphaGo, 350
    Deep Blue, 350
Artificial neural network, 350
    Activation network, 351
    Back propagation, 353
    Sigmoid function, 351
    Weighting, 351
Axiomatic, 39

Bell's theorem, 84
Big Bang, 98
Biology
    Fractal, 167
Boltzmann constant, 277

Cause
    Uncaused, 94
Cellular Automata, 70
Cellular automata
    Convergence, 163
Centromere, 182
Chaotic system, 54
Chaotic systems, 54, 92

Sensitivity to initial
    conditions, 59
Closure, 293
Coacervate, 296
Complexity
    Convective flow, 13
    In nature, 12
    Snowflake, 12
Conditional independence
    condition, 215
Constraint, 291
Constructal law, 168
Control theory, 123
Cosmic microwave background
    radiation, 100
Covalent bond, 286
Creationism, 185
CRISPR, 357

Dark energy, 258
Darwin, Charles, 137
Deep learning, 354
Delta function, 94
Design inference, 207, 215
Deterministic systems, 77
Discovery Institute, 187
Discrete functions, 61
Doppler effect, 98
Dynamic simulation, 252

Edwin Hubble, 98
Emile Borel, 219
Entanglement, 83

Enthalpy, 280
Entropy, 277
    Expected value, 308
    Macrostate, 277
    Microstate, 277
EPR paradox, 83
Ergodic, 160
Evolution, 135
    Artificial selection, 356
    Complexity, 158
    Convergence, 151, 158
    Evolutionary pressure, 173
    Extending function space, 169
    Macroevolution, 141
    Microevolution, 141
    Natural selection, 137
    Phyletic gradualism, 174
    Punctuated equilibrium, 175
    Stasis, 173
    Superior forms, 155
    Theory of, 137
    Transitional forms, 153
    Unnatural selection, 356
Exponential growth, 32

Faith
    Definition, 45
Faraday's law of induction, 50
Feigenbaum, 61
Fine-tuning, 252
Flagellar motor, 145, 196, 199, 201
Fractal
    Applications to technology, 69
    Benoit Mandelbrot, 66
    Box, 68
    Dimension, 67
    Dual pendulum, 60
    Koch curve, 67
    Sierpinski triangle, 69
Free will, 341
    Libertarian, 345
Fundamental thermodynamic
    relation, 280

Gene editing, 357

Gene mutation, 138
Genetic algorithm, 129
    Flowchart, 130
Genotype, 137
Gibbs free energy, 280

Heisenberg uncertainty relations, 82
Hidden variables, 83, 84
Hydrophilic, 289
Hydrophobic, 289

Information, 75
    Kolmogorov measure, 78
Intelligence, 80
Intelligent design, 185
    Not a theory, 188
Irreducible Complexity, 195, 199

Kalam cosmological argument, 93
Kurt Gödel, 92

Laws of logic, 46, 191
Least squares fit, 97
Libet, Benjamin, 345
Lissajous figure, 198
Logistic equation, 61, 62
    Period doubling into chaos, 63
Lorentz force, 51
Lotka-Volterra model, 155
Louis de Broglie, 18

Mandelbrot
    Benoit, 64
Mandelbrot set, 64
Max Planck, 16
Maxwell, 52
Mean, 230
Michelson-Morley experiment, 40
Molecular biology
    Affinity, 159
    Lock and key, 159
Motor, 196
Mutation
    Neutral, 164

Negative feedback, 140

INDEX

Neuralink, 359
Neuron
    Axon, 328
    Dendrite, 327
    Synapse, 328
Neuroplasticity, 171
Newton's law of gravitation, 49
Newton's laws of motion, 37

Olbers' paradox, 100
Optimization, 104
    Cost function, 110
    Gradient ascent, 107
    Random perturbation, 108
    Random search, 105
    Using derivative, 108

Panspermia, 263
Persistent excitation, 123
Phenotype, 137
    Constant networks, 152
Planck's law, 16
Positive feedback, 140
Probabilistic resources, 211
Probability, 25, 192
    Event space, 194
    Independent variables, 142, 147
    Law of large numbers, 192
    Let's make a deal, 25
    Sample space, 194
Probability resources, 219
Proteins, 139
Pulsar binary, 97

Quantum computing, 355
    Qubit, 355
Quantum mechancis, 14
Quantum mechanics
    Copenhagen interpretation, 21
    Double-slit experiment, 17
    Harmonic oscillator, 15
    Superposition, 355
    Wave function, 19, 231
    Wave function collapse, 19
    Wave-particle duality, 18, 231

Ramachandran, V. S., 343
Randomness, 75
    Kolmogorov measure, 78
    Radioactive decay, 76
Rayleigh-Jeans law, 14
Red shift, 98
Reductionism, 89
Religion, 373
Replicational resources, 211
RNA world, 295

Schrodinger equation, 19
Science
    Evolution, 366
    Modeling reality, 48
Second law of thermodynamics, 278
Sets, 28
    Cantor's binary proof, 29
    Infinite, 28
Simulated reality, 360
Simulation
    Double pendulum, 57
    Single pendulum, 55
Special relativity, 22
    Length contraction, 24
    Mass-energy equivalence, 24
    Time dilation, 22
Specified Complexity, 207
Specified complexity, 158
Standard deviation, 230
Supernatural, 257

Technology
    Evolution, 366
Telomere, 182
Theory, 36
    What is a theory?, 36
Thermodynamics
    Equipartition theorem, 309
    Intensive variable, 306
    Maxwell's demon, 315
Tiktaalik, 177
Tractability condition, 216
Traveling salesman problem, 129
Turing complete, 152

Turing test, 354

Ultraviolet catastrophe, 15
Unification, 53
  Of electric and magnetic fields, 53
Universal probability bound, 218
Universe
  Expansion, 98, 258
  Universe design intuition, 10

Wallace, Alfred Russel, 138
Weasel problem, 131
  Genetic algorithm, 131
Wedge document, 187
Wolfram, Stephen, 74

www.ingramcontent.com/pod-product-compliance
Lightning Source LLC
Chambersburg PA
CBHW052307220526
45472CB00001B/13